Smart Cities as a Solution for Reducing Urban Waste and Pollution

Goh Bee Hua
National University of Singapore, Singapore

A volume in the Advances in Environmental Engineering and Green Technologies (AEEGT) Book Series

Information Science REFERENCE
An Imprint of IGI Global

Published in the United States of America by
Information Science Reference (an imprint of IGI Global)
701 E. Chocolate Avenue
Hershey PA 17033
Tel: 717-533-8845
Fax: 717-533-8661
E-mail: cust@igi-global.com
Web site: http://www.igi-global.com

Library of Congress Cataloging-in-Publication Data

Names: Hua, Goh Bee, 1964- editor.
Title: Smart cities as a solution for reducing urban waste and pollution / Goh Bee Hua, editor.
Description: Hershey : Information Science Reference, 2016. | Includes bibliographical references and index.
Identifiers: LCCN 2016006150| ISBN 9781522503026 (hardcover) | ISBN 9781522503033 (ebook)
Subjects: LCSH: City planning--Technological innovations. | Transportation--Energy conservation. | Renewable energy sources. | Greenhouse gas mitigation. | Urban ecology (Sociology)
Classification: LCC HT166 .S58777 2016 | DDC 307.1/216--dc23 LC record available at https://lccn.loc.gov/2016006150

This book is published in the IGI Global book series Advances in Environmental Engineering and Green Technologies (AEEGT) (ISSN: 2326-9162; eISSN: 2326-9170)

British Cataloguing in Publication Data
A Cataloguing in Publication record for this book is available from the British Library.

Advances in Environmental Engineering and Green Technologies (AEEGT) Book Series

ISSN: 2326-9162
EISSN: 2326-9170

Mission

Growing awareness and an increased focus on environmental issues such as climate change, energy use, and loss of non-renewable resources have brought about a greater need for research that provides potential solutions to these problems. Research in environmental science and engineering continues to play a vital role in uncovering new opportunities for a "green" future.

The **Advances in Environmental Engineering and Green Technologies (AEEGT)** book series is a mouthpiece for research in all aspects of environmental science, earth science, and green initiatives. This series supports the ongoing research in this field through publishing books that discuss topics within environmental engineering or that deal with the interdisciplinary field of green technologies.

Coverage

- Radioactive Waste Treatment
- Waste Management
- Biofilters and Biofiltration
- Green Transportation
- Water Supply and Treatment
- Policies Involving Green Technologies and Environmental Engineering
- Cleantech
- Contaminated Site Remediation
- Industrial Waste Management and Minimization
- Pollution Management

IGI Global is currently accepting manuscripts for publication within this series. To submit a proposal for a volume in this series, please contact our Acquisition Editors at Acquisitions@igi-global.com or visit: http://www.igi-global.com/publish/.

The Advances in Environmental Engineering and Green Technologies (AEEGT) Book Series (ISSN 2326-9162) is published by IGI Global, 701 E. Chocolate Avenue, Hershey, PA 17033-1240, USA, www.igi-global.com. This series is composed of titles available for purchase individually; each title is edited to be contextually exclusive from any other title within the series. For pricing and ordering information please visit http://www.igi-global.com/book-series/advances-environmental-engineering-green-technologies/73679. Postmaster: Send all address changes to above address.

Titles in this Series

For a list of additional titles in this series, please visit: www.igi-global.com

Food Science, Production, and Engineering in Contemporary Economies
Andrei Jean-Vasile (Petroleum-Gas University of Ploiesti, Romania)
Information Science Reference • copyright 2016 • 473pp • H/C (ISBN: 9781522503415) • US $240.00 (our price)

Smart Grid as a Solution for Renewable and Efficient Energy
Ayaz Ahmad (COMSATS Institute of Information Technology, Pakistan) and Naveed Ul Hassan (LUMS School of Science & Engineering, Pakistan)
Information Science Reference • copyright 2016 • 415pp • H/C (ISBN: 9781522500728) • US $220.00 (our price)

Biologically-Inspired Energy Harvesting through Wireless Sensor Technologies
Vasaki Ponnusamy (Universiti Tunku Abdul Rahman, Malaysia) Noor Zaman (King Faisal University, Saudi Arabia) Tang Jung Low (Universiti Teknologi Petronas, Malaysia) and Anang Hudaya Muhamad Amin (Multimedia University, Malaysia)
Information Science Reference • copyright 2016 • 318pp • H/C (ISBN: 9781466697928) • US $200.00 (our price)

Handbook of Research on Waste Management Techniques for Sustainability
Ulas Akkucuk (Bogazici University, Turkey)
Information Science Reference • copyright 2016 • 438pp • H/C (ISBN: 9781466697232) • US $240.00 (our price)

Control and Treatment of Landfill Leachate for Sanitary Waste Disposal
Hamidi Abdul Aziz (Universiti Sains Malaysia, Malaysia) and Salem Abu Amr (Universiti Sains Malaysia, Malaysia)
Information Science Reference • copyright 2016 • 459pp • H/C (ISBN: 9781466696105) • US $220.00 (our price)

Toxicity and Waste Management Using Bioremediation
Ashok K. Rathoure (Vardan Environet Guargaon, India) and Vinod K. Dhatwalia (Uttaranchal University, India)
Engineering Science Reference • copyright 2016 • 421pp • H/C (ISBN: 9781466697348) • US $205.00 (our price)

Editorial Advisory Board

Table of Contents

Detailed Table of Contents

Section 1
Smart Building Components and Systems

The section targets smart city solutions on the micro-scale of building components and systems. Chapter 1 presents a study that examined the reuse of agro-industrial by-products and concluded that it is a feasible solution for smart city development owing to the ever increasing demand for conventional construction materials, as well as an increase in agro-industrial by-products, which necessitates the reuse of those materials.

R. V. Ralegaonkar, Visvesvaraya National Institute of Technology, India
M. V. Madurwar, Visvesvaraya National Institute of Technology, India
V. V. Sakhare, Visvesvaraya National Institute of Technology, India

Due to ever increasing demand for the conventional construction materials as well as an increase in agro-industrial by-products it is essential to reuse these materials. As a smart city solution this chapter briefs an overview for the application of alternate raw materials as a principal source for the development of sustainable construction materials. The potential application of the discussed raw materials is elaborated as cementitious material, the aggregates as well as alternative reinforcement material. To understand the process of application, sustainable masonry product development is discussed in detail. In order to evaluate the feasibility of the raw material, the necessary physico-chemical test evaluation methods are also briefed. The developed end product performance evaluation is also discussed by desired tests as recommended by standards. The chapter concludes with a positive note that reuse of agro-industrial by-products is a feasible solution for the smart city development.

Section 2
Smart Innovations for Buildings

The section covers innovative solutions for smart buildings as a whole. Chapter 2 details an experiment of passive designs of residential buildings located in 16 cities in the severe cold regions of Northeast China and the results has shown the independent and integrated effectiveness of nine kinds of passive design strategy. In chapter 3, a study of the factors affecting the possibility of introducing smart homes, in developing countries like Nigeria, that would conserve electricity, reduce a need for reliance on the national grid and lower energy costs has provided a better understanding of the advantages and disadvantages, as well as the impact on society and development in general.

With a strong dependency on regions and climate, passive design is the most economically effective strategy to reduce energy consumption. Using Climate Consultant 5.5 software, data analysis is conducted to analyze 16 cities in Northeast China. The independent and integrated effectiveness of nine kinds of passive design strategy are studied like sun shading of windows, high thermal mass, high thermal mass night flushed, direct evaporative cooling, two-stage evaporative cooling, natural ventilation cooling, passive solar direct gain low mass, passive solar direct gain high mass, wind protection of outdoor spaces. The results show that passive solar heat gains are the most important passive design strategy, whether in the heating or non-heating season, with an average contribution rate of 14.64% and 28.38%, respectively. This study provides suggestions on passive design in severe cold regions in Northeast China. The effectiveness-evaluation tables proposed can be referred to by architects in their passive designs.

In many developing countries, there is limited access to electricity with the populace typically resorting to purchasing personal electricity generating units. This research explores the factors affecting the possibility of introducing smart homes that would conserve electricity, reduce a need for reliance on the national grid and lower energy costs. Advantages and disadvantages, as well as the impact on society and development in general are examined in addition to future perspectives on smart homes and sustainable energy.

Section 3
Smart Innovations for Infrastructures

The section concerns smart solutions for creating and operating a city's infrastructures for the provision of public transportation, as well as all the basic utilities and amenities. Chapter 4 deals with the efficiency improvements in the transportation sector, with the aim of reducing pollution from engine fuel emissions, and concluded that it would necessitate behavior and urban planning changes as well rather than financial input alone. In chapter 5, a study of how smart transportation system can be adopted in Harare, the capital city of Zimbabwe, has found that besides the need to design and implement air pollution control measures in the urban areas, the integration of transportation, land use and decision making is also critical in the achievement of smart transport.

On a worldwide scale population is increasing, necessitating efficiencies in transportation essential for a modern and healthy environment. Most cities have already reached their limitations on population and vehicle movement. Street layouts and cycling lanes, which were initially planned for a certain number of inhabitants, are now not able to cope with increasing demand. The same applies for fresh water supply and effluent treatment which have limitations also. With an increasing population in urban areas, restriction levels in these regions can be reached very quickly, creating a problem which needs to be addressed. A smarter city can only be developed when there is pre-planning on all of these issues, when there is an understanding of newly developed human behaviours, and when new inventions and technologies in urban areas can adapt to these changes.

The concept of smart transportation systems is increasingly becoming critical in addressing the challenges posed by an increasing number of mega cities in both developed and developing regions in maintaining safety, smooth traffic flow, and an environmentally friendly and sustainable urban environment. The Government of Zimbabwe deregulated its transport sector in the early 1990s. This development ushered in the informal public transport operators, locally referred to as 'kombis'. Major cities such as Harare are characterised by a disjointed and chaotic urban public transport system. The major problems are the impacts this has on the quality of the environment. Currently, the urban public transport system is contributing greatly to both air and noise pollution within the confines of the city, especially in the Central Business Districts. The problem is further exacerbated by the massive importation of used vehicles from outside the country. These developments have resulted in high emission rates of major air pollutants resulting in a deterioration of the ambient air quality especially in the major cities such as Harare. Transportation is a major source of air pollutants. Vehicles are probably the largest single source of pollutants such as hydrocarbons, nitrogen dioxide and carbon monoxide. Other harmful emissions include as lead, benzene, arsenic, aldehydes, sulphates, particulate matter and the secondary creation of ozone. In Harare the number of registered vehicles increased from 192 901 in 1994 to 292 862 and by August 1999 showing that the increase in the number of vehicles is mostly in the cities. This chapter seeks to explore how smart transportation system can be adopted in Harare, the capital city of Zimbabwe. Harare was purposefully selected as it is experiencing rapid urbanization and motorization in the country. Using documentary analysis, discourse analysis and textual analysis, the chapter also describes and examines the challenges, constraints and opportunities of adopting smart urban transportation system in Harare. From this chapter the major conclusions are that the main problem associated with this rapid growth in vehicle population in the major cities is increase concentration of line and area emission sources due to traffic congestion at peak times. The stock of vehicles is quite old and they lack emission control equipment. Major constraints and limitations are observed in the current pieces of legislation. For example, the current Atmospheric Pollution Prevention Act (1971) does not require vehicles to be fitted with emission control equipment. Most of the vehicles use leaded fuel resulting in emission of the dangerous lead particulate matter in the urban areas. There is therefore an urgent need to design and implement air pollution control measures in the urban areas of the country. It is also critical to develop smart and eco-friendly transportation infrastructures so as to achieve sustainable urban communities. The integration of transportation, land use and decision making is also critical in the achievement of smart transport.

Section 4
Smart Innovations for Cities and Communities

The section covers innovative concepts and solutions developed on the macro-scale of cities and communities. Chapters 6, 7 and 8 present possible urban smart solutions to help reduce noise pollution, as well as minimize waste generation. In one study, accounting for noise pollution by incorporating appropriate mathematical optimization models for designing the layout of facilities during the planning stages of smart cities was found to be effective. The issue of managing and treating construction waste and organic fraction of municipal solid waste (OFMSW) in urban areas has been examined in another study, which included a case-study analysis of the related Dutch, German and Polish guidelines for the sustainable siting of waste processing infrastructure within an urban unit. Concerning urban medical waste, a study has introduced a statistical learning method to estimate the rate of medical waste generation to allow for a more effective way of storing, transporting and disposing the waste. The topic of building smart resilience in cities is next. In chapter 9, the proposed study aims to document the potential and need of developing smart cities in India, as well as understand smart city principles and various dimensions of smart city adopted in various parts of the world which will further help in deriving recommendations and lessons for India's future smart cities. In chapter 10, the building of smart resilience in cities that allows various stakeholders of urban infrastructures, including the local community, to adapt to changes and uncertainties, via a facility information platform for managing the needs of different user groups, was explained through a new concept of building resilience information modeling. Finally, chapter 11 presents an investigation of the role of total quality management (TQM) in smart city development which culminated in the design of an integrative framework with building information modeling and TQM serving as the information architecture and the ideological premise, respectively.

A.W.A. Hammad, University of New South Wales, Australia
A. Akbarnezhad, University of New South Wales, Australia
D. Rey, University of New South Wales, Australia

The incorporation of sustainable design measures in urban planning and development has been steadily increasing in the recent years. Achieving a sustainable urban environment requires accounting for the economic, environmental and social impacts of the development involved. An important factor affecting the social and environmental sustainability of urbanised areas which is commonly overlooked in urban planning is the noise pollution level. Despite the proven impacts of noise pollution on the general wellbeing of individuals within an urban setting, there remains a lack of systematic methods to integrate the impact of noise within the design of urban areas. This chapter seeks to raise awareness of the issue of noise pollution in urban settings while proposing novel approaches for its incorporation as a design parameter in planning the layout of smart cities.

This chapter focuses on the metabolic concept for the management and treatment of construction waste and organic fraction of municipal solid waste in urban areas. Analysis of related Dutch, German and Polish guidelines for environmental zoning of industrial plants, allows formulation of conditions for an optimal siting of waste infrastructure within urban unit. Protection zones are defined in accordance with specific requirements for waste facilities, which treat and recycle both municipal and construction waste. Distances from inhabited areas are related to environmental burdens generated by such facilities (incl. parameters such as odours, noise level, explosion impacts and emissions of other substances). Moreover, this chapter provides the analysis of a selected case studies of waste facilities processing. A comparison of European guidelines and implementation of practical solutions is described in the case study analysis, including the issues open for the discussion about sustainable siting for waste processing infrastructure within an urban unit.

The prediction of wastes generated in the hospital will help their management for several activities like storage, transport and disposing. This chapter adopts Support Vector Machine (SVM), Least Square Support Vector Machine (LSSVM) and Genetic Programming (GP) in order to estimate the rate of medical waste generation. In the event of predicting the rate, type of hospital, capacity and bed occupancy has been used as inputs of SVM, LSSVM and GP. SVM is based on statistical learning theory, which provides an elegant tool for nonlinear system modeling. LSSVM is the re-formulation to the general SVM. GP, a best part of evolutionary algorithm and also the specification of Genetic Algorithm (GA). These SVM, LSSVM and GP have been used as the regression techniques. The results show the performance of the developed SVM, LSSVM and GP models were elegant and outstanding.

Preface

The concept of 'Smart Cities' arises to draw a distinction from terms like 'Digital Cities' or 'Intelligent Cities' so that it looks at the process of urbanization more holistically by involving the social and environmental aspects as well in order to balance the drive for global competitiveness with sustainable development. In other words, cities in future should be well functioning organisms that are environmentally, socially and economically sustainable. A city has to be created by minimizing its ecological impact, where landscape and built form are balanced and where buildings and infrastructures are safe and resource efficient. In essence, cities and individual buildings can focus on achieving resource conservation by enhancing waste minimization and recycling and improving energy efficiency. In addition, new materials, construction methods and architectural designs can greatly improve the environmental performance of urban buildings. These aspects indeed relate to the effective design, construction and management of the built environment. With today's technology, the increased use of information technology (IT) is key to enabling cities and individual buildings arrive at a more 'ideal' state of sustainability through innovation. The much-needed information and knowledge have to be captured, utilized and managed appropriately to allow companies to exploit them as strategic resources. Innovative tools are available and used by various stakeholders for analyzing data to arrive at better decisions, anticipating problems so as to resolve them proactively and coordinating resources for more effective operations. Overall, smarter cities have to drive sustainable economic growth and prosperity for their citizens. At the end of the day, improving lives is central to implementing a smart city given that the main focus of all objectives is about satisfying the population's needs and priorities. However, the benefits provided by a smart city can be different in terms of context as each would have its own distinct needs and priorities which are based on what the population needs and values at a certain point in time. Therefore, in order for smart city planning to be effective and purposeful, the systems created have to be flexible to change, scalable to size and customizable to requirement because population needs are constantly evolving through time.

The realization of smart building, infrastructure and city will increasingly become widespread with the rapid advancement of innovative technologies that allow the automatic, intelligent (or context-aware) and virtual manipulation of real-time information in the new era of Big Data. A smart city is where the development of infocomm-based integrated networks, capabilities and solutions for the urban environments through a systems-of-systems approach is one of the keys. As such, the building of intelligent enterprises in the various economic sectors has to take place simultaneously to complement the social and environmental targets of achieving the 'smart' status. In the more recent years, strategic and smart applications of information technologies are increasingly becoming important as more innovative organizations have witnessed improving their efficiency, effectiveness and performance through better aligning of their technologies with business goals. Moving up to the societal level, people have benefited from the smart systems as they can pre-empt their needs and help them carry out basic activities. They can also empower people by providing them with the information to learn and do the activities themselves in the most efficient way.

Clearly, the consideration of sustainability is highly important in ensuring that the smart city concept works holistically. It is simply because the cities use more technology and as a result they consume more power and other resources which raises their impact on the environment through an increased carbon footprint and waste generation. However, technology-driven solutions have emerged to counter the effects of any potential damage to the environment as cities are beginning to adopt more efficient management of energy use. Take for instance the application of smart grid technology which comprises an electricity system that incorporates IT to determine the levels of demand for power and adjusts the levels of supply accordingly to only generate as much power the city needs. Such systems can be also integrated with waste collection systems to burn biomass as fuel, which adds to the reduction of the city's dependency on fossil fuels and re-using some of the waste it produces on a daily basis. Other smart solutions for better waste management include plasma gasification whereby organic materials are more efficiently converted to synthetic gas (as harvested energy), as well as using IT to capture data on harvested energy extracted from various stages of waste's life cycle and making it readily available to businesses for continuous improvement of the design of materials that are biodegradable or recyclable.

The primary reason for presenting a Handbook of Research on the emerging area of smart innovations in built environments for reducing urban waste and pollution is to bring together relevant concepts and technologies, including their applications, that aim to address mounting issues concerning the need to capture and manage huge volumes of data, information and knowledge at different levels ranging from the individual component and system to an entire building and even city. The collection of works from academics, researchers, scientists, practitioners, policymakers and other experts will present the latest in concept, technology and application of innovations, including those of smart and knowledge-based systems. The unique feature of the book is the structure or, specifically, the arrangement of the parts and chapters in each part. The flow will allow the reader to appreciate the development of the issues, concepts and applications from the micro scales of building components and systems to macro scales of buildings, infrastructures and cities. The target audience will be graduate students, academics, researchers and scientists, engineers, builders, architects, facility managers and government officials who are concerned with advancing information management through the application of smart innovations to create, manage and redevelop the built environment.

Within the Handbook, there are four distinct sections and a total of 11 chapters. Section 1 covers smart city solutions on the micro-scale of building components and systems. In chapter 1, it presents a study that examined the reuse of agro-industrial by-products and concluded that it is a feasible solution for smart city development owing to the ever increasing demand for conventional construction materials, as well as an increase in agro-industrial by-products, which necessitates the reuse of those materials. The focus of section 2 is on innovative solutions for smart buildings as a whole. Chapter 2 details an experiment of passive designs of residential buildings located in 16 cities in the severe cold regions of Northeast China and the results has shown the independent and integrated effectiveness of nine kinds of passive design strategy. In chapter 3, a study of the factors affecting the possibility of introducing smart homes, in developing countries like Nigeria, that would conserve electricity, reduce a need for reliance on the national grid and lower energy costs has provided a better understanding of the advantages and disadvantages, as well as the impact on society and development in general. Moving on to section 3, it offers smart solutions for creating and operating a city's infrastructures for the provision of public transportation, as well as all the basic utilities and amenities. Chapter 4 deals with the efficiency improvements in the transportation sector, with the aim of reducing pollution from engine fuel emissions, and concluded that it would necessitate behavior and urban planning changes as well rather than financial input alone. In chapter 5, a study of how smart transportation system can be adopted in Harare, the capital city of Zimbabwe, has found that besides the need to design and implement air pollution control measures in the urban areas, the integration of transportation, land use and decision making is also critical in the achievement of smart transport. The last section covers innovative concepts and solutions developed on the macro-scale of cities and communities. Chapters 6, 7 and 8 present possible urban smart solutions to help reduce noise pollution, as well as minimize waste generation. In one study, accounting for noise pollution by incorporating appropriate mathematical optimization models for designing the layout of facilities during the planning stages of smart cities was found to be effective. The issue of managing and treating construction waste and organic fraction of municipal solid waste (OFMSW) in urban areas has been examined in another study, which included a case-study analysis of the related Dutch, German and Polish guidelines for the sustainable siting of waste processing infrastructure within an urban unit. Concerning urban medical waste, a study has introduced a statistical learning method to estimate the rate of medical waste generation to allow for a more effective way of storing, transporting and disposing the waste. The topic of building smart resilience in cities is next. In chapter 9, the proposed study aims to document the potential and need of developing smart cities in India, as well as understand smart city principles and various dimensions of smart city adopted in various parts of the world which will further help in deriving recommendations and lessons for India's future smart cities. In chapter 10, the building of smart resilience in cities that allows various stakeholders of urban infrastructures, including the local community, to adapt to changes and uncertainties, via a facility information platform for managing the needs of different user groups, was explained through a new concept of building resilience information modeling. Finally, chapter 11 presents an investigation of the role of total quality management (TQM) in smart city development which culminated in the design of an integrative framework with building information modeling and TQM serving as the information architecture and the ideological premise, respectively.

The scope of the Handbook is broad. The proposed solutions for smart cities to reduce the urban waste and pollution consist of ideas, concepts and/or applications from nine countries, namely, Australia, China, Germany, India, Nigeria, Poland, South Africa, USA and Zimbabwe. They also consist of contributions, many in the form of case studies, from both the developing and developed parts of the world and, as such, have brought together some of the more critical issues of smart city development faced by countries at different stages of economic development for a useful comparison if need be. There are aspects of urban planning and land use, building and infrastructure designs, information and knowledge management, construction project management and IT covered in the 11 chapters. It provides a valuable text for students reading urban planning, architecture and civil engineering, and construction IT and project management subjects at both the undergraduate and graduate levels. It also offers useful insights into the complex workings of smart cities for researchers, policy-makers and practitioners who have a special interest in strategic technology-enabled change. In particular, the case studies in the chapters on passive residential building designs, smart transportation systems, construction waste management in urban areas and smart designs of resilient neighborhoods will provide practitioners, educators and students with important examples of real-life considerations and applications, as well as lessons that can be learnt from the problems encountered in some of the implementations.

Acknowledgment

First and foremost, I wish to thank IGI Global for giving me the opportunity to complete my second book. I thank the development editors, Meghan Lamb, Jacqueline Sternberg and Janine Haughton, for their guidance, advice and patience. I also wish to thank all the members of my editorial advisory board who had contributed their time and expertise, especially those who also served as anonymous reviewers and provided invaluable feedback to authors on how to improve their chapters. To them, I own my sincere gratitude. Last but not least, I am grateful to the authors for having contributed their work and making this achievement possible.

This book is dedicated to my husband, Kevin, to thank him for having given me the much-needed support and encouragement throughout the duration of my project.

Section 1
Smart Building Components and Systems

The section targets smart city solutions on the micro-scale of building components and systems. Chapter 1 presents a study that examined the reuse of agro-industrial by-products and concluded that it is a feasible solution for smart city development owing to the ever increasing demand for conventional construction materials, as well as an increase in agro-industrial by-products, which necessitates the reuse of those materials.

Chapter 1
Sustainable Construction Materials

R. V. Ralegaonkar
Visvesvaraya National Institute of Technology, India

M. V. Madurwar
Visvesvaraya National Institute of Technology, India

V. V. Sakhare
Visvesvaraya National Institute of Technology, India

ABSTRACT

Due to ever increasing demand for the conventional construction materials as well as an increase in agro-industrial by-products it is essential to reuse these materials. As a smart city solution this chapter briefs an overview for the application of alternate raw materials as a principal source for the development of sustainable construction materials. The potential application of the discussed raw materials is elaborated as cementitious material, the aggregates as well as alternative reinforcement material. To understand the process of application, sustainable masonry product development is discussed in detail. In order to evaluate the feasibility of the raw material, the necessary physico-chemical test evaluation methods are also briefed. The developed end product performance evaluation is also discussed by desired tests as recommended by standards. The chapter concludes with a positive note that reuse of agro-industrial by-products is a feasible solution for the smart city development.

DOI: 10.4018/978-1-5225-0302-6.ch001

INTRODUCTION

Globally, due to the rapid urbanization, industrialization and increasing population, there's a large demand for the building materials over a few decades. That resulted in a chronic shortage of building construction materials. The manufacturing of these conventional building materials consumes a lot of thermal and electrical energy which in-turn pollutes air, water and land. The world's total cement production has increased from (rounded) 4.08 x10^4 to 4.18 x10^4 thousand metric tons during the year 2013-2014 (U.S. Geological Survey, 2015). The number of bricks produced in the UK during the first quarter of 2015 is 464 million bricks (provisional), a rise of 22 million bricks or 5% from the same period in 2014 (Brick Development Association). While the global market for construction aggregates is expected to increase 5.2% per year through 2015 to 48.3 billion metric tons (Concrete construction, 2015). The solid waste generation worldwide is 1.86 billion tons (Waste Atlas Report, 2013). According to the report (2013) of the American Coal Ash Association, the total production of coal combustion process (fly ash, bottom ash, boiler slag, gypsum, etc.) is 114 million tons, while the utilization of the applications like concrete/concrete products/grout, blended cement/feed for clinker, structural fills/embankments etc. is 51 million tons, which is around 55% of the total production. Disposal of solid waste generated from agricultural and industrial production activity is another serious problem. The accumulation of these by-products is not only a burden to the industry, but also affects the environment adversely. In many countries, due to the increasing cost of raw materials and the continuous reduction of natural resources, the use of agro-industrial by-products as a potential alternative construction material is practiced in the construction industry. Industrial by-product, when properly processed, has shown to be selected as construction materials and readily meet the design specifications. In the countries where abundant agricultural and industrial by-products are discharged, these by-products can be used as potential material or replacement material in the construction industry. This will have the double advantage of reduction in the cost of construction material and also as a means of disposing by-products. In view of the utilization of agro-industrial by-products for developing sustainable construction material, the present study reviewed various by-products that were added as a principal raw material to develop the sustainable end products and also relevant tests.

BACKGROUND

Natural resources that are being used especially in process industries leave a by-product called industrial by-products. It normally includes ash and slag coming out of boilers. To manage this generated by-products which normally go for land-filling needs additional resources like transportation of these materials as well as a piece of land. The global industrial by-product management market is segmented on the basis of their services (collection, recycling, incineration, and landfill) and further on the basis of regions (Asia-Pacific, Europe, Middle East & Africa, and America). It is expected that the by-products generation in a 5-year span may get doubled with the current practices of production industry. The aspiring countries under rapid development are also eying for better and better infrastructure which is also putting a lot of pressure on conventional construction material manufacturers (cement, bricks, concrete, steel, etc.). In the era of smart city development to manage the supply of the end products and also to have the free access to piece of land it is projected to reuse the potential by-products for the development of sustainable construction materials. The present chapter shall briefly discuss the design procedure for sustainable material development, especially where raw materials can be used as partial or full substitute of cement or fine aggregates.

RESEARCH STATEMENT AND METHODOLOGY

The solid waste generated by industrial activity (e.g. Boiler ash) is generally dumped. However, if local manufacturers like cement, bricks, ready mix concrete plant, etc. are aware about the application and advantages of the use of such by-products they not only will save the quantity and cost on conventional products which are energy intensive but also shall help to reduce the solid waste management problem.

In order to cater the ever increasing demand for the construction materials as well as to reuse the industrial by-products which are otherwise waste for the society, globally researchers made significant contributions to identify the potential applications. The objective of the present chapter is to brief the reader about these novel construction materials that are energy efficient and green products which is essentially the need for smart city infrastructure. In order to investigate the potential of specific raw material the desired methodology for physico-chemical tests recommended by the standards are elaborated further. Also, the desired end product laboratory tests (physical, mechanical and chemical) are also briefed to ascertain the conformance of novel products.

APPLICATION OF LOCALLY AVAILABLE BY-PRODUCTS FOR CONSTRUCTION MATERIAL DEVELOPMENT

Particle Boards

Zhou et al. (2010) manufactured environment friendly binderless cotton stalk fibreboard (BCSF) made from cotton stalks with no chemical additives. The results showed that the thermal conductivity values had a strong linear correlation with the board density. The internal bonding strength (IBS) of boards was good at the relatively low density level. As an environment friendly and renewable material, the BCSF is particularly suitable for ceiling and wall application to save energy. Quintana et al. (2009) produced the fiber board by performing the pre-treatment on a banana bunch using non synthetic binders. The lignocellulosic material was steam exploded with thermo-mechanical aqueous vapor process in a batch reactor. Lertsutthiwong et al. (2008) developed composite particle boards with low thermal conductivity produced from a mixture of by-products from tissue paper manufacturing and corn peel. Khedari et al. (2004) developed low thermal conductivity particleboards with optimized durian peel and coconut coir mixture ratio. Two parameters were investigated, namely, the mixture ratio of durian peel and coconut coir and board density. It was observed that in comparison with either durian or coconut-based particle boards, mixture of them had better thermal and mechanical properties. Sampathrajan et al. (1992) manufactured particle boards from five different by-products (Maize Husk, Paddy Straw, Maize Cob, Coconut Coir/Pith and Groundnut Shell) using urea formaldehyde as the binding material. The mechanical and thermal properties of the boards were evaluated. Several researchers had developed various by-products based particle boards and evaluated physico-mechanical properties like density, thickness (Thk.), modulus of elasticity (MOE), modulus of rupture (MOR), water absorption and thermal conductivity. Table 1 highlights the studied potential agro by-products as a principal raw material and physico-mechanical properties as well.

Cementitious / Pozzolana / Binder Material

Amin (2011) described the by-products from sugar industries (Bagasse ash) as a cement replacement in concrete. The results indicated that Sugarcane bagasse ash (SBA) is an effective mineral admixture and pozzolana with the original ratio of 20% cement which reduced the chloride diffusion by more than 50% without any adverse effects on other properties of the hardened concrete. Suvimol and Daungruedee (2008) used the ground ash, with percentages retained on sieve no. 325 in the range of 3-28, as a cement replacement in the range of 20- 40%. According to Jaturapitakkul et al. (2007) the expansion and loss in compressive strength were

Table 1. Physico-mechanical properties of particle boards manufactured from various agro-wastes

Agro-Waste	Density (Kg/ m³)	Thk. (mm)	MOE (MPa)	MOR (MPa)	Water Absorption (%)	Thermal Conductivity (W/mK)	Source
Cotton Stalk	150 - 450	25	~75	~0.55	13	0.0585 - 0.0815	Zhou et al., 2010
Banana Bunch	1000	3	3361.95	22.30	-	N.A.	Quintana et al., 2009
TPM/Corn Peel	789 + 16	3.5	21.3 + 5.4	5.6 + 2.2	53.8 + 3.2	0.1470 + 0.0082	Lertsutthiwong et al., 2008
Durian peel and Coconut Coir	311 - 856	10	146.413 -2239.152	2.934 - 36.161	227.382- 32.291	0.0764 - 0.1254	Khedari et al., 2004
Maize Husk	310	16	427	5.2	11 - 14	0.000348	Sampathrajan et al., 1992
Paddy Straw	190	16	930	6.5	11 - 14	0.000229	
Coconut Pith	290	16	282	5.8	11 - 14	0.000314	
Groundnut Shell	540	16	523	6.3	11 - 14	0.000548	
Kenaf Board	150 - 200	-	-	-	-	0.051 - 0.058	Xu et al., 2004

reduced when the palm oil fuel ash (POFA) fineness was increased. The results suggested that ground POFA could be used as a pozzolanic material and could also improve the sulfate resistance of concrete. Ramasamy and Biswas (2008) utilized rice husk ash (RHA) as a cement replacement material. The test results indicated that the optimum quantity of RHA increases the mechanical properties of concrete. Xu et al. (2012) studied the microstructure of RHA from the X-ray diffraction (XRD) analysis and compressive strength test results. It was observed that the optimum combustion temperature for obtaining highly reactive RHA is 600 °C. Thus, RHA can partially replace cement as a mineral admixture for construction purpose. Tashima et al. (2004) in their research reported that the addition of RHA enhances the strength and reduces the water absorption of concrete. Adesanya and Raheem (2009) utilized corn cob ash (CCA) as a pozzolanic material for cement production. Factory production of the CCA blended cement was carried out by replacing ordinary Portland cement clinker with CCA. The developed blended cement satisfies both the NIS 439:2000 and ASTM C 150 requirements. Ganesan et al. (2007) investigated the effects of Bagasse ash (BA) content as partial replacement of cement on physical and mechanical properties of hardened concrete. The investigated properties of concrete were compressive strength, splitting tensile strength, water absorption, permeability characteristic, chloride diffusion and resistance to chloride ion penetration. The test results indicated that BA is an effective mineral admixture, with 20% as the optimal replacement ratio of cement. Chusilp et al. (2009) investigated the physical properties of concrete containing ground BA including compressive strength, water permeability, and heat evolution. Bagasse ash from a sugar factory was ground using a ball mill until the particles retained on a No. 325 sieve were less than 5wt%. They were then used as a replacement for Type I Portland cement at 10, 20 and 30 wt% of the binder. The water to binder (W/B) ratio and binder content of the concrete were held constant at 0.50 and 350 kg/m^3, respectively. The results showed that, at the age of 28 days, the concrete samples containing 10-30% ground bagasse ash by weight of binder had greater compressive strengths than the controlled concrete (concrete without ground bagasse ash), while, the water permeability was lower than the controlled concrete. Concrete containing 20% ground bagasse ash had the highest compressive strength at 113% of the controlled concrete. The water permeability of concrete decreased as the fractional replacement of ground bagasse ash was increased. For the heat evolution, the maximum temperature rise of concrete containing ground bagasse ash was lower than the controlled concrete. It was also found that the maximum temperature rise of the concrete was reduced by 13, 23, and 33% as compared with the controlled concrete when the cement was replaced by ground bagasse ash in 10, 20, and 30 wt% of the binder, respectively. According to Cordeiro et al. (2009) incorporation of an ultra finely-ground ash in a high-performance concrete for partial replacement of Portland cement resulted in no measurable change in mechanical behavior, but improved rheology and resistance to penetration of chloride ions. Sata et al. (2007) investigated the effects of agro-industrial by-product pozzolana on mechanical properties of high-strength concrete. The results suggested that the agro by-products can be utilized as a substitute for Portland cement to produce high strength concrete without altering the mechanical properties of concrete. Rukzon and Chindaprasirt (2012) utilized bagasse ash (BA) as a pozzolanic material for producing high strength concrete. The concrete mixtures, in part, were replaced with 10%, 20% and 30% of BA respectively. The compressive strength, porosity, the coefficient of water absorption, the rapid chloride penetration and chloride diffusion of concrete were determined. The test results indicated that the incorporation of BA up to 30% replacement level not only increases the resistance to chloride penetration but also increases the compressive strength reasonably. Nair et al. (2006) examined the long-term strength of different rice husk ash samples as a pozzolana with lime or cement to suggest a sustainable affordable option for rural housing in India. According to Akram et al. (2009) the compressive strengths developed by the self compacting concrete (SCC) mixes with bagasse ash at 28 days were comparable to the controlled concrete. Cost analysis showed that the cost of the ingredients of the specific SCC mix was 35.63% less than that of controlled concrete, both having a compressive strength above 34 MPa. Memon et al. (2011) utilized rice husk ash as viscosity modifying agent in SCC. Based on the experimental results it was observed that the possibility of developing low cost SCC using RHA is feasible. Several researchers had utilized various agro-industrial by-products as pozzolanic/cementitious material in various classifications of concretes (i.e. Ordinary concrete, Table 2 and high strength concrete, Table 3), and evaluated the physico-mechanical properties like compressive strength and water absorption.

Aggregate

Mannan and Ganapathy (2004) utilized oil palm shell (OPS) as a coarse aggregate for structural concrete. The results indicated that the bulk density of OPS concrete was about 1850 kg/m^3 and the compressive strength was in between 20-24 N/mm^2 for 28 days; that satisfied the strength requirement of structural lightweight concrete.

Table 2. Cement replacement, water to binder ratio (w/b) and compressive strength of various agro-wastes for ordinary concrete

Agro-waste	Cement Replacement (%)	w/b ratio (%)	Compressive Strength, 28 Days (MPa)	Source
Bagasse Ash	30	0.53	32.00	Amin, 2011
Palm Oil Fuel Ash	30	0.71	30.10	Jaturapitakkul et al., 2007
Rice Husk Ash	30	0.504	49.74	Nair et al., 2006

Table 3. Cement replacement, water to cement ratio (w/c) and compressive strength of various agro-wastes for high strength concrete

Agro-waste	Cement Replacement (%)	w/c ratio (%)	Compressive strength, 28 Days (MPa)	Design Strength, (MPa)	Source
Rice Husk Barak Ash	10	0.28	86	78	**Sata et al., 2007**
	20	0.28	88		
	30	0.28	85		
Palm Oil Fuel Ash	10	0.28	81	78	
	20	0.28	86		
	30	0.28	80		
Bagasse Ash	10	0.30	65	65	**Rukzon & Chindaprasirt, 2012**
	20	0.30	67		
	30	0.30	68		

Sales and Lima (2010) prepared mortar and concrete with sugarcane bagasse ash (SBA) as a sand replacement and performance tests were carried out. The results indicated that the SBA samples presented physical properties similar to those of natural sand. The mortars produced with SBA in place of sand showed better mechanical results than the conventional mortar.

Fiber Reinforcement

As reported by Coutts (1992), vegetable fibers contain cellulose, a natural polymer, as the main reinforcement material. The chains of cellulose form micro-fibers, which are held together by amorphous hemicelluloses and lignin that form fibrils. The fibrils are assembled in various layers to build up the structure of the fiber. Fibers or cells are cemented together in the plant by lignin, which can be dissolved by the alkalinity of the cement matrix (Gram, 1983). According to Ghavami (2005) vegetable fibers can be used either alone or as reinforcement in different types of matrices such as soil and cement composites. Li et al. (2000) recommended use of sisal fiber as a promising reinforcement for use in composites on account of its low cost, low density, high specific strength and modulus, no health risk, easy availability in some countries and renewability. Ramakrishna and Sundararajan (2005) studied the variations in chemical composition and tensile strength of coir, sisal, jute and hibiscus fibers, when they were subjected to alternate wetting and drying and continuous immersion for 60 days in three different mediums, water, saturated lime and sodium hydroxide respectively. From the results it was observed that there's a substantial reduction in the salient chemical composition of all four fibers, after exposure in the various mediums. Coir fibers were found to retain higher percentages of their initial strength than all other fibers, after the specified period of exposure in the various mediums. The compressive and flexural strengths of all natural fiber reinforced with mortar specimens using corroded fibers (i.e. The fibers subjected to continuous immersion/alternate wetting and drying in the above mediums) were found to have lesser strength than that of the reference mortar (i.e. Without fibers) and fiber reinforced mortar specimen reinforced with dry natural fibers. Ashour et al. (2010) utilized straw as a reinforcing fiber in plaster material for wall plaster. Three types of straw were used such as wheat straw, barley straw and wood shavings. The results indicated that the thermal conductivity of materials decreased with increasing straw fiber content and decreased with increasing sand content. The straw fibers showed a greater effect on the change of thermal conductivity than the effect of sand. The result also revealed that plaster reinforced with barley straw fibers had the highest values of thermal insulation. The studied physico-chemical characteristics of the natural fibers are presented in Table 4.

Table 4. Physico-chemical properties of natural fibres

Fibre Type	Fibre Length (mm)	Fibre Diameter (mm)	Tensile Strength (N/mm²)	Elongation (%)	Hemi-Cellulose (%)	Cellulose (%)	Lignin (%)	Source
Coir	60 - 250	0.40 - 0.10	15 - 327	75.0	31.1	33.2	**20.5**	**Ramakrishna & Sundararajan, 2005**
Sisal	180 - 600	0.10 - 0.50	31 - 221	14.8	26	38.2	26.0	
Jute	128 - 1525	0.04 - 0.35	29 - 312	19.0	22.7	33.4	28.0	
Hibiscus	163 - 1527	0.04 - 0.16	18 - 180	12.4	25	28	22.7	

Reinforcement

Ghavami (1995) studied the ultimate load behavior of bamboo reinforced lightweight concrete beams and the results indicated that the tensile strength of bamboo was relatively high and could reach 370 MPa that makes bamboo an attractive alternative to steel in tensile loading applications. As reported, the ratio of tensile strength to the specific weight of bamboo is six times greater than that of steel. Rahman et al. (2011) carried out the performance evaluation of bamboo reinforced concrete beam and studied stress-strain behavior of bamboo. It was observed that bamboo possesses low modulus of elasticity compared to steel and it cannot prevent cracking of concrete under ultimate load. But, from the flexural test of bamboo reinforced beam, it was observed that using bamboo as reinforcement in concrete can increase the load carrying capacity of beam having the same dimensions. For singly reinforced bamboo concrete beam, the load carrying capacity increased about 2 times and that for doubly reinforced bamboo concrete beam about 2.5 times than that of a plain concrete beam having the same dimensions. The maximum deflection of singly reinforced beam and doubly reinforced beam were about 4.5 and 8 times respectively than that of plain concrete. Mahzuz et al. (2011) studied the strength properties of composite members made up of bamboo. They determined the compressive strength of bamboo, the strength of a composite column made up of bamboo and mortar in its hollow area, and the strength of a composite column of bamboo and concrete. Different mixing ratios and cross-sectional areas were used for conducting the study. From the results of the study it was observed that samples constructed as an aid of bamboo can offer respectable amount of strength that can be safely used for low-cost housing. Mark and Russell (2011) carried out a comparative study of bamboo reinforced concrete beams with shear links made of different materials. The different stirrup materials considered were bamboo, rattan cane and steel. From the results it was recommended that bamboo reinforced concrete beams were reinforced with steel stirrups to improve on its load carrying behavior.

Masonry Composites / Bricks

Brick is one of the most accommodating masonry units as a building material due to its properties. Attempts were made to incorporate various industrial by-products in the production of bricks; for instance, petroleum effluent treatment plant sludge, billet scale, red mud, fly ash, granulated blast furnace slag, steel industry dust, and sewage sludge were used to develop bricks and other construction materials (Sengupta et al., 2002; Shakir et al., 2013; Amrilphale & Patel, 1987; Malhotra & Tehri, 1995; Dominguez & Ullmann, 1996; Tay, 1987). Several researchers investigated the effect of adding agro-industrial by-products in the production of bricks. Rajput et al. (2012) utilized recycled paper mill and cotton by-products to manufacture waste-create bricks (WCB). WCB with varying content of cotton by-products from 1-5 percentages by weight (1-5 wt. %), recycled paper mill by-products from 89-85 percentage by weight (89-85 wt. %) and fixed content of Portland cement (10 wt. %) were prepared and tested as per IS 3495 (Part 1-3): 1992 standards. Results indicated that bricks were thermally stable and meet the standard (BIS 3495 (Part I-III). Alami (2010) investigated the effect of adding unburnt olive husk to clay bricks on its thermo-mechanical properties. The amount of husk added was varied, along with the drying temperature. The specimen scored the lower thermal conductivity resulted than the conventional clay bricks. Chiang et al. (2009) produced the lightweight bricks from the sintering mixes of dried water treatment sludge and rice husk. Samples containing up to 20 wt. % rice husks were fired using a heating schedule that allowed effective organic burn-out. It was observed that addition of rice husk below 15 wt. % and sintered at 1100°C produced the low density and relatively high strength bricks compliant with relevant Taiwan standards for lightweight bricks. Ling and Teo (2011) had developed the bricks from rice husk ash (RHA) and expanded polystyrene (EPS) beads. RHA was used as partial replacement for cement while, EPS was used as a partial aggregate replacement in the mixes. The engineering properties of the bricks were investigated. It was found that the properties of the bricks were mainly influenced by the content of EPS and RHA in the mix and also the used curing conditions.

Lertsatitthanakorn et al. (2009) had developed rice husk ash based sand-cement brick. Its performance was compared with that of a standard commercial clay brick. It was resulted that the RHA based sand-cement brick reduces solar heat transfer by 46 W. Demir (2008) utilized several organic residues having long cellulose fibers like sawdust, tobacco residues in clay bricks. The results indicated that the insulation capacity of brick was increased with the increasing porosity of clay body. It was observed that the fibrous nature of residue did not create extrusion problems. As a result, sawdust, tobacco residue and grass can be utilized in an environmentally safe way as organic pore-forming agents in clay bricks. Algin & Turgat (2008) and Turgat & Algin (2007) used a combination of cotton waste (CW) – limestone powder waste (LPW) and a combination wood sawdust waste (WSW) and limestone powder waste (LPW) for producing low cost and lightweight composite as a building material. The obtained compressive strength, flexural strength, ultrasonic pulse velocity, unit weight and water absorption values satisfied the relevant international standards. It resulted in lighter weight composite having potential to be used for walls, wooden board substitute, economical alternative to the concrete blocks, ceiling panels, sound barrier panels, etc. Rahman (1987) developed bricks from clay-sand mixes with different percentages of RHA. The effects of RHA contents on workable mixing water content, Atterberg limits, linear shrinkage, density, compressive strength and water absorption of the bricks were investigated. Test results indicated that the compressive strength of the bricks was higher with rice husk ash contents. Lima et al. (2012) analyzed the effect of incorporating SBA in compressed earth bricks. The results showed that the average compressive strength, water absorption and specific density of the brick incorporating SBA was increased as compared to the control brick without incorporating SBA. Faria et al. (2012) utilized SBA (up to 20 weight percentage) to provide raw material for clay brick bodies. The result showed that the linear shrinkage of bricks was decreased with the increase of the SBA. On the contrary, to earlier research the specific density and the tensile strength were also decreased with increase of SBA weight percentage. Performance of various agro-industrial by-product materials evaluated from the common tests of compressive strength, water absorption and specific weight (Sp. Wt.) are studied further (Table 5).

Thermal Insulator

Buildings are less able to control the internal environment to comfortable conditions without mechanical air conditioning (Panyakaew & Fotios, 2008). An only technique for reducing the scale of air conditioning is application of thermal insulation in walls and roofs. Pinto et al. (2011) studied Tabique construction (Portuguese traditional building techniques). An experimental work concluded that there were significant similarities between the corn cob and the extruded polystyrene material in terms of microstructure and chemical composition. Furthermore, the results indicated that the corn cob may have adequate thermal properties for building purpose. Korjenic et al. (2011) carried out the research to develop new insulating material from renewable resources (jute, flax and hemp) with comparable building physics and mechanical properties to commonly used insulation materials. The test results showed that the

Table 5. Physico-mechanical properties of bricks made from various agro-waste materials

Agro-waste	Compressive Strength (MPa)	Water Absorption (%)	Sp. Wt. (10^3 x kg/m^3)	Source
Paper – Cotton Waste – Cement	22.27 + 0.5	100	0.56 + 0.02	**Rajput et al., 2012**
Tobacco Residue	11.55	21.85	1.53	**Demir, 2008**
Grass	12.35	21.65	1.55	
Cotton Waste - Cement	7.0 + 0.3	17.4 + 1.4	1.51	**Algin & Turgat 2008**
Sawdust	13.60	21.40	1.56	**Demir, 2008 ; Turgat & Algin, 2007**
RHA	12.65	20.20	1.42	**Rahman, 1987**

correct combination of natural materials was absolutely comparable with convectional materials. Onesippe et al. (2010) examined the thermal properties of cement composites reinforced with vegetable bagasse fibers. Experimental investigations revealed that adding ratified bagasse fibers reduces thermal conductivity of composites and yields a weaker specific heat. There are several possible materials like bagasse, rice husk, coconut coir, corn stalk, durian peel, and palm oil leaves which are studied further on the basis of physico-mechanical results (Table 6). Among the studied by-product bagasse, rice husk and coconut coir observed to have lower thermal conductivity.

Table 6. Physico-mechanical properties of the particle insulation board made from agro-waste materials

Agro-Waste	Density (Kg/m^3)	Thermal Conductivity (W/mK)	Source
Rice Husk	150-175	0.046-0.056	**Panyakaew & Fotios, 2008 ; Yarbrough 2005**
Coconut Coir	300-350	0.047-0.085	**Panyakaew & Fotios, 2008**
Bagasse	90-140	0.047-0.050	**Panyakaew & Fotios, 2008 ; Onésippe et al., 2010 ; Krishpersad 2006**
Corn Cob	300-330	0.097-0.098	**Panyakaew & Fotios, 2008 ; Pinto et al., 2011**
Durian Peel	400-875	0.065-0.150	**Khedari et al., 2004 ; Panyakaew & Fotios, 2008**
Oil Palm Leaves	800-1000	0.118-0.240	**Panyakaew & Fotios, 2008 ; Al-Juruf et al., 1988**

It is observed from the study (Table 1- 6) that researchers used various by-products in different proportions and also adopted various methodologies to produce different building materials and composites like particle boards, thermal insulated wall and ceiling panels, bricks, cementitious pozzolana for concrete. Several thermo-mechanical tests were conducted on different materials and composites as per the various available standards. The common parameters calculated for the particle and insulating boards were density, MOE, MOR and thermal conductivity. It is observed that densities of particle boards manufactured from incorporating agro by-products are meeting the minimum requirement of standards (S-20 grade of JIS A 5905-1994). Thermal conductivity values of these materials and composites lies in the standard range 0.054 – 0.143 W/mK and satisfies the requirement of insulation materials as per BS874-2. The common test parameters calculated by the various researchers for the brick or masonry are water absorption and compressive strength. The common parameter required for utilizing any material as a mineral admixture or pozzolana depends on the proportion of silica in its by-product. Burnt residue by-product of quality conforming BIS 456:2000 can be used as pozzolana in the manufacturing of concrete provided that there are satisfactory data on their suitability such as performance test on concrete containing that the residue (BIS 465, 2000). The performance test conducted on concrete compositions containing various proportions of different by-product residues indicated that the characteristic compressive strength of concrete is enhanced whereas, effect of chemical attacks like chloride and sulfate are diminished. The overall performance and durability of the concrete is enhanced with the addition of burnt residue. The purpose of fiber reinforcement is to improve the properties of a building material, but the main drawback to the use of natural fibers is the durability of these fibers in a cementitious matrix. The alkaline media weaken the most natural fibers, especially the vegetable ones which are actually strands of individual filaments liable to be separated from each other. The severe degradation of exposed composites can also be attributed to interfacial damages due to continuous volume changes of the porous vegetable fibers inside the cement matrix. There are two approaches to improve the durability of vegetable fibers. One is based on the protection of the fibers by coating them or sealing the dry composite to avoid the effect of water, mainly alkalinity. The other approach is the reduction of free alkalis in the matrix by developing low alkaline binders based on industrial and agricultural by-products. The consumption of building components made of fiber reinforced cement is increasing rapidly, especially in developing countries because with this type of material it is possible to produce light weight building components, with good mechanical performance, suitable thermal-acoustic insulation and its economic viability. Traditionally, steel is used as reinforcement in concrete. But because of cost and availability, replacement of steel with some other suitable materials as reinforcement is now a major concern. Bamboo is natural, cheap, widely available and most importantly strong in both tension and compression. Maximum tensile strength of bamboo is up to 370 MPa whereas, compression tests of bamboo give an average stress of 41.02 MPa. These are the good values to resist moderate loads, especially for low cost, low rise buildings.

CONCERNS IN THE BRICK SECTOR

Stack emissions are the major source of air pollution from brick industry. The brick production industry has been identified as a highly resource, energy intensive and polluting industry owing to the prevalence of obsolete production technologies. While, the clusters are the source of local air pollution affecting local population, agriculture and vegetation; at a global scale, they also contribute to climate change. The brick industry also uses better quality of top soil from agricultural fields. The unplanned and unregulated exploitation of good quality agriculture soil for brick making is a major area of concern. The fast depletion of arable land, thus caused due to brick making is a matter of concern regarding food security (www.ecobrick.in). The large coal consumption of the brick industry is the cause of significant air pollution in terms of carbon dioxide (CO_2), carbon monoxide (CO), sulfur dioxide (SO_2), nitrogen oxides (NOx) and suspended particulate matter (SPM). The large amount of coal used for brick firing also leaves behind bottom ash as a residue. The air pollution and the generated bottom ash cause considerable health problems especially related to respiratory health and also causes damage to property and crops (Skinder, 2013).

Brick Manufacturing Process

Traditionally for manufacturing the clay bricks, clay is mined and stored in the open (http://www.ecobrick.in). This makes the clay soft and removes unwanted oxides. Soft clay is then mixed with water to get the right consistency for molding. Mixing is done and lump of mix is taken, rolled in sand and drive into the mold. Sand is used so the brick does not stick to the mold. The mold is emptied onto the drying area, where the bricks are arranged in a herringbone pattern to dry in the sun. Every two days they are turned over to facilitate uniform drying and prevent warping. After two weeks they are ready to be burnt. The green bricks are arranged in a kiln and insulation is provided with a mud pack. Fire holes left to ignite the kiln are later sealed to keep the heat inside. This is maintained for a week. After the kiln is disassembled, the bricks are sorted according to color. Color is an indication of the level of burning. Over burnt bricks are used for paving or covering the kiln, while, slightly under burnt bricks are used for building inner walls or burnt once again in the next kiln.

Wherein, for fly ash brick manufacturing [NTPC guidelines] raw materials required are fly ash, sand and cement. They are mixed thoroughly in pan mixer and made as a semi dry mix by adding a sufficient amount of water. This mix is fed in hydraulic press and molded into bricks. The green bricks are air dried for 1-2 days and the water cured for 20 days period.

USE OF LOCALLY AVAILABLE INDUSTRIAL WASTE FOR NOVEL CONSTRUCTION MATERIALS

In order to investigate the potential locally available industrial by-products it is very much essential to understand the chemical composition and the behavior of the material. In general, it is required to conduct following important tests for acceptability of the desired raw material.

Physical Properties

Particle Size Distribution (Hydrometer Method)

The percentage of sand, silt and clay in the inorganic fraction of the soil is measured in this method. The method is based on Stoke's law governing the rate of sedimentation of particles suspended in water. The sample is to be treated with sodium hexametaphosphate to complex Ca^{++}, Al^{3+}, Fe^{3+}, and other cations that bind clay and silt particles into aggregates. The organic matter is to be suspended in this solution. The density of the soil suspension is to be determined with a hydrometer calibrated to read in grams of solids per liter after the sand settles out and again after the silt settles. Corrections are made for the density and temperature of the dispersing solution. Results are to be reported as percentages of the mineral fraction, sand, silt, and clay (Bouyoucos, 1992).

Specific Gravity

Specific gravity, as it is a ratio of densities, is a dimensionless quantity. Specific gravity varies with temperature and pressure; reference and sample must be compared at the same temperature and pressure, or corrected to a standard reference temperature and pressure. In scientific work the relationship of mass to volume is usually expressed directly in terms of the density (mass per unit volume) of the substance under study. Specific gravity is the ratio of the mass of a unit volume of soil at a stated temperature to the mass of the same volume of gas-free distilled water at the stated temperature. The specific gravity of a soil is used in the phase relationship of air, water, and solids in a given volume of the soil.

Chemical Properties

X-Ray Fluorescence (XRF)

XRF is an analytical technique that can be used to determine the chemical composition of a wide variety of sample types including solids, liquids, slurries and loose powder. XRF is also used to determine the thickness and composition of layers and coatings. It can analyze elements from beryllium (Be) to uranium (U) in concentration ranges from 100 weight percentage to sub-PPM levels. XRF is a robust technique, combining high precision and accuracy with straightforward, fast sample preparation. XRF provides both qualitative and quantitative types of information on a sample. XRF is the emission of characteristic "secondary" (or fluorescent) X-rays from a material that has been excited by bombarding with high-energy X-rays or gamma rays. The phenomenon is widely used for elemental analysis and chemical analysis, particularly in the investigation of metals, glass, ceramics and building materials (Williams & Carter, 2009).

X-Ray Diffraction (XRD)

X-ray diffraction is an analytical technique for looking at X-ray scattering from crystalline materials. Each material produces a unique X-ray "fingerprint" of X-ray intensity versus scattering angle that is characteristic of its crystalline atomic structure. Qualitative analysis is possible by comparing the XRD pattern of an unknown material to a library of known patterns. When X-rays interact with a crystalline substance (phase), one gets a diffraction pattern. The X-ray diffraction pattern of a pure substance is, therefore, like a fingerprint of the substance. The powder diffraction method is thus ideally suited for the characterization and identification of polycrystalline phases. The main use of powder diffraction is to identify components in a sample by a search/match procedure. Furthermore, the areas under the peak are related to the amount of each phase present in the sample (Williams & Carter, 2009).

Thermogravimetric Analysis (TGA)

Thermogravimetric analysis or thermal gravimetric analysis (TGA) is a method of thermal analysis in which changes in physical and chemical properties of materials are measured as a function of increasing temperature (with constant heating rate), or as a function of time (with constant temperature and/or constant mass loss). TGA can provide information about physical phenomena, such as second-order phase transitions, including vaporization, sublimation, absorption, adsorption, and desorption. Likewise, TGA can provide information about chemical phenomena, including chemisorptions, desolvation (especially dehydration), decomposition, and solid-gas reactions (e.g., oxidation or reduction). TGA is commonly used to determine selected characteristics of materials that exhibit either mass loss or gain due to decomposition, oxidation, or loss of volatiles (such as moisture). Common applications of TGA are (1) material characterization through analysis of characteristic decomposition patterns, (2) studies of degradation mechanisms and reaction kinetics, (3) determination of organic content in a sample, and (4) determination of inorganic (e.g. ash) content in a sample, which may be useful for corroborating predicted material structures or simply used as a chemical analysis (Coats &, Redfern, 1963).

Scanning Electron Microscope (SEM)

Mode of detection is by secondary electrons emitted A scanning electron microscope is a type of electron microscope that produces images of a sample by scanning it with a focused beam of electrons. The electrons interact with the atoms in the sample, producing various signals that can be detected and that contain information about the sample's surface topography and composition. The electron beam is generally scanned in a raster scan pattern and the beam's position combines with the detected signal to produce an image. SEM can achieve resolution better than 1 nanometer. Specimens can be observed in high vacuum, in low vacuum, and (in environmental SEM) in wet conditions. The most common by atoms excited by the electron beam. The number of secondary electrons is a function of the angle between the surface and the beam. On a flat surface, the plume of secondary electrons is mostly contained in the sample, but on a tilted surface, the plume is partially exposed and more electrons are emitted. By scanning the sample and detecting the secondary electrons, an image displaying the tilt of the surface is created (McMullan, 1995).

Proximate Analysis

Proximate analysis is a partitioning of compounds in a feed into six categories based on the chemical properties of the compounds. The six categories are moisture, ash, crude protein (or Kjeldahl protein), crude lipid, crude fiber, nitrogen-free extracts (digestible carbohydrates). The standards which describe the proximate analysis are (ASTM D7582-15, BIS 1350-1, 1984).

With a concern about development of building blocks (bricks) using locally available industrial by-products following necessary tests must be performed to make use of such novel products.

PRODUCT TESTING

Physico-Mechanical Properties

Compressive Strength (BIS 3495 (part 1): 1992)

Place the specimen with flat faces horizontal and mortar filled face facing upwards between two 3-ply plywood sheets each of 3 mm thickness and carefully centered between the plates of the testing machine. Apply load axially at a uniform rate of 14 N/mm^2 (140 kg/cm^2) per minute till failure occurs and note the maximum load at failure. The load at failure shall be the maximum load at which the specimen fails to produce any further increase in the indicator reading on the testing machine.

The compressive strength of the test specimen is expressed as:

$$\text{Compressive strength}\left(\text{N / mm}^2\right) = \frac{\text{Maximum load at failure}\left(\text{N}\right)}{\text{Average area of the bed faces}\left(\text{mm}^2\right)} \quad (1)$$

Water Absorption (BIS 3495 (part 2): 1992)

Dry the specimen in a ventilated oven at a temperature of 105 to 115°C till it attains substantially constant mass. Cool the specimen to room temperature and obtain its weight (M_1). Specimen warm to touch shall not be used for the purpose. Immerse completely the dried specimen in clean water at a temperature of 27 ± 2°C for 24 hours. Remove the specimen and wipe out any traces of water with a damp cloth and weigh the specimen. Complete the weighing in 3 minutes after the specimen has been removed from the water (M_2).

Water absorption, percent by mass, after 24-hour immersion in cold water is given by the following formula:

$$\text{Water absorption}\left(\%\right) = \frac{M_2 - M_1}{M_1} \times 100 \quad (2)$$

Efflorescence (BIS 3495 (part 3): 1992)

Place the end of the bricks in the dish, the depth of immersion in water being 25 mm. Place the whole arrangement in a warm (for example, 20 to 30°C) well ventilated room until all the water in the dish is absorbed by the specimens and the surplus water evaporates. Cover the dish containing the brick with suitable glass cylinder so that excessive evaporation from the dish may not occur. When the water has been absorbed and bricks appear to be dry, place a similar quantity of water in the dish and allow it to evaporate as before. Examine the bricks for efflorescence after the second evaporation and the results are reported. The report shall be as given below:

The liability to efflorescence shall be reported as 'nil', 'slight', 'moderate', 'heavy' or 'serious' in accordance with the following definitions:

1. **Nil:** When there is no perceptible deposit of efflorescence.
2. **Sight:** When not more than 10 percent of the exposed area of the brick is covered with a thin deposit of salts.
3. **Moderate:** When there is a heavier deposit than under 'slight' and covering up to 50 percent of the exposed area of the brick surface but unaccompanied by powdering or flaking of the surface.
4. **Heavy:** When there is a heavy deposit of salts covering 50 percent or more of the exposed area of the brick surface but unaccompanied by powdering or flaking of the surface.
5. **Serious:** When there is a heavy deposit of salts accompanied by powdering and/or flaking of the exposed surfaces.

Brick Density (BIS 2185 (Part I): 1979)

Three bricks taken at random shall be dried to a constant mass in a suitable oven heated to approximately 100°C. After cooling the bricks to room temperature, the dimensions of each brick shall be measured in millimeters and the overall volume is computed in cubic millimeters. The bricks shall then be weighed in kilograms and the density of each brick calculated as follows.

$$\text{Density} = \frac{\text{mass of brick}\left(\text{kg}\right)}{\text{volume of specimen}(\text{m}^3)} \tag{3}$$

Flexural Strength (BIS 4860: 1968)

The test specimen shall be whole brick, the dimension of which shall be measured to the nearest 1 mm. A minimum of five bricks shall be separately tested. The test specimens shall be immersed in water at 27± 2°C for 24 hours. The test specimen shall be placed centrally on self-aligning bearers A, B and C as shown in Figure 1. The bearers shall be of mild steel 40 mm in diameter and shall be in the same horizontal plane and parallel to each other. The distance between the bearers A and B at the lines of contact with the specimen shall be three-fourths of the length dimension of the brick. Bearer C shall be midway between bearers A and B measured horizontally and rests upon the surface of the specimen. The load shall be applied at uniform rate not greater than 30 kg/min through bearer C.

Compressive Strength of Masonry Prism (ASTM C1314)

The following test method covers procedure for masonry prism construction and testing to determine the compressive strength of masonry.

At the time of construction, the surfaces of units shall be free of moisture and dirt. Place the brick on perfectly leveled surface; lay a full bed of mortar on it of thickness 10 to 12 mm. Then place next brick perfectly on its top. Build prism of height of minimum two-unit height with height to thickness ratio between 1.3 and 5.0. Construction of prisms is done as shown in Figure 2. Do not disturb prisms for the first 48 hours after construction. Curing of prisms is done by maintaining prisms

Figure 1. Arrangement of loading for flexural strength test

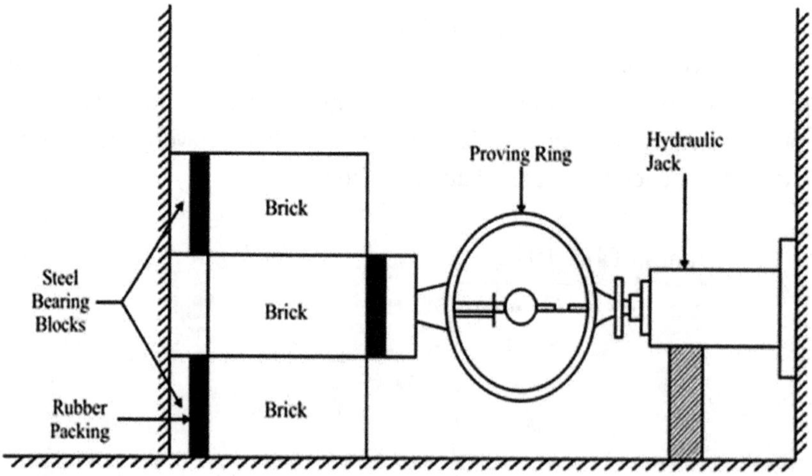

Figure 2. Arrangement of prism compressive strength

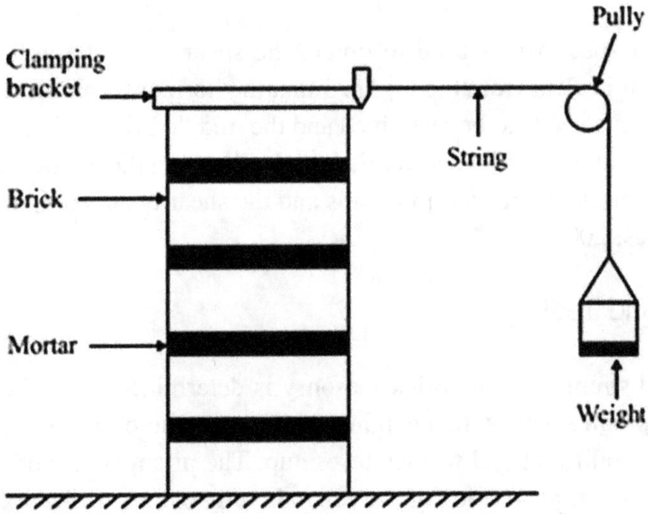

in the area of 24±8°C. Capping of prism is done by fine slurry of cement to fill all irregularities of brick and spirit level is used to bring the surface of brick in level. Measure dimensions of prism accurately. Test the prisms at the age of 7, 28 days or at the designated test age. The surface of bearing plate of compression testing machine must be cleaned before testing. Place the prism on plate of the machine, aligning centroidal axis of prism with that of machine. Apply load to prism to one-half of the expected total load at any convenient rate. Apply remaining load at uniform rate in not less than one or more than two minutes. Calculate the masonry prism strength by dividing each prism's compressive load sustained by net cross sectional area of that prism. Calculate h_p/t_p ratio for each prism (h_p- Height of prism, t_p-Least lateral dimension of prism) and determine the correction factor from Table 7.

$$Permissible\,compressive\,stress\,in\,masonry = \sigma = correction\,factor\,x\,f_m\left(MPa\right) \tag{4}$$

where, f_m = Masonry compressive strength from prism test

Table 7. Height to thickness ratio for compressive strength of masonry prism

hp/tp	**1.3**	**1.5**	**2**	**2.5**	**3**	**4**	**5**
Correction factor	0.75	0.86	1	1.04	1.07	1.15	1.22

Shear Bond Test

A brick triplet specimen is used to obtain the shear bond strength of the brick mortar joints. It is clear from the Figure 3 that the horizontal movement of the top and bottom bricks needs to be restrained and the middle brick is free to move horizontally. Horizontal shear load is applied gradually using the hydraulic jack till the bond between brick and mortar joint fails and the shear bond strength is estimated (Reddy & Vyas, 2008).

Flexural Bond Test

Flexural bond strength of the brick masonry is determined by testing five-brick-high bonded prism stacks with a height to thickness ratio of 4. Figure 4 shows the details of the modified bond wrench test setup. The prism is supported on a rigid bottom. The bottom most brick of the prism is fully clamped. The load is applied to the topmost brick of the prism through a pulley arrangement. This load causes a moment in the prism, which will further cause a flexure failure between the masonry unit and the mortar. In this test setup, only one value of flexural bond strength is obtained for each prism, because once one joint fails the other joints of the prism are not tested (ASTM C-1072).

Figure 3. Arrangement of loading for shear bond test

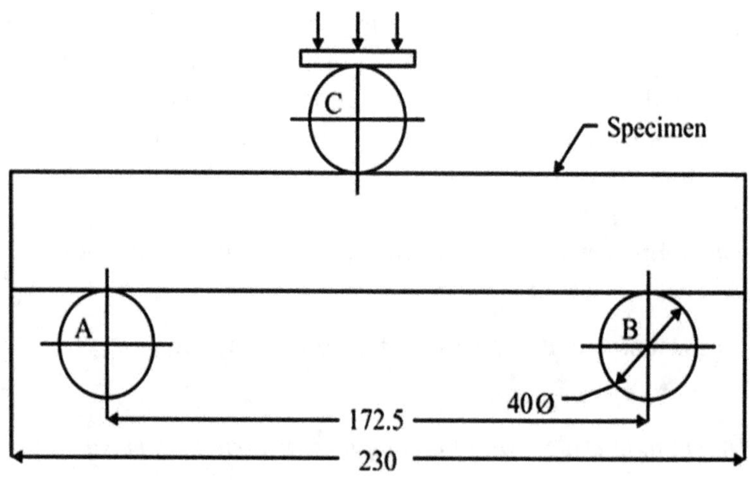

Figure 4. Arrangement of loading for flexural bond strength test

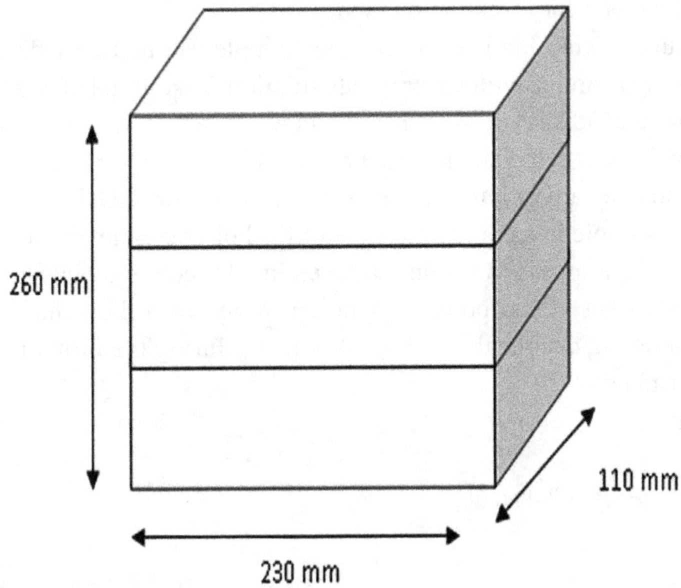

Durability Properties

For masonry/concrete to be durable, careful selection of the mix and materials is necessary, so that deleterious constituents do not exceed the limits. The durability tests like chloride sulfate and effect of carbonation needs to be carried out on the test samples.

Chloride

Whenever there is chloride in concrete there is an increased risk of corrosion of embedded metal. Higher the chloride content or, if subsequently exposed to warm, moist conditions, the greater is the risk of corrosion. All constituents may contain chlorides and concrete may be contaminated with chlorides from the external environment. To minimize the chances of deterioration of concrete from harmful chemical salts, the levels of such harmful salts in concrete coming from constituent materials, that is cement, aggregates, water and admixtures, as well as by diffusion from the environment should be limited. The total acid soluble chloride content is calculated from the mix proportions and the measured chloride content of each of the constituents. Wherever possible, the total chloride content of the concrete is determined. The developed brick needs to be crushed to 125 micron size and suspended in double distilled water. The mix is then allowed to roll at room temperature for 6 hours. The mixtures are filtered and the filtrates sample is tested for chloride content by the laboratory titration method.

The amount of chloride present in water sample can be easily determined by titrating the water sample with silver nitrate solution. Take 10 ml of NaCl reagent in a conical flask and add 2-3 drops of K_2CrO_4 indicator to it. Titrate till turbid pinkish yellow end point and calculate the exact normality of $AgNO_3$. Check the pH of the sample solution and adjust it to near neutral with H_2SO_4 or NaOH as required. Take 100 ml of the sample in a conical flask. Add 1 ml of potassium chromate solution to it and titrate against $AgNO_3$ solution taken in a burette to a pinkish yellow end point. Be consistent in end point recognition. Note down the volume of $AgNO_3$ required. Following chemical reaction takes place during the titration of chloride content determination.

Reaction:

$$Cl^- + AgNO_3 = AgCl + NO_{3^-} \tag{5}$$

$$K_2CrO_4 + 2AgNO_3 = AgCrO_3 + 2KNO_3 \tag{6}$$

Formula:

$$Chloride\ (mg/l) = \frac{\left[(A-B)\, x\, N\, x\, 35.45\, x\, 1000\right]}{ml\, of\, sample\, taken} \tag{7}$$

Where,

A = ml of $AgNO_3$ required for sample
B = ml of $AgNO_3$ required for distilled water
N = Normality of $AgNO_3$

Sulfate

Sulfates are present in majority of cement samples and in some aggregates. Excessive amount of water-soluble sulfate from these or other mix constituents can cause expansion and disruption of the concrete. To prevent this, the total water-soluble sulfate content of the concrete mix (expressed as SO_3 and SO_4) should not exceed 4 percent by mass of the cement in the mix in terms of SO_3. The sulfate content should be calculated as the total from the various constituents of the mix. The developed bricks have to be crushed to 125 microns and suspended in double distilled water. The mix was then allowed to roll at room temperature for 6 hours. The mixtures were filtered and the filtrates sample was tested for sulfate concentration by laboratory spectrometer test.

Take six 100 ml glass stopper standard flasks (four for standards, one for the sample and one for the blank). Add 10 ml of the standard sulfate solution to the first standard flask, 20 ml to the second, 30 ml to the third and 40 ml to the fourth. To the fifth standard flask, add 20 ml of the sample water. The sixth standard flask is for the blank. To this standard flask, add distilled water alone. Add 5 ml of conditioning reagent to all the standard flasks. Then make up the volume to the 100 ml mark using distilled water. The ultraviolet (UV) visible spectrometer is used to measure the sulfate content of the given sample. The fully automatic UV Spectrometer has the capabilities to display the absorbance reading of the solution placed in the chamber of the spectrometer.

Carbonation

Carbonation of concrete is a process by which carbon dioxide from the air penetrates into concrete and reacts with calcium hydroxide to form calcium carbonates. In the presence of moisture, CO_2 changes into dilute carbonic acid, which attacks the concrete and also reduces alkalinity of concrete. A common and simple method for establishing the extent of carbonation is to treat the freshly broken surface of concrete with a solution of phenolphthalein in diluted alcohol. If the Ca $(OH)_2$ is unaffected by CO_2 the color turns out to be pink. If the concrete is carbonated, it remains uncolored. It should be noted that the pink color indicates that enough Ca $(OH)_2$ is present; but it may have been carbonated to a lesser extent (Shetty, 2013).

Thermal Properties

Thermal Conductivity (k)

Thermal conductivity, k, is the property of a material that indicates its ability to conduct heat. Conduction will take place if there exists a temperature gradient in a solid (or stationary fluid) medium. Energy is transferred from more energetic to less energetic molecules when neighboring molecules collide. Conductive heat flow occurs in the direction of the decreasing temperature because higher temperature is associated with higher molecular energy.

The procedure to evaluate thermal conductivity by Lee's disc apparatus (Figure 5) is elaborated further. Place a disc made by the poor conductor, radius r and thickness x, between a steam chamber and two good conductivity metal discs (of the same metal) and allow the setup to come to equilibrium, so that the heat lost by the lower disc to convection is the same as the heat flow through the poorly conducting disc. The upper disc temperature T_2 and the lower disc temperature T_1 are recorded. The poor conductor is removed and the lower metal disc is allowed to heat up to the upper disc temperature T_2. Finally, the steam chamber and upper disc are removed and replaced by a disc made of a good insulator. The metal disc is then allowed to cool through $T_1 < T_2$ and towards room temperature T_0. The temperature of the metal disc is recorded as it cools so a cooling curve can be plotted. Then the slope s1 $=\Delta T/\Delta t$ of the cooling curve is measured graphically where the curve passes through temperature T_1 (NISER manual).

The thermal conductivity can be obtained by the equation

$$k = \frac{mc\frac{dT}{dt}x}{A\left(T_2 - T_1\right)} \tag{8}$$

Where,

k = Coefficient of thermal conductivity of the sample,
A = Area of the sample in contact with the metallic disc,
x = Thickness of the sample,
$(T_2 - T_1)$ = Temperature difference across the sample thickness,

Figure 5. Schematic diagram of Lee's disc apparatus

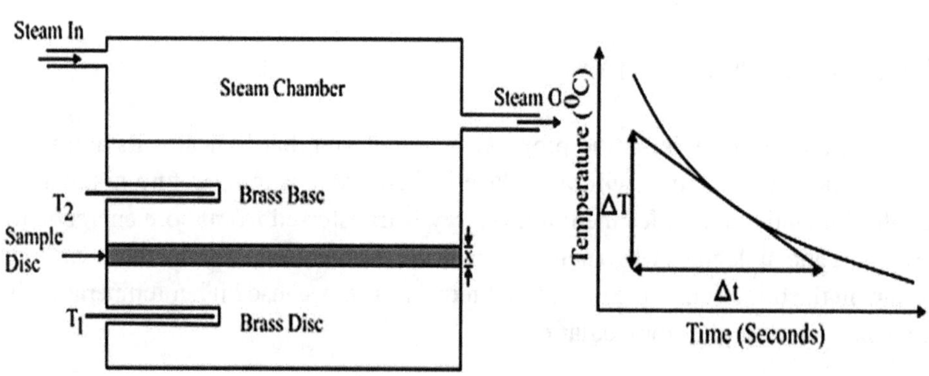

m = Mass of the metallic disc,
c = Heat capacity of the metallic disc,
dT/dt = Rate of cooling of the metallic disc at T_2

Specific Heat Capacity

The Specific Heat is the amount of heat required to change a unit mass of a substance by one degree in temperature (The engineering toolbox). The heat supplied to a unit mass can be expressed as

$$dQ = mc\,dt \quad (9)$$

Where,

dQ = heat supplied (kJ, Btu)
m = unit mass (kg, lb)
c = specific heat (kJ/kg °C, kJ/kg °K, Btu/lb °F)
dt = temperature change (K, °C, °F)

Expressing Specific Heat using

$$c = {dQ}/{m\,dt} \quad (10)$$

Environmental Properties

Toxicity Characteristics of Leaching Protocol (TCLP)

To check the possibility of leaching of heavy metals, toxicity characteristics of the leaching protocol (TCLP) was carried out. The TCLP test, as modified by Zhao et al. (1992) was used for determination of leaching characteristics. The developed bricks were crushed to 125 micron size and suspended in distilled water and TCLP fluid (acetic acid solution at pH 4.9 ± 0.5) maintaining 1:20 (wt/vol) sample-to-fluid ratio. The mixtures were then allowed to roll at rolling speed of 26 rpm in a roller oven, at room temperature for 24 hours to extract the trace metals. The mixtures were filtered and the filtrates were acidified with 6M HNO_3 to about pH 2.5 and then preserved in a plastic container until the determination of the trace metals.

PROMOTION OF THE NOVEL PRODUCTS

Although a lot of innovative products have been developed by the eminent researchers worldwide key challenge is to transfer these ideas from laboratory to field. All the concern stakeholder, especially policy makers, certifying agencies, builders, consumers, etc. should understand these developments and promote them in the right way for sustainable development of the society that aims to be a smart city. Locally available raw material utilization shall not only help to reduce the problem of industrial by-product management, but also help to develop the cost effective solution for the smarter cities. A local governing body needs to make such efforts with material research institutions to lay the guidelines for the application of the materials. Builders and other buyers should be facilitated such information and promote incentives and better schemes to use them. Owners should be made aware so that newer construction products shall reach to the masses. At national level similar attempts are to be made by bureau of standards to bring out suitable national codes for the use of products with regular updates. The government should come up with novel incentive schemes for promotion of newer products like evaluating the greenness of the products, allowing the promoters additional floor space index if permissible in the local jurisdiction, tax subsidies, etc. In order to manage the resource effectively web based graphical user interface needs to be developed wherein raw material producers, manufacturers, builders and owners can be connected together for better dissemination of novel construction products. The outcome of such integration can be evaluated in terms of environmental impact assessment. Based on the performed study following recommendations are suggested.

Appropriate framework (Figure 6) needs to be designed for the potential application of novel products.

Locally available non-hazardous agro-industrial by products has significant potential to be used as alternate raw material.

Central Government should proactively frame the policy and standards to promote the use of such raw materials.

In order to make the designed framework functional more graphical user interface tools needs to be designed and developed. It is proposed to have city level planning related to agro-industrial waste generation and utilization for novel construction material development. Newer admixtures and advanced technologies (from conventional bricks to cellular light weight concrete blocks, autoclaved aerated concrete blocks, pre-fabricated wall material, etc.) needs to be tested and certified. Proper calculators related to total product cost, carbon footprint estimation, etc. needs to be designed. The geographical information System or Building Information Modeling tools shall be used to organize the stakeholders in the smart cities.

Figure 6. Framework for promotion of novel construction materials

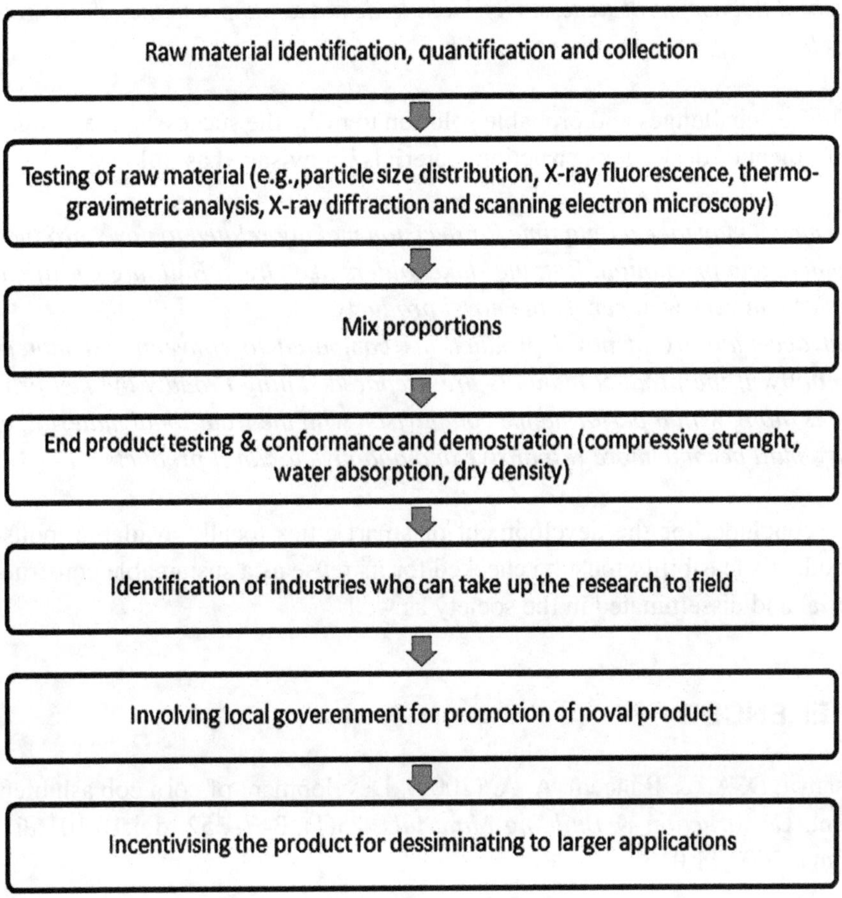

Local governance should establish a sustainability index of the upcoming infrastructure to access the eco-sensitiveness of the buildings and data shall be available in the public domain.
State Government should promote the novel product manufacturing as well as extend necessary incentives for its application on the field with the help of builders and developers.
Product information and knowledge dissemination amongst stakeholders (designers, contractors, quality auditors, etc.) as well as end users must be made with local professional bodies for the betterment of the society.
Smart tools (e-resource) related to mapping of important stakeholders related to construction material design, development and application must be enabled at lo-

cal level for easy sharing of information and the potential application as well. The relevant database shall necessarily include demand, supply management related key fields

The key challenges and probable solution to make the success in the design and development of desired construction materials is envisaged as follows:

Government may take a long time for decision making related to novel product development and promotion. But, the stakeholders like Green Building Certification agencies can very well certify the novel products.
Social acceptability of novel products as compared to conventional materials. Essentially, if the product manuals are prepared stating broadly the key product features along with a performance comparison with the traditional material, community shall become more responsive and adaptive to novel products.

To conclude, for the development of smart cities locally available industrial by-product's feasibility must be checked for its reuse as a sustainable construction material and disseminated in the society as well.

REFERENCES

Adesanya, D. A., & Raheem, A. A. (2009). Development of corn cob ash blended cement. *Construction & Building Materials*, *23*(1), 347–352. doi:10.1016/j.conbuildmat.2007.11.013

Akram, T., Menon, S. A., & Obaid, H. (2009). Production of low cost self compacting concrete using bagasse ash. *Construction & Building Materials*, *23*(2), 703–712. doi:10.1016/j.conbuildmat.2008.02.012

Al-Juruf, R. S., Ahmed, F. A., & Alam, I. A. (1988). Development of heat insulating materials using date palm leaves. *Thermal insulation*, 11, 158-68.

Alami, A. H. (2010). Experiments on olive husk-addition to masonry clay bricks on their mechanical properties, and their application and manufacturability as an insulating material. *Advanced Materials Research*, *83-86*, 874–880. doi:10.4028/www.scientific.net/AMR.83-86.874

Algin, H. M., & Turgut, P. (2008). Cotton and limestone powder wastes as brick material. *Construction & Building Materials*, *22*(6), 1078–1080. doi:10.1016/j.conbuildmat.2007.03.006

American Coal Ash Association. (2013). Coal Combustion Product (CCP) Production & Use Survey.

American Society for Testing and Materials International. (n. d.). ASTM C-1072, *Standard test methods for measurement of masonry flexural bond strength.*

American Society for Testing and Materials International. (n. d.). ASTM D7582 – 15, *Standard Test Methods for Proximate Analysis of Coal and Coke by Macro Thermo-gravimetric Analysis.*

American Society for Testing and Materials International. (n. d.). *ASTM C* 1314-12. *Standard test method for compressive strength of masonry prisms.*

Amin, N. (2011). Use of Bagasse Ash in Concrete and Its Impact on the Strength and Chloride Resistivity. *Journal of Materials in Civil Engineering, 23*(5), 717–720. doi:10.1061/(ASCE)MT.1943-5533.0000227

Amrilphale, S. S., & Patel, M. (1987). Utilization of red mud, fly ash for manufacturing bricks with pyrophyllite. *Silic. Ind, 52*(3-4), 31–35.

Ashour, T., Wieland, H., Georg, H., Bockisch, F. J., & Wu, W. (2010). The influence of natural reinforcement fibres on insulation values of earth plaster for straw bale buildings. *Materials & Design, 31*(10), 4676–4685. doi:10.1016/j.matdes.2010.05.026

Bouyoucos, G. J. (1962). Hydrometer method improved for making particle size analysis of soils. *Agronomy Journal, 54*(5), 464–465. doi:10.2134/agronj1962.00021962005400050028x

Brick Development Association. (2015). Brick manufacturing rises to meet housebuilding challenge. Retrieved from http://www.brick.org.uk/2015/05/brick-manufacturing-rises-to-meet-housebuilding-challenge/

British Standard. BS 874: Part 2. (1986). Methods for determining thermal insulating properties. London.

Bureau of Indian Standard. IS 2185 Part-I. (1979). *Specifications for concrete masonry - hollow and solid concrete blocks.* BIS, New Delhi.

Bureau of Indian Standard. IS 1350-1. (1984). *Methods of Test for Coal and Coke*, Part I: Proximate Analysis [PCD 7: Solid Mineral Fuels]

Bureau of Indian Standard. IS 3495 Part-I. (1992). *Methods of tests of burnt clay building bricks – determination of compressive strength.* BIS, New Delhi.

Bureau of Indian Standard. IS 3495 Part-II. (1992). *Methods of tests of burnt clay building bricks – determination of water absorption.* BIS, New Delhi.

Bureau of Indian Standard. IS 3495 Part-III. (1992). *Methods of tests of burnt clay building bricks – determination of efflorescence.* BIS, New Delhi.

Bureau of Indian Standard. IS 4860. (1996). Specifications for acid resistant bricks. BIS, New Delhi.

Bureau of Indian Standard. IS 456. (2000). Plain and reinforced concrete – code of practice. BIS, New Delhi.

Chiang, K. Y., Chou, H., Hua, C. R., Chien, K. L., & Cheeseman, C. (2009). Lightweight bricks manufactured from water treatment sludge and rise husks. *Hazardous Materials, 171*(1-3), 76–82. doi:10.1016/j.jhazmat.2009.05.144 PMID:19596512

Chusilp, N., Jaturapitakkul, C., & Krattikomol, K. (2009). Utilization of bagasse ash as a pozzolanic material in concrete. *Construction & Building Materials, 23*(11), 3352–3358. doi:10.1016/j.conbuildmat.2009.06.030

Coats, A. W., & Redfern, J. P. (1963). Thermogravimetric analysis: A review. *Analyst (London), 88*(1053), 906–924. doi:10.1039/an9638800906

Concrete construction.net. (2015). Global demand for construction aggregates to exceed 48 billion metric tons in 2015. Retrieved from http://www.concreteconstruction.net/aggregates/global-demand-for-construction-aggregates-to-exceed-48-billion-metric-tons-in-2015.aspx

Cordeiro, G. C., Filho, R. D. T., Tavares, L. M., & Fairbairn, E. R. (2009). Ultrafine grinding of sugar cane bagasse ash for application as pozzolanic admixtures in concrete. *Cement and Concrete Research, 39*(2), 110–115. doi:10.1016/j.cemconres.2008.11.005

Coutts, R. P. (1992). From forest to factory to fabrication. In R.N. Swamy (Ed.), *Proceedings of the 4th International Symposium Fibre Reinforced Cement and Concrete* (pp. 31-47).

Demir, I. (2008). Effect of organic residues addition on the technological properties of clay bricks. *Waste Management (New York, N.Y.), 28*(3), 622–627. doi:10.1016/j.wasman.2007.03.019 PMID:17512183

Dominguez, E. A., & Ullmann, R. (1996). Ecological bricks made with clays and steel dust pollutant. *Applied Clay Science, 11*(2-4), 237–249. doi:10.1016/S0169-1317(96)00020-8

Ecobricks. (n. d.). Issues in the Indian Brick Sector. Retrieved from http://www.ecobrick.in/challenges_Issues_in_the_Indian_Brick_Sector.aspx

Faria, K. P., Gurgel, R. F., & Holanda, J. F. (2012). Recycling of sugarcane bagasse ash waste in the production of clay bricks. *Journal of Environmental Management*, *101*, 7–12. doi:10.1016/j.jenvman.2012.01.032 PMID:22387325

Ganesan, K., Rajagopal, K., & Thangavel, K. (2007). Evaluation of bagasse ash as supplementary cementitious material. *Cement and Concrete Composites*, *29*(6), 515–524. doi:10.1016/j.cemconcomp.2007.03.001

Ghavami, K. (1995). Ultimate load behavior of bamboo reinforced lightweight concrete beams. *Cement and Concrete Composites*, *17*(4), 281–288. doi:10.1016/0958-9465(95)00018-8

Ghavami, K. (2005). Bamboo as reinforcement in structural concrete elements. *Cement and Concrete Composites*, *27*(6), 637–649. doi:10.1016/j.cemconcomp.2004.06.002

Gram, H. E. (1983). Durability of natural fibres in concrete. Stockholm: Swedish Cement and Concrete Research Institute. Retrieved from http://www.engineeringtoolbox.com/specific-heat-capacity-d_339.html

Japanese Industrial Standard. (1994). *JIS A 5905*. Tokyo, Japan: Fibreboards.

Jaturapitakkul, C., Kiattikomol, K., Tangchirapat, W., & Saeting, T. (2007). Evaluation of the sulfate resistance of concrete containing palm oil fuel ash. *Construction & Building Materials*, *21*(7), 1399–1405. doi:10.1016/j.conbuildmat.2006.07.005

Khedari, J., Nankongnab, N., Hirunlabh, J., & Teekasap, S. (2004). New low-cost insulating particleboards from mixture of durian peel and coconut coir. *Building and Environment*, *39*(1), 59–65. doi:10.1016/j.buildenv.2003.08.001

Korjenic, A., Petranek, V., Zach, J., & Hroudova, J. (2011). Development and Performance evaluation of natural thermal insulation materials composed of renewable resources. *Energy and Building*, *43*(9), 2518–2523. doi:10.1016/j.enbuild.2011.06.012

Krishpersad, M. (2006). Biodegradable fibrous thermal insulation. *Journal of the Brazilian Society of Mechanical Sciences and Engineering*, *28*, 45–47.

Lertsatitthanakorn, C., Atthajariyakul, S., & Soponornnarit, S. (2009). Techno-economical evaluation of a rice husk ash (RHA) based sand-cement block for reducing solar conduction heat gain to a building. *Construction & Building Materials*, *23*(1), 364–369. doi:10.1016/j.conbuildmat.2007.11.017

Lertsutthiwong, P., Khunthon, S., Siralertmukul, K., Noomum, K., & Chandrachang, S. (2008). New insulating particleboards prepared from mixture of solid wastes from tissue paper manufacturing and corn peel. *Bioresource Technology*, *99*(11), 4841–4845. doi:10.1016/j.biortech.2007.09.051 PMID:17977719

Li, Y., Mai, Y. W., & Ye, L. (2000). Sisal fibre and its composites: A review of recent developments. *Composites Science and Technology*, *60*(11), 2037–2055. doi:10.1016/S0266-3538(00)00101-9

Lima, S. A., Varum, H., Sales, A., & Neto, V. F. (2012). Analysis of the mechanical properties of compressed earth block masonry using the sugarcane bagasse ash. *Construction & Building Materials*, *35*, 829–837. doi:10.1016/j.conbuildmat.2012.04.127

Ling, I. H., & Teo, D. L. (2011). Properties of EPS RHA lightweight concrete bricks under different curing conditions. *Construction & Building Materials*, *25*(8), 3648–3655. doi:10.1016/j.conbuildmat.2011.03.061

Mahzuz, H. A., Ahmed, M., Ashrafuzzaman, M., Karim, R., & Ahmed, R. (2011). Performance evaluation of bamboo with mortar and concrete. *Journal of Engineering and Technology Research*, *3*(12), 342–350.

Malhotra, S. K., & Tehri, S. P. (1995). Building materials from granulated blast furnace slag – some new prospects. *Indian Journal of Engineering and Materials Sciences*, *2*, 80–82.

Mannan, M. A., & Ganapathy, C. (2004). Concrete from an agricultural waste-oil palm shell (OPS). *Building and Environment*, *39*(4), 44–448. doi:10.1016/j.buildenv.2003.10.007

Mark, A. A., & Russell, A. O. (2011). A comparative study of Bamboo reinforced concrete beams using different stirrup materials for rural construction. *International Journal of Civil and Structural Engineering*, *2*(1), 407–423.

McMullan, D. (1995). Scanning electron microscopy 1928–1965. *Scanning*, *17*(3), 175–185. doi:10.1002/sca.4950170309

Memon, S. A., Shaikh, M. A., & Akbar, H. (2011). Utilization of rice husk ash as viscosity modifying agent in self compacting concrete. *Construction & Building Materials*, *25*(2), 1044–1048. doi:10.1016/j.conbuildmat.2010.06.074

Nair, D. G., Jagadish, K. S., & Fraaij, A. (2006). Reactive pozzolanas from rice husk ash, An alternative to cement for rural housing. *Cement and Concrete Research*, *36*(6), 1062–1071. doi:10.1016/j.cemconres.2006.03.012

Niser, M. (n. d.). Lee's Method. Retrieved from http://physics.niser.ac.in/labmanuals/sem1/Lee's-Method.pdf

NTPC guidelines. (n. d.). Ch. 6 Coal Ash. Retrieved from https://www.ntpc.co.in/ash-download/1664/6/ntpc-guide-users-coal-ash-chapter-6

Onésippe, C., Passe-Coutrin, N., Toro, F., Delvasto, S., Bilba, K., & Arsène, M. A. (2010). Sugar cane bagasse fibres reinforced cement composites, Thermal considerations. *Composites*, 41(A), 549–556.

Panyakaew, S., & Fotios, S. (2008, October 22-24). Agricultural waste materials as thermal insulation for dwellings in Thailand: Preliminary results. *Proceedings of the 25th conference on Passive and Low Energy Architecture PLEA '08*, Durbin.

Pinto, J., Paiva, A., Varum, H., Costa, A., Cruz, D., Pereira, S., & Agarwal, J. et al. (2011). Corn's cob as a potential ecological thermal insulation material. *Energy and Building*, *43*(8), 1985–1990. doi:10.1016/j.enbuild.2011.04.004

Quintana, G., Velasquez, J., Betancourt, S., & Ganan, P. (2009). Binderless fibre board from steam exploded babana bunch. *Industrial Crops and Products*, *29*(1), 60–66. doi:10.1016/j.indcrop.2008.04.007

Rahman, M. A. (1987). Properties of clay-sand-rice husk ash mixed bricks. *Cement Composites and Lightweight Concrete*, *9*(2), 105–108. doi:10.1016/0262-5075(87)90026-1

Rahman, M. M., Rashid, M.H., Hossain, M. A., Hasan, M. T., & Hasan, M. K. (2011). Performance Evaluation of Bamboo Reinforced Concrete Beam. *International Journal of Engineering and Technology*, 11(4), 142-46.

Rajput, D., Bhagade, S. S., Raut, S. P., Ralegaonakar, R. V., & Mandavgane, S. A. (2012). Reuse of cotton and recycle paper mill waste as building material. *Construction & Building Materials*, *32*, 470–475. doi:10.1016/j.conbuildmat.2012.02.035

Ramakrishna, G., & Sundararajan, T. (2005). Studies on the durability of natural fibres and the effect of corroded fibres on the strength of mortar. *Cement and Concrete Composites*, *27*(5), 575–582. doi:10.1016/j.cemconcomp.2004.09.008

Reddy, B. V. V., & Vyas, U. C. V. (2008). Influence of shear bond strength on compressive strength and stress-strain characteristics of masonry. *Materials and Structures*, *41*(10), 1697–1712. doi:10.1617/s11527-008-9358-x

Rmasamy, V., & Biswas, S. (2008). Performance of Rice Husk Ash Concrete with Super-plasticizers. *ICI Journal*, *4*, 27–34.

Rukzon, S., & Chindaprasirt, P. (2012). Utilization of bagasse ash in high-strength concrete. *Materials & Design*, *34*, 45–50. doi:10.1016/j.matdes.2011.07.045

Sales, A., & Lima, S. A. (2010). Use of Brazilian sugarcane ash in concrete as sand replacement. *Waste Management (New York, N.Y.)*, *30*(6), 1114–1122. doi:10.1016/j.wasman.2010.01.026 PMID:20163947

Sampathrajan, A., Vijayaraghavan, N. C., & Swaminathan, K. R. (1992). Mechanical and thermal properties of particle boards made from farm residues. *Bioresource Technology*, *40*(3), 249–251. doi:10.1016/0960-8524(92)90151-M

Sata, V., Jaturapitakkul, C., & Kiattikomol, K. (2007). Influence of pozzolana from various by-product materials on mechanical properties of high-strength concrete. *Construction & Building Materials*, *21*(7), 1589–1598. doi:10.1016/j.conbuildmat.2005.09.011

Sengupta, P., Saikia, N., & Borthakur, P. C. (2002). Bricks from petroleum effluent treatment plant sludge: Properties and environmental characteristics. *Journal of Environmental Engineering*, *128*(11), 1090–1094. doi:10.1061/(ASCE)0733-9372(2002)128:11(1090)

Shakir, A. A., Naganathan, S., & Mustapha, N. K. (2013). Properties of bricks made using fly ash, quarry dust and billet scale. *Construction & Building Materials*, *41*, 131–138. doi:10.1016/j.conbuildmat.2012.11.077

Shetty, M. S. (2013). *Concrete technology – theory and practice. S.* New Delhi, India: Chand Publishing.

Skinder, B. M., Sheikh, A. Q., Ashok, K. P., & Ganai, B. A. (2013). Brick kiln emissions and its environmental impact: A review. *Journal of Ecology and the Natural Environment*, *6*(1), 1–11.

Suvimol, S., & Daungruedee, C. (2008). Bagasse ash, Effect of pozzolanic activity and application in cement use aspect. *Proceedings of the3rd ACF International Conference*.

Tashima, M.M., Da-Silva, C. R., & Akasaki, J.L. (2004). The possibility of adding RHA to the concrete. *Proceedings of the International RILEM conference on the use of recycled materials in building structures* (pp. 778-786).

Tay, J. H. (1987). Bricks manufactured from sludge. *Journal of Environmental Engineering*, *113*(2), 278–283. doi:10.1061/(ASCE)0733-9372(1987)113:2(278)

Turgut, P., & Algin, H. M. (2007). Limestone dust and wood sawdust as brick material. *Building and Environment*, *42*(9), 3399–3403. doi:10.1016/j.buildenv.2006.08.012

U.S. Geological Survey. (2015). Mineral Commodity Summaries, January 2015.

Waste atlas 2013 report. (2013). Retrieved from http://www.atlas.d-waste.com/

Williams, D. B., & Carter, C. B. (2009). Transmission electron microscopy: a textbook for materials science (Vol. 2, pp. 559-560). Springer. doi:10.1007/978-0-387-76501-3

Xu, J., Sugawara, R., Widyorini, R., Han, G., & Kawai, S. (2004). Manufacture and thermal properties of low-density binderless particleboard from kenaf core. *Journal of Wood Science*, *50*(1), 62–67. doi:10.1007/s10086-003-0522-1

Xu, W., Lo, T. Y., & Menon, S. A. (2012). Microstructure and reactivity of rice husk ash. *Construction & Building Materials*, *29*, 541–547. doi:10.1016/j.conbuildmat.2011.11.005

Yarbrough, D. W. (2005). Apparent thermal conductivity data and related information for rice hulls and crushed pecan shells. *Thermal Conductivity*, *27*, 222–230.

Zhao, B., Daughetry, K. E., & Zhang, L. (1992). In R. W. Piggot (Ed.), *Ingram KD, Hill R. Portland cement doped with lead compounds." Proc., 1st Int. Sump. Cement Industry Solution to Waste Management* (pp. 359–372). Calgary Alta., Canada.

Zhou, X.-, Zheng, F., Li, H.-, & Lu, C.-. (2010). An environment-friendly thermal insulation material from cotton stalk fibers. *Energy and Building*, *42*(7), 1070–1074. doi:10.1016/j.enbuild.2010.01.020

Section 2
Smart Innovations for Buildings

The section covers innovative solutions for smart buildings as a whole. Chapter 2 details an experiment of passive designs of residential buildings located in 16 cities in the severe cold regions of Northeast China and the results has shown the independent and integrated effectiveness of nine kinds of passive design strategy. In chapter 3, a study of the factors affecting the possibility of introducing smart homes, in developing countries like Nigeria, that would conserve electricity, reduce a need for reliance on the national grid and lower energy costs has provided a better understanding of the advantages and disadvantages, as well as the impact on society and development in general.

Chapter 2
Simulating Passive Design Strategies of Rural Residential Buildings in Severe Cold Regions of Northeast China

Meng Zhen
Harbin Institute of Technology, People's Republic of China

Cheng Sun
Harbin Institute of Technology, People's Republic of China

Bee Hua Goh
National University of Singapore, Singapore

ABSTRACT

With a strong dependency on regions and climate, passive design is the most economically effective strategy to reduce energy consumption. Using Climate Consultant 5.5 software, data analysis is conducted to analyze 16 cities in Northeast China. The independent and integrated effectiveness of nine kinds of passive design strategy are studied like sun shading of windows, high thermal mass, high thermal mass night flushed, direct evaporative cooling, two-stage evaporative cooling, natural ventilation cooling, passive solar direct gain low mass, passive solar direct gain high mass, wind protection of outdoor spaces. The results show that passive solar heat gains are the most important passive design strategy, whether in the heating or non-heating season, with an average contribution rate of 14.64% and 28.38%, respectively. This study provides suggestions on passive design in severe cold regions in Northeast China. The effectiveness-evaluation tables proposed can be referred to by architects in their passive designs.

DOI: 10.4018/978-1-5225-0302-6.ch002

INTRODUCTION

The "Twelfth Five-year" Building Energy Saving Special Planning program of The Housing and Urban-rural Development of China set the goal to save 116 million tons of standard coal by the end of "12th five-year" through building energy conservation. According to the data of second national agricultural census data of the three northeast province, there are 14.658 million rural households and 1.17 billion square meters of building area, and the trend is rising. The main source of energy consumption in severe cold regions is the heating in buildings. Most rural residential buildings are self-built without the use of scientific design and construction methods and as such they consume 2-2.5 times more than those in the developed countries located in the same latitude (Jiang et al. 2009). This means there is a great potential for energy saving if the buildings, as mentioned, are designed and constructed to a new performance standard. With the improvement of rural residents' living standards and the increase in commodity energy use, building energy consumption is rising year by year. The surveyed annual household income of rural residents in the severe cold regions is about 3200 dollars and by current world economic standard is comparatively low. It puts pressure on the rural residents in terms of the high energy costs.

Passive design concerns the planning, designing and environment configuration construction of buildings to improve and create a comfortable indoor environment that conforms to the natural principles of the sun, wind, temperature and humidity, in order to reduce the consumption of conventional energy (Mao, 2006). Scholars have carried out extensive research of passive design. Lam et al. (2006) used the bioclimatic method to study passive design strategies in China's five climate zones. The paper identified nine passive design strategy zones and proposed design strategies for both summer and winter. Gong et al. (2012) studied the interactive effects between passive design parameters in different Chinese areas and put forward the optimization strategies, and summarized seven passive design zones in a map and the neighbor cities can use the same optimal combinations. Morrissey et al. (2011) studied passive solar heat gain for residential buildings and concluded that passive solar design should be considered in the design phase. Results showed that higher building energy efficiency standards are easier to apply to smaller dwellings, and costs are lower. Qiu and Liu (2011) researched passive solar design for rural residential buildings in the Northeast region of China, expecting to reduce the construction cost of farmhouses as well as reduce building energy consumption.

However, the studies mentioned have not taken the difference in effect into consideration between heating season and non-heating season. In this study, the whole year will be divided into the heating season and non-heating season, and passive design will be studied from the point of effectiveness of 9 kinds of passive design strategies and design potential in rural residential buildings located in the severe cold regions of Northeast China. Above all, in the view of the different seasons, the research aims to establish that the application of passive design strategies that would make rural residential buildings green energy saving, which reduces the economic burden of energy consumption by the rural residents. It would also effectively improve the indoor thermally comfortable environment in order to realize the sustainable development of rural residential buildings in severe cold regions.

Burning coal is a universal phenomenon in the rural areas for heating, but it can cause global warming and pollution. Cristóbal et al. (2012) developed a systematic method to retrofit coal-fired plants, and the paper found that environmental indicators tend to behave in a non-conflictive manner in the pollution control. Zhang et al. (2012) studied how to reduce Chinese emissions of SO_2, NOx and fine particles, and finally, the paper presented a policy recommendation to make China`s air more clean. Nema et al. (2012) considered that burning coal can intensify climate change, and the paper illustrated the current situation of climate changing and causes. Barnes et al. (2013) studied the climate change and its impact on the environment in the view of respiratory and allergic disease, and the results showed that increased energy costs may lead to human source pollution.

BACKGROUND OF CLIMATE DIVISION OF CHINA

According to the "Thermal Design Code for Civil Buildings" (GB50176-93), China was divided into five Building thermal design partitions: severe cold region, cold region, hot summer and cold winter region, warm region and hot summer and warm winter region. The severe cold region is shown in Table 1. In this study, the severe cold region refers to the most area of Heilongjiang, Jilin, Liaoning province and Northeast of Inner Mongolia Autonomous Region, shown in Figure 1. According to the "Severe Cold and Cold Area Residential Building Energy Efficiency Design Standard" (JGJ26-2010), Chinese severe cold region subdivided into three climate zones under the heating Degree-days, as shown in Table 2.

Table 1. Building thermal design partition and design requirements (GB50176-93, 1993)

Name of the Partition	Partition Standard		Design Requirements
	Main Indicators	Auxiliary Indicators	
Severe Cold Region	The average temperature of the coldest month≤-10°C	The number of days when the average daily temperature is less than 5 degrees≥145d	Severe cold area

Figure 1. Map of building thermal design partition

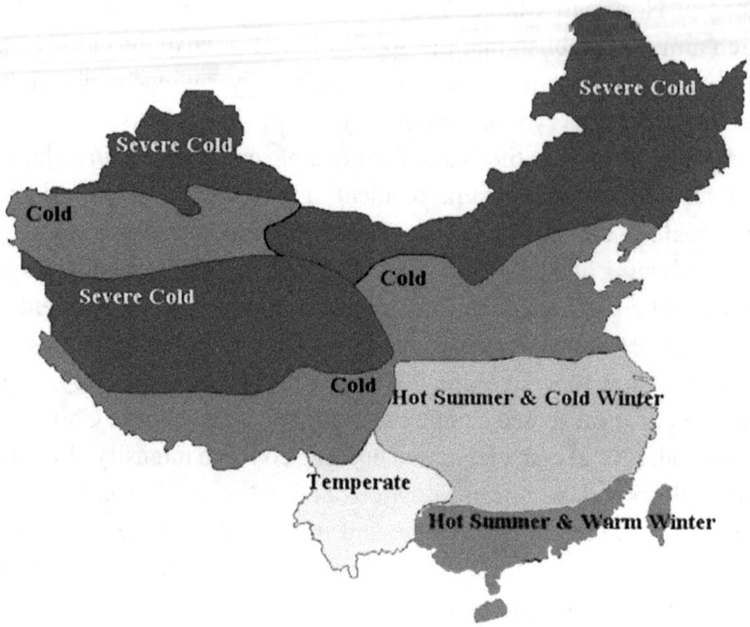

Table 2. The Climate sub-region of residential building energy saving design in severe cold region (JGJ26-2010, 2010)

Climate Sub-Region		Criterion of Partition
Severe Cold Region (Region I)	Region A	6000≤HDD18
	Region B	5000≤HDD18<6000
	Region C	3800≤HDD18<5000

Note: HDD18 refers to heating number (heating degree days based on 18 °C). In a year, when the daily outdoor average temperature below 18 degrees, multiplying the difference between the daily average temperature and 18 degrees by 1 day, then the cumulative value of all the year round comes to heating Degree-days of one year.

RESEARCH METHODOLOGY

Data Collection

The China meteorological administration monitors each city`s environment every year, and the Chinese National Standards use the data to guide architectural design in the area of building energy efficiency. The standard is named "Design code for heating ventilation and air conditioning of civil buildings" (GB50736-2012). As such, there will be accuracy in using authoritative data collected by the government.

Method of Data Analysis

The Climate Consultant 5.5 which is a kind of software used for passive building design strategy validity analysis developed by the University of California at Los Angeles (UCLA). It can express climate data graphically, such as temperature, humidity, wind speed, and it is used to analyze the effectiveness of the passive design, such as building heat storage, solar heat gain and natural ventilation through psychrometric chart to put forward the guidance on design. Climate Consultant 5.5 can read Climate data in CSWD, SWEAR, and in this study, the data will be presented in CSWD format, that is, the Chinese building thermal environment analysis special meteorological data set developed by the China meteorological administration and Tsinghua University. It based the meteorological data collected from 270 weather stations across the country from 1971 to 2003 on this typical meteorological data set.

Key Assumptions Used in the Analysis

In this study, there are three key assumptions:

1. The temperature and humidity of surrounding areas are same as typical city.
2. Thermal comfort of all rural residents is accord with ideal condition.
3. There is no cold air infiltration in the simulation model.

The National Weather Service set up 270 ground meteorological stations in China to collect over meteorological information such as temperature, humidity, wind speed, wind direction, rainfall, and each weather station can represent the main climatic features of the region and its surroundings. A village uses meteorological data of the nearest weather station because there is no weather station in each rural county.

In their daily lives, when the weather gets hotter, the residents open the windows. But when the weather gets colder, they wear more clothes. They have obvious adaptive thermal comfort behaviors. Furthermore, the perception of thermal comfort of rural residents may be impacted by age, gender, income and other factors, and different individual has different perception of thermal comfort. So the study use standard comfort data as the basis for judging in holistic approach without considering individual differences.

To study the effectiveness of each passive design accurately and quantify, Climate Consultant 5.5 use the ideal model without considering cold air infiltration. But building has cold air infiltration problems just as slit windows, doors, and other human activities switching behavior in actual environment. Therefore, it will be added to cold air infiltration coefficient as a model in future studies based on ideal model to make it closer to the actual situation.

Typical Cities

According to the three climate sub-areas in the severe cold region of Northeast China, four representative provinces of Heilongjiang, Jilin, Liaoning and the northeast of Inner Mongolia autonomous region have been selected with a total of 16 towns to represent the whole countryside of Northeast China. The purpose is to study the applicability and effectiveness of the nine kinds of passive design strategies as shown in Table 3. They are:

1. Sun shading of windows
2. High thermal mass
3. High thermal mass night flushed
4. Direct evaporative cooling
5. Two-stage evaporative cooling
6. Natural ventilation cooling
7. Passive solar direct gain low mass
8. Passive solar direct gain high mass
9. Wind protection of outdoor spaces

Thermal Comfort Zone

The simulation analysis used the California Energy Code Comfort Model which is a thermal comfort evaluation model of Climate Consultant 5.5. This model is based on the study of Auliciem, an Australian scholar, who devised the neutral temperature determination method and the equation is stated as: Tn = 17.6 + 0.314 Tm, where Tn is neutral temperature; and Tm is the average outdoor temperature. It has been revised to make it more conforming to the climatic characteristics of China, and the equation is changed to Tn = 19.7 + 0.3 Tm. The modified equation can be used to calculate the regional neutral temperature and thermal comfort range. The thermal comfort zones` boundary temperature is above or below the neutral temperature by 2-2.5°C. The maximum relative humidity of the thermal comfort range is not more than 84.6% (Milne et al., 1979). In this study, the thermal comfort correction model developed by Yang (2003), as shown in Table 4 is adopted.

Table 3. Typical cities in the four selected provinces in Northeast China

No.	Climatic Regions	Province	City Site	Latitude(N)	Longitude(E)	Altitude(m)	Heating degree-days(HDD18)	Air conditioning degree-days(CDD26)
17.	Severe cold region A	Hei Long Jiang	Mo He	52.13	122.52	433	7994	0
18.	Severe cold region A		Nen Jiang	49.17	125.23	242	6352	5
19.	Severe cold region B		Qiqihar	47.38	123.92	147	5259	23
20.	Severe cold region B		Harbin	45.75	126.77	142	5032	14
21.	Severe cold region B	Ji Lin	Dun Hua	43.37	128.20	524	5221	1
22.	Severe cold region C		Chang Chun	43.90	125.22	236	4642	12
23.	Severe cold region C		The former Guo Er Ross	45.08	124.87	136	4800	17
24.	Severe cold region C	Liao Ning	Lin Jiang	41.72	126.92	332	4736	4
25.	Severe cold region C		Zhang Wu	42.42	122.53	79	4134	13
26.	Severe cold region C		Shen Yang	41.77	123.43	44	3929	25
27.	Severe cold region C		Ben Xi	41.32	123.78	185	4046	16
28.	Severe cold region C		Kuan Dian	40.72	124.78	260	4095	4
29.	Severe cold region A	The Northeast of Inner Mongolia autonomous region	Hailar	49.22	119.75	610	6713	3
30.	Severe cold region A		Arshan	47.17	119.93	997	7364	0
31.	Severe cold region B		West Ujimqin	44.58	117.60	995	5812	4
32.	Severe cold region C		Tong Liao	43.60	122.27	178	4376	22

Table 4. Thermal comfort zones of typical cities in severe cold region of the Northeast China (°C)

No.	City Site	The Annual Average Temperature	The Neutral Temperature	Thermal Comfort Range
17.	Mo He	-3.5	18.7	16.2-21.2
18.	Nen Jiang	0.17	19.8	17.3-22.3
19.	Qiqihar	4.25	21	18.5-23.5
20.	Harbin	4	20.9	18.4-23.4
21.	Dun Hua	3.25	20.7	18.2-23.2
22.	Chang Chun	5.3	21.3	18.8-23.8
23.	The former Guo Er Ross	5.5	21.4	18.9-23.9
24.	Lin Jiang	3.58	20.8	18.3-23.3
25.	Zhang Wu	7.42	21.9	19.4-24.4
26.	Shen Yang	8.33	22.2	19.7-24.7
27.	Ben Xi	7.58	22	19.5-24.5
28.	Kuan Dian	7	21.8	19.3-24.3
29.	Hailar	-0.58	19.5	17-22
30.	Arshan	-2.25	19	16.5-21.5
31.	West Ujimqin	1.67	20.2	17.7-22.7
32.	Tong Liao	6.42	21.6	19.1-24.1

PASSIVE DESIGN STRATEGIES ANALYSIS

Passive design strategies contribute to improve the interior comfort conditions, increasing the energy efficiency in buildings and reducing their energy consumption (Rodriguez-Ubinas et al., 2014). Passive design is a sustainable design approach, and it can reduce the burning of non-renewable energy just as burning coal etc. and slow down greenhouse gas emissions using sun shading of windows, natural ventilation cooling and passive solar etc. The study analyzed 9 passive design strategies using Climate Consultant 5.5. The purpose is to find the passive design potential of buildings.

The Case of Heating and Non-Heating Seasons

According to the climate characteristics of severe cold regions of the Northeast China, the effectiveness of passive design strategies has been studied in the heating season and Non-heating season, respectively. The heating season lasts from October to April and the non-heating season from May to September.

From Table 5, the results showed that in the heating season, the comprehensive effectiveness and the passive design potential of the nine kinds of passive design strategies were both between 9% and 20%. Based on the mean of the representative cities of the three sub climatic zones, it was concluded that the comprehensive effectiveness and potential order of passive design strategies of the three zones were severe cold area C, severe cold area B and severe cold area A, as shown in Figure 2. In the heating season where there was passive solar direct gain with low/high thermal mass, the average contribution rate of passive solar direct gain with low mass is 8.78% and the passive solar direct gain with high mass is 5.86%. Therefore, strengthening the passive solar direct gain in the heating period should be stressed. It is worth noting that because the default model of Climate Consultant is a standard model, there has been no problem of doors and windows closed lax. Therefore, it means that the contribution rate is very low, only 1.3%, but in the actual case of the rural residence, due to the limited of economic conditions and low standard of design and construction of the buildings, cold air infiltration is very serious. As such, the prevention of infiltration is an important passive design strategy in those rural residential buildings.

From Table 6, the results showed that in the non-heating season, comprehensive effectiveness and the passive design potential of the nine kinds of passive design strategies were between 40% and 80%, which were far higher than that in the heating season. Based on the calculated mean of the representative cities of the three sub climate zones, the order of passive design potential were severe cold area A, severe cold area B and severe cold area C, as shown in Figure 3. Based on the calculated mean of the nine kinds of passive strategies in the 16 representative cities, respectively, the order of the single strategy contribution rate is sun shading of window (20.2%) > high passive solar direct gain with thermal mass (19.7%) > natural ventilation (17.7%) > high thermal mass with night ventilation (11.9%) > indirect evaporative cooling (9.4%) > passive solar direct gain with low thermal mass (8.7%) > direct evaporative cooling (7.8%) > wind protection of outdoor spaces (0.1%). From the analysis, it was concluded that whether in the heating season or in the non-heating season, passive solar direct gain is an important passive design strategy and priority should be given to it for the rural residential buildings.

The Case of 16 Cities

The effectiveness of passive design strategies was analyzed for 16 cities as city representatives of the three sub climate zones.

Table 5. Effectiveness evaluation of passive design strategies (%) in heating season (October - April)

	Sun Shading of Windows	High Thermal Mass	High Thermal Mass Night Flushed	Direct Evaporative Cooling	Two-Stage Evaporative Cooling	Natural Ventilation Cooling	Passive Solar Direct Gain Low Mass	Passive Solar Direct Gain High Mass	Wind Protection of Outdoor Spaces	Comprehensive Passive Design Measures	No Passive Design Measure	Passive Design Potential
Mo He	0.2	0	0	0	0	0	7.5	2.8	0.1	10.2	0.1	10.1
Nen Jiang	0.3	0.1	0.1	0.4	0.4	0.1	7.2	4.4	2.7	12.2	0.3	11.9
Qiqihar	0.5	0.1	0.1	0.3	0.3	0.1	9.5	6.3	0.2	16.4	0.4	16.0
Harbin	0.8	0.4	0.4	0.7	0.7	0.3	7.5	5.7	0.2	14.7	1.0	13.7
Dun Hua	0.6	0.1	0.1	0.4	0.4	0	6.5	4.0	0.5	11.4	0.6	10.8
Chang Chun	0.6	0	0	0.3	0.3	0	8.5	6.3	2.8	15.6	0.5	15.1
The former Guo Er Ross	0.3	0	0	0	0	0	10.7	7.4	0	18.6	0.8	17.8
Lin Jiang	0.6	0	0	0.1	0.1	0	8.8	6.0	0	15.4	0.8	14.6
Zhang Wu	0.6	0	0	0.1	0.1	0	8.8	6.6	3.7	16.3	1.2	15.1
Shen Yang	0.6	0	0	0.1	0.1	0	9.7	7.7	1.7	18.0	1.0	17.0
Ben Xi	0.5	0	0	0.1	0.1	0	9.0	7.2	0.1	16.7	0.8	15.9
Kuan Dian	0.2	0	0	0	0	0	9.2	5.2	0.1	14.5	0.3	14.2
Hailar	0.3	0	0	0	0	0	8.6	6.2	2.9	14.7	0.3	14.4
Arshan	0.3	0	0	0.1	0.1	0	6.1	3.2	2.2	9.3	0.2	9.1
West Ujimqin	0.3	0	0	0.2	0.2	0	11.4	7.2	1.3	18.3	0	18.3
Tong Liao	0.4	0	0	0	0	0	11.5	7.6	2.1	19.3	0.6	18.7

Figure 2. Comprehensive effectiveness and potential order of passive design strategies in the heating season

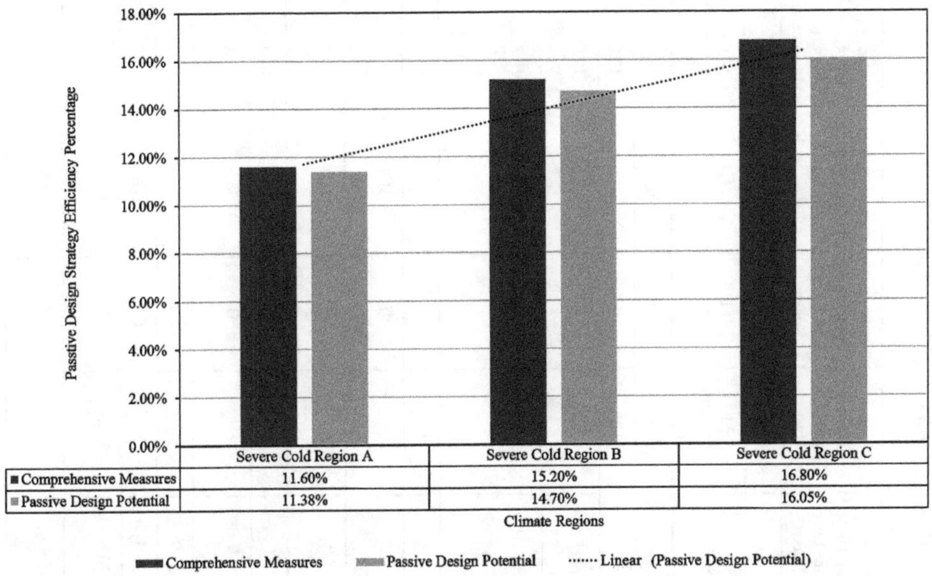

	Severe Cold Region A	Severe Cold Region B	Severe Cold Region C
■ Comprehensive Measures	11.60%	15.20%	16.80%
■ Passive Design Potential	11.38%	14.70%	16.05%

Figure 3. Comprehensive effectiveness and potential order of passive design strategies in the non-heating season

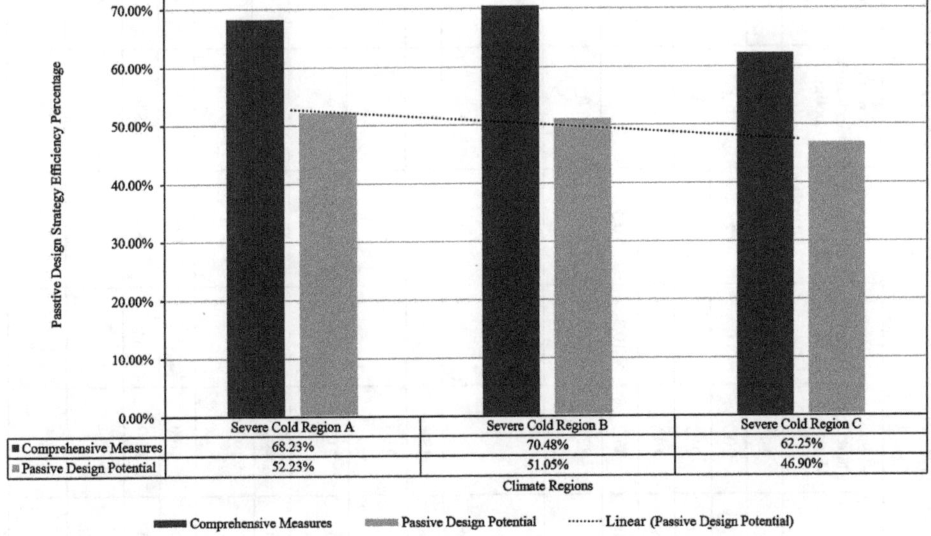

	Severe Cold Region A	Severe Cold Region B	Severe Cold Region C
■ Comprehensive Measures	68.23%	70.48%	62.25%
■ Passive Design Potential	52.23%	51.05%	46.90%

Table 6. Effectiveness evaluation of passive design strategies (%) in non-heating season (May-September)

	Sun Shading of Windows	High Thermal Mass	High Thermal Mass Night Flushed	Direct Evaporative Cooling	Two-Stage Evaporative Cooling	Natural Ventilation Cooling	Passive Solar Direct Gain Low Mass	Passive Solar Direct Gain High Mass	Wind Protection of Outdoor Spaces	Comprehensive Passive Design Measures	No Passive Design Measure	Passive Design Potential
Mo He	19.2	15.0	15.9	12.0	13.3	12.5	8.0	20.6	0	58.1	10.3	47.8
Nen Jiang	20.7	11.7	12.6	10.2	11.4	15.4	11.5	25.1	0.1	71.4	16.6	54.8
Qiqihar	21.2	11.4	13.7	9.2	11.2	22.5	8.9	17.7	0.2	72.8	20.2	52.6
Harbin	19.7	9.1	10.3	7.3	8.7	20.2	8.4	17.4	0.1	65.7	18.8	46.9
Dun Hua	17.7	10.3	10.3	7.3	8.3	14.2	8.2	19.9	0	61.0	17.0	44.0
Chang Chun	23.1	10.7	12.8	7.0	9.2	23.2	4.5	13.2	0.1	62.5	19.0	43.5
The former Guo Er Ross	21.1	9.4	11.0	6.6	8.7	21.4	9.5	18.3	0	68.7	18.2	50.5
Lin Jiang	18.6	9.9	11.1	5.0	6.5	13.7	6.5	16.5	0	55.8	14.2	41.6
Zhang Wu	21.5	12.6	13.6	7.0	9.5	21.9	4.4	15.9	0	61.7	15.4	46.3
Shen Yang	21.5	8.6	9.8	3.9	5.1	24.9	5.4	15.0	0	61.3	14.7	46.6
Ben Xi	20.1	10.8	12.8	4.9	7.7	22.6	5.0	14.5	0	63.8	18.5	45.3
Kuan Dian	19.2	8.2	8.2	2.6	4.0	19.8	6.0	18.1	0	58.3	14.4	43.9
Hailar	18.6	11.2	12.9	11.0	12.3	12.4	13.9	26.0	0.1	73.5	18.9	54.6
Arshan	18.0	8.0	8.1	8.9	9.2	8.1	16.3	28.2	0.2	69.9	18.2	51.7
West Ujimqin	20.3	10.7	12.0	11.7	12.5	10.2	16.8	32.2	0	82.4	21.7	50.7
Tong Liao	22.4	13.6	15.9	9.4	12.4	19.9	5.4	16.8	0.1	65.9	18.7	47.2

Mohe is located in severe cold area A and the annual average temperature is -3.5°C, the reading for passive design strategies in the psychrometric chart is in the form of hourly. Figure 4 shows that 86% of the outdoor dry bulb temperature is below 0°C in heating season, it can take 10.2% of the time to reach thermal comfort using passive design strategies, including 10.3% in contribution rate from passive solar direct gain low/high mass, 0.1% in contribution rate from wind protection of outdoor spaces. From Figure 5, 20% of the time in the range of 23~38°C, people feel uncomfortable. It can take 57.7% of the time to reach thermal comfort using passive design strategies. There was 19.2% in contribution rate from sun shading of windows, 14.1% in contribution rate from high thermal mass, 15% in contribution rate from high thermal mass night flushed, 11.7% in contribution rate from direct evaporative cooling, 12.8% in contribution rate from two-stage evaporative cooling, 14% in contribution rate from natural ventilation cooling, and 28.6% in contribution rate from passive solar direct gain low/high mass.

Nenjiang is located in severe cold area A and the annual average temperature is 0.17°C. Figure 5 shows that 79% of the outdoor dry bulb temperature is below 0°C in heating season, it can take 12.2% of the time to reach thermal comfort using pas-

Figure 4. Psychrometric chart of mohe in the heating and non-heating seasons

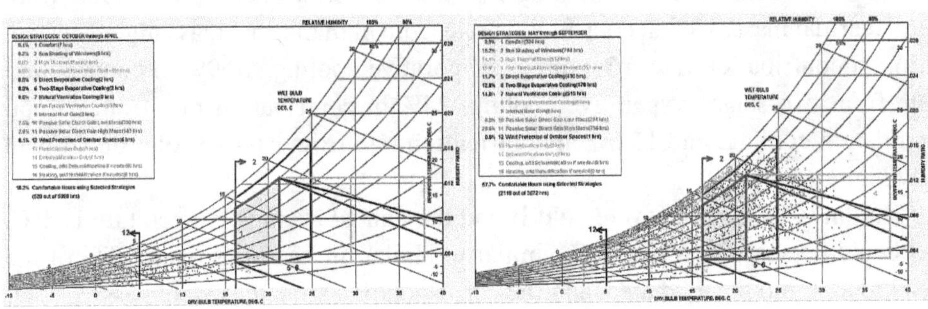

Figure 5. Psychrometric chart of Nenjiang in the heating and non-heating seasons

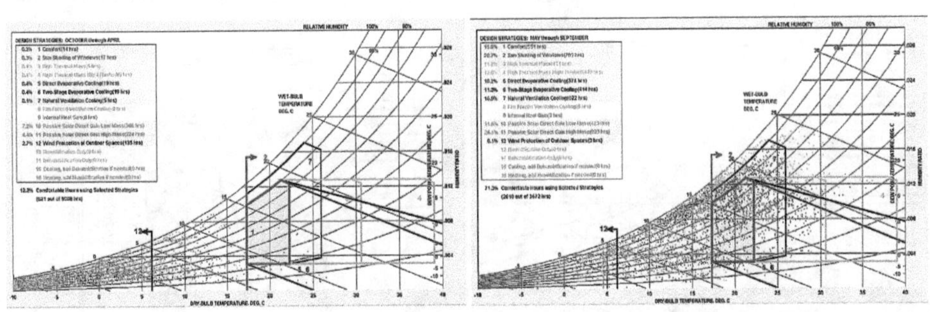

sive design strategies, including 11.6% in contribution rate from passive solar direct gain low/high mass, 0.1% in contribution rate from high thermal mass, 2.7% in contribution rate from wind protection of outdoor spaces. From Figure 5, 19% of the time in the range of 23~38°C, people feel uncomfortable. It can take 71.3% of the time to reach thermal comfort using passive design strategies. There was 20.7% in contribution rate from sun shading of windows, 11.2% in contribution rate from high thermal mass, 12% in contribution rate from high thermal mass night flushed, 10.2% in contribution rate from direct evaporative cooling, 11.3% in contribution rate from two-stage evaporative cooling, 16.9% in contribution rate from natural ventilation cooling, and 36.6% in contribution rate from passive solar direct gain low/high mass.

Qiqihar is located in severe cold area B and the annual average temperature is 4.25°C. Figure 6 shows that 69% of the outdoor dry bulb temperature is below 0°C in heating season, it can take 16.4% of the time to reach thermal comfort using passive design strategies, including 15.8% in contribution rate from passive solar direct gain low/high mass, 0.1% in contribution rate from high thermal mass, 0.2% in contribution rate from wind protection of outdoor spaces. From Figure 6, 24% of the time in the range of 24~38°C, people feel uncomfortable. It can take 72.8% of the time to reach thermal comfort using passive design strategies. There was 21.2% in contribution rate from sun shading of windows, 10.9% in contribution rate from high thermal mass, 13% in contribution rate from high thermal mass night flushed, 9.2% in contribution rate from direct evaporative cooling, 10.9% in contribution rate from two-stage evaporative cooling, 24% in contribution rate from natural ventilation cooling, and 16.6% in contribution rate from passive solar direct gain low/high mass.

Harbin is located in severe cold B and the annual average temperature is 4°C. As could be seen from Figure 7, similar to Hailar, in heating season, 85% of the reading fell into the cold area, and the corresponding strategies were passive solar direct gain with high/low thermal mass, wind protection of outdoor spaces, includ-

Figure 6. Psychrometric chart of Qiqihar in the heating and non-heating seasons

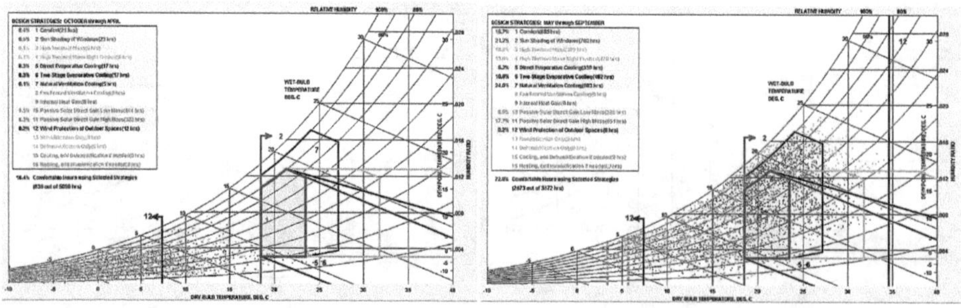

Figure 7. Psychrometric chart of Harbin in the heating and non-heating seasons

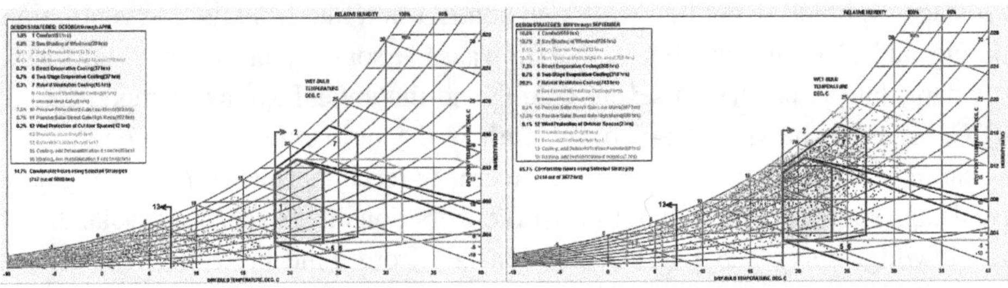

ing 13.2% in contribution rate of passive solar direct gain. From Figure 7, the passive design strategies applied in the heating season could achieve thermal comfort, giving 66% comfort hours without energy consuming. And 34% of non-comfort hours was scattered in cold area, wet area and hot area. Thus, the potential application of natural ventilation to improve comfortable hours was 20.2%, and the application potential of high thermal mass and high thermal mass with night ventilation to improve comfortable hours were 9.1% and 10.3%, respectively.

Dunhua is located in severe cold B and the annual average temperature is 3.25°C. Figure 8 shows that 70% of the outdoor dry bulb temperature is below 0°C in heating season, it can take 11.4% of the time to reach thermal comfort using passive design strategies, including 10.5% in contribution rate from passive solar direct gain low/high mass, 0.1% in contribution rate from high thermal mass, 0.5% in contribution rate from wind protection of outdoor spaces. From Figure 8, 12% of the time in the range of 23~38°C, people feel uncomfortable. It can take 60.9% of the time to reach thermal comfort using passive design strategies. There was 17.7% in contribution rate from sun shading of windows, 9.6% in contribution rate from high thermal mass, 9.6% in contribution rate from high thermal mass night flushed, 7.3%

Figure 8. Psychrometric chart of Dunhua in the heating and non-heating seasons

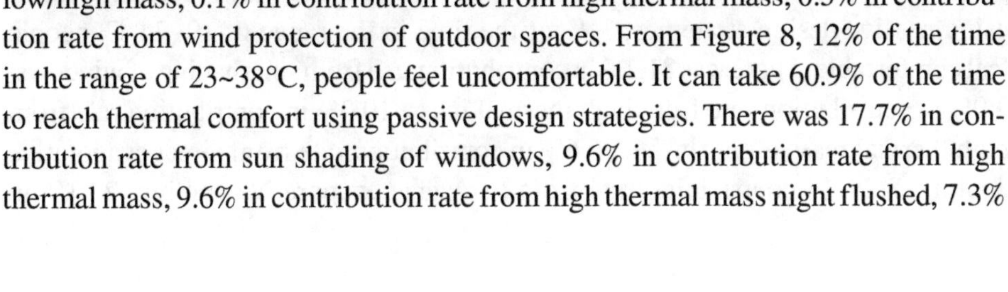

in contribution rate from direct evaporative cooling, 8% in contribution rate from two-stage evaporative cooling, 15.8% in contribution rate from natural ventilation cooling, and 28.1% in contribution rate from passive solar direct gain low/high mass.

Changchun is located in severe cold C and the annual average temperature is 5.3°C. Figure 9 shows that 64% of the outdoor dry bulb temperature is below 0°C in heating season, it can take 15.6% of the time to reach thermal comfort using passive design strategies, including 14.8% in contribution rate from passive solar direct gain low/high mass, 2.8% in contribution rate from wind protection of outdoor spaces. From Figure 9, 20% of the time in the range of 23~38°C, people feel uncomfortable. It can take 62.3% of the time to reach thermal comfort using passive design strategies. There was 23.1% in contribution rate from sun shading of windows, 9.6% in contribution rate from high thermal mass, 11.5% in contribution rate from high thermal mass night flushed, 7% in contribution rate from direct evaporative cooling, 8.7% in contribution rate from two-stage evaporative cooling, 24.6% in contribution rate from natural ventilation cooling, and 17.7% in contribution rate from passive solar direct gain low/high mass.

Qian Gorlos is located in severe cold C and the annual average temperature is 5.5°C. Figure 10 shows that 65% of the outdoor dry bulb temperature is below 0°C in heating season, it can take 18.6% of the time to reach thermal comfort using passive design strategies, including 18.1% in contribution rate from passive solar direct gain low/high mass. From Figure 10, 23% of the time in the range of 24~38°C, people feel uncomfortable. It can take 68.6% of the time to reach thermal comfort using passive design strategies, 31% of the time in the range of 19~24°C, people feel comfortable. There was 21.1% in contribution rate from sun shading of windows, 8.9% in contribution rate from high thermal mass, 10.3% in contribution rate from high thermal mass night flushed, 6.6% in contribution rate from direct evaporative cooling, 8.2% in contribution rate from two-stage evaporative cooling, 22.1% in

Figure 9. Psychrometric chart of Changchun in the heating and non-heating seasons

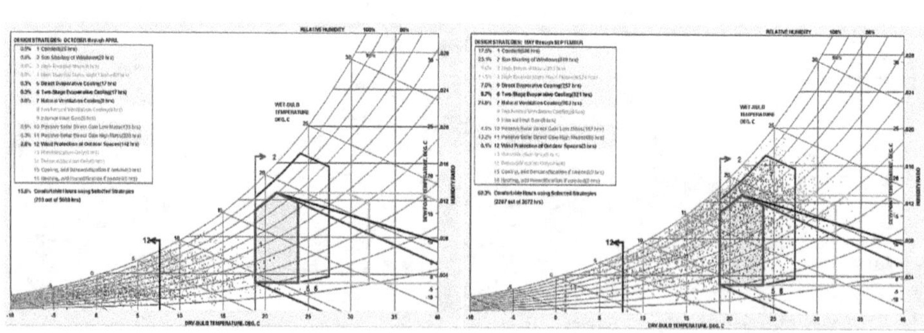

Figure 10. Psychrometric chart of Qian Gorlos in the heating and non-heating seasons

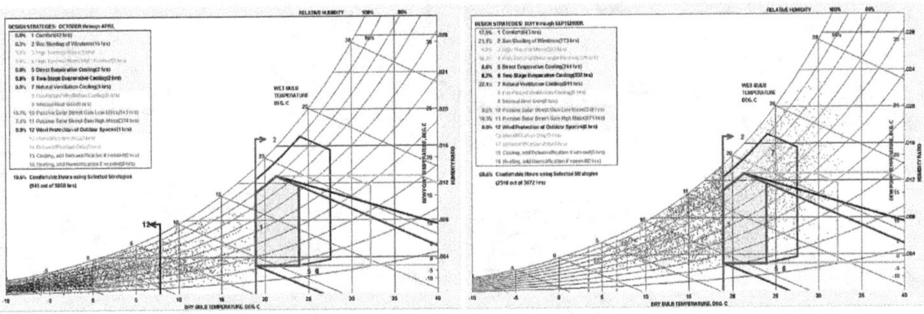

contribution rate from natural ventilation cooling, and 27.8% in contribution rate from passive solar direct gain low/high mass.

Linjiang is located in severe cold C and the annual average temperature is 3.58°C. Figure 11 shows that 60% of the outdoor dry bulb temperature is below 0°C in heating season, it can take 15.4% of the time to reach thermal comfort using passive design strategies, including 14.8% in contribution rate from passive solar direct gain low/high mass. From Figure 11, 19% of the time in the range of 23~38°C, people feel uncomfortable. It can take 55.3% of the time to reach thermal comfort using passive design strategies. There was 18.6% in contribution rate from sun shading of windows, 9.1% in contribution rate from high thermal mass, 9.9% in contribution rate from high thermal mass night flushed, 5% in contribution rate from direct evaporative cooling, 6.3% in contribution rate from two-stage evaporative cooling, 15.1% in contribution rate from natural ventilation cooling, and 23% in contribution rate from passive solar direct gain low/high mass.

Zhangwu is located in severe cold C and the annual average temperature is 7.42°C. Figure 12 shows that 57% of the outdoor dry bulb temperature is below

Figure 11. Psychrometric chart of Linjiang in the heating and non-heating seasons

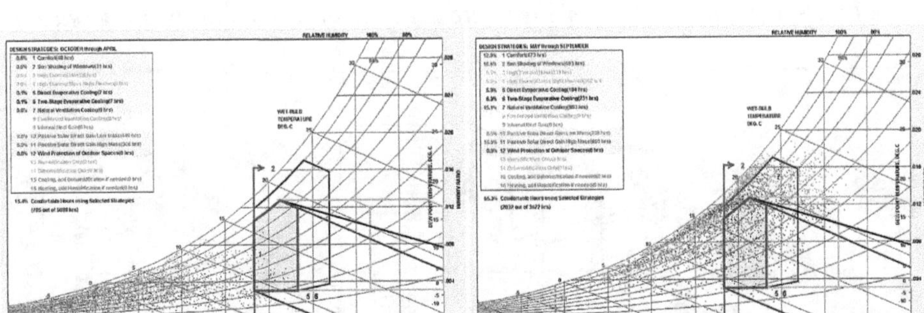

Figure 12. Psychrometric chart of Zhangwu in the heating and non-heating seasons

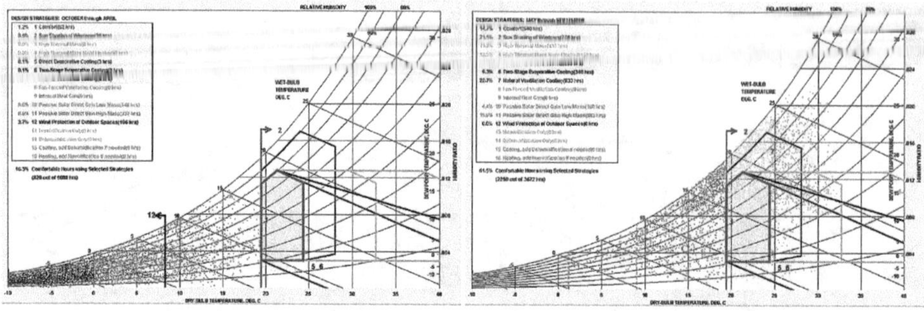

0°C, 41% of the outdoor dry bulb temperature between 0~20°C in heating season, it can take 16.3% of the time to reach thermal comfort using passive design strategies, including 15.4% in contribution rate from passive solar direct gain low/high mass, 3.7% in contribution rate from wind protection of outdoor spaces. From Figure 12, 24% of the time in the range of 25~38°C, people feel uncomfortable. It can take 61.5% of the time to reach thermal comfort using passive design strategies, 36% of the time in the range of 20~25°C, people feel comfortable. There was 21.5% in contribution rate from sun shading of windows, 11.8% in contribution rate from high thermal mass, 12.5% in contribution rate from high thermal mass night flushed, 7% in contribution rate from direct evaporative cooling, 9.3% in contribution rate from two-stage evaporative cooling, 22.7% in contribution rate from natural ventilation cooling, and 20.3% in contribution rate from passive solar direct gain low/high mass.

Shen Yang is located in severe cold area C. The annual average temperature is 8.33°C. From Figure 13, it could be seen that in the heating season, 82% of the reading fell within the cold area, and the application potential of passive solar direct gain with low/high thermal mass were 9.7% and 7.7%, respectively. From Figure

Figure 13. Psychrometric chart of Shenyang in the heating and non-heating seasons

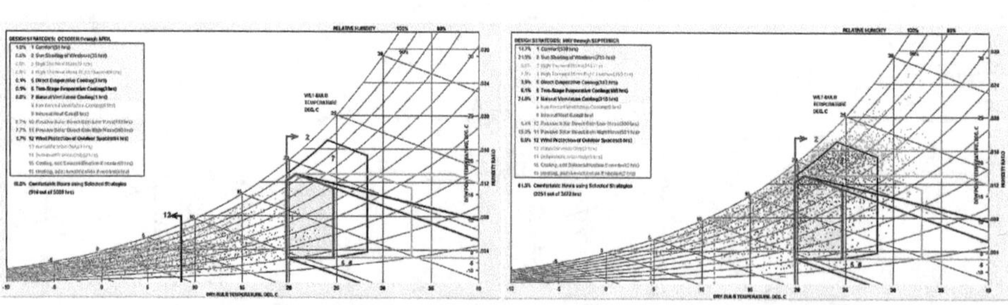

13, the results also showed that the passive design strategies applied could achieve 61% thermal comfort hours in the heating season while sun shading of window accounted for 21.5%, and the contribution rate of natural ventilation was 24.9%.

Benxi is located in severe cold C and the annual average temperature is 7.58°C. Figure 14 shows that 54% of the outdoor dry bulb temperature is below 0°C, 45% of the outdoor dry bulb temperature between 0~20°C in heating season, it can take 16.7% of the time to reach thermal comfort using passive design strategies, including 16.2% in contribution rate from passive solar direct gain low/high mass, 0.1% in contribution rate from wind protection of outdoor spaces. From Figure 14, 22% of the time in the range of 25~38°C, people feel uncomfortable. It can take 63.5% of the time to reach thermal comfort using passive design strategies, 35% of the time in the range of 20~25°C, people feel comfortable. There was 20.1% in contribution rate from sun shading of windows, 9.7% in contribution rate from high thermal mass, 11.5% in contribution rate from high thermal mass night flushed, 4.9% in contribution rate from direct evaporative cooling, 7.2% in contribution rate from two-stage evaporative cooling, 23.3% in contribution rate from natural ventilation cooling, and 19.5% in contribution rate from passive solar direct gain low/ high mass.

Kuandian is located in severe cold C and the annual average temperature is 7°C. Figure 15 shows that 56% of the outdoor dry bulb temperature is below 0°C, 43% of the outdoor dry bulb temperature between 0~20°C in heating season, it can take 14.5% of the time to reach thermal comfort using passive design strategies, including 14.4% in contribution rate from passive solar direct gain low/high mass, 0.1% in contribution rate from wind protection of outdoor spaces. From Figure 15, 15% of the time in the range of 25~38°C, people feel uncomfortable. It can take 58% of the time to reach thermal comfort using passive design strategies, 37% of the time in the range of 20~25°C, people feel comfortable. There was 19.2% in contribution rate from sun shading of windows, 6.8% in contribution rate from high thermal

Figure 14. Psychrometric chart of Benxi in the heating and non-heating seasons

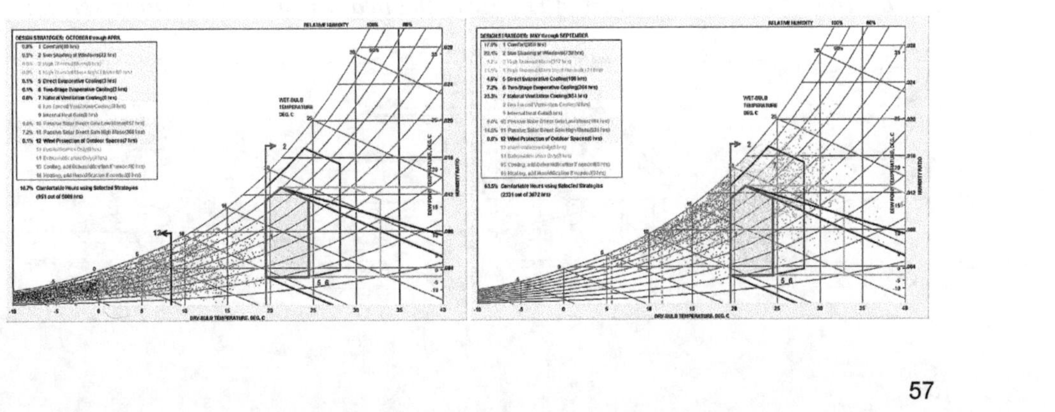

Figure 15. Psychrometric chart of Kuandian in the heating and non-heating seasons

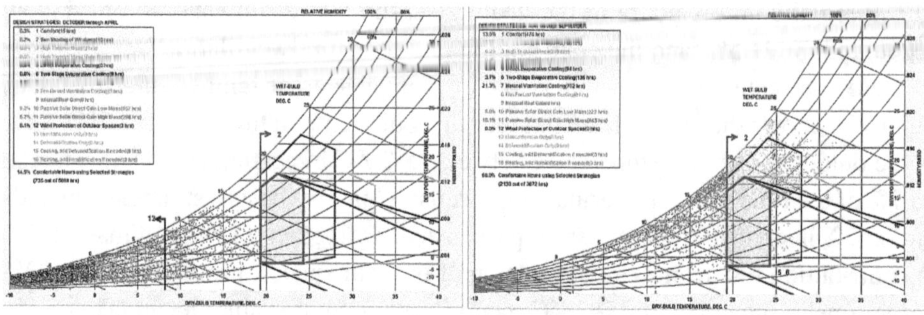

mass, 6.8% in contribution rate from high thermal mass night flushed, 2.6% in contribution rate from direct evaporative cooling, 3.7% in contribution rate from two-stage evaporative cooling, 21.3% in contribution rate from natural ventilation cooling, and 24.1% in contribution rate from passive solar direct gain low/high mass.

Hailar is located in severe cold area A. The annual average temperature is 0.58°C, it could be seen from Figure 16 that in the heating season, 85% of the reading fell within the cold area, and the corresponding strategies were solar direct gain with high/low thermal mass, wind protection of outdoor spaces, including 14.8% in contribution rate from passive solar direct gain. As could be seen also from Figure 16, the passive design strategies applied in heating season could achieve 74% of comfort hours without energy consumption. In addition, 26% of non-comfort hours was scattered in cold area, wet area and hot area. Wet area and hot area can be achieved thermal comfort by natural ventilation, direct evaporative cooling and two-stage evaporative cooling, sun shading of window, high thermal mass and high thermal mass with night ventilation. Among them, the natural ventilation accounted for 12.4%. In the cold area, it could be improved by passive solar direct gain with low/high thermal mass and their contribution rates were 13.9% and 26%, respectively.

Figure 16. Psychrometric chart of Hailar in the heating and non-heating seasons

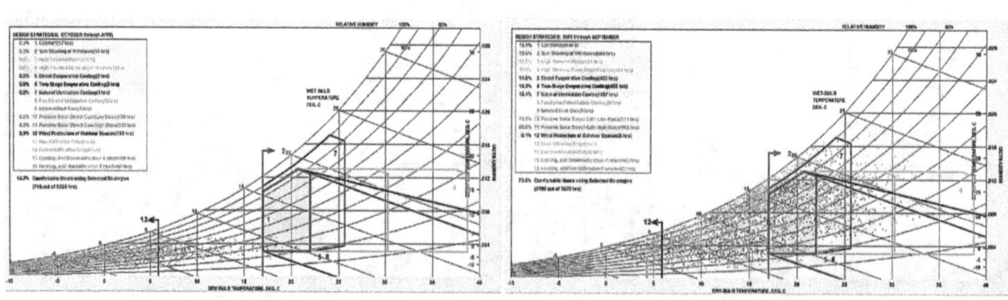

Arxan is located in severe cold A and the annual average temperature is -2.25°C. Figure 17 shows that 86% of the outdoor dry bulb temperature is below 0°C, 14% of the outdoor dry bulb temperature is between 0~17°C in heating season, it can take 9.3% of the time to reach thermal comfort using passive design strategies, including 9.3% in contribution rate from passive solar direct gain low/high mass, 2.2% in contribution rate from wind protection of outdoor spaces. From Figure 17, 9% of the time in the range of 22~38°C, people feel uncomfortable. It can take 69.9% of the time to reach thermal comfort using passive design strategies, 22% of the time in the range of 17~22°C, people feel comfortable. There was 18% in contribution rate from sun shading of windows, 8% in contribution rate from high thermal mass, 8.1% in contribution rate from high thermal mass night flushed, 8.9% in contribution rate from direct evaporative cooling, 9.2% in contribution rate from two-stage evaporative cooling, 9.4% in contribution rate from natural ventilation cooling, and 44.5% in contribution rate from passive solar direct gain low/high mass.

Xiwuzhumuqin Banner is located in severe cold A and the annual average temperature is 1.67°C. Figure 18 shows that 77% of the outdoor dry bulb temperature is below 0°C, 23% of the outdoor dry bulb temperature is between 0~18°C in heating season, it can take 18.3% of the time to reach thermal comfort using passive design strategies, including 18.6% in contribution rate from passive solar direct gain low/high mass, 1.3% in contribution rate from wind protection of outdoor spaces. From Figure 18, 13% of the time in the range of 23~38°C, people feel uncomfortable. It can take 82.4% of the time to reach thermal comfort using passive design strategies, 26% of the time in the range of 18~23°C, people feel comfortable. There was 20.3% in contribution rate from sun shading of windows, 10.5% in contribution rate from high thermal mass, 11.8% in contribution rate from high thermal mass night flushed, 11.7% in contribution rate from direct evaporative cooling, 12.3% in contribution rate from two-stage evaporative cooling, 11.6% in contribution rate

Figure 17. Psychrometric chart of Arxan in the heating and non-heating seasons

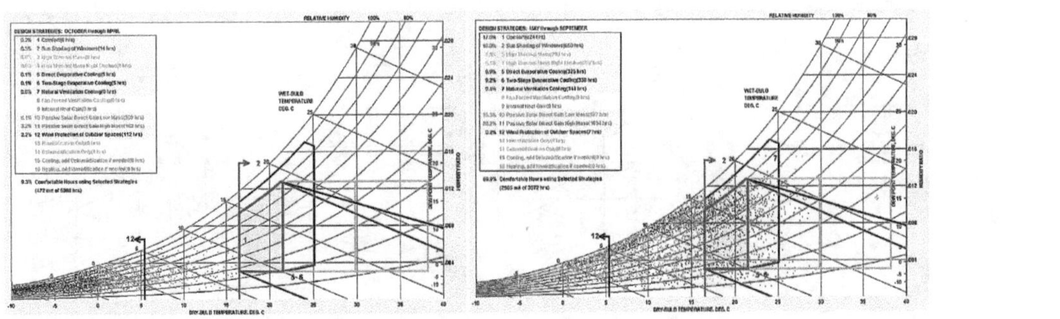

Figure 18. Psychrometric chart of Xiwuzhumuqin Banner in the heating and non-heating seasons

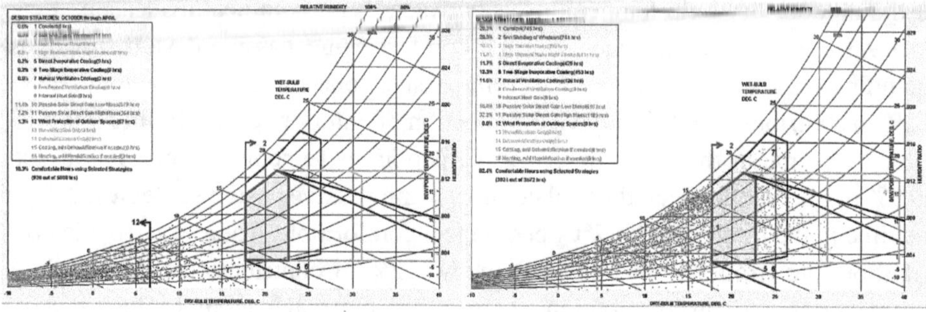

from natural ventilation cooling, and 49% in contribution rate from passive solar direct gain low/high mass.

Tongliao is located in severe cold c and the annual average temperature is 6.42°c. figure 19 shows that 61% of the outdoor dry bulb temperature is below 0°C, 38% of the outdoor dry bulb temperature is between 0~20°C in heating season, it can take 19.3% of the time to reach thermal comfort using passive design strategies, including 19.1% in contribution rate from passive solar direct gain low/high mass, 2.1% in contribution rate from wind protection of outdoor spaces. From Figure 19, 25% of the time in the range of 25~38°C, people feel uncomfortable. It can take 65.8% of the time to reach thermal comfort using passive design strategies, 33% of the time in the range of 20~25°C, people feel comfortable. There was 22.4% in contribution rate from sun shading of windows, 12.8% in contribution rate from high thermal mass, 14.9% in contribution rate from high thermal mass night flushed, 9.4% in contribution rate from direct evaporative cooling, 12% in contribution rate

Figure 19. Psychrometric chart of Tongliao in the heating and non-heating seasons

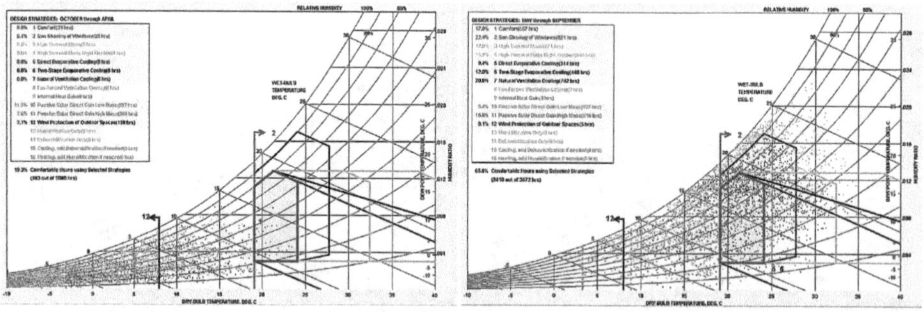

from two-stage evaporative cooling, 20.8% in contribution rate from natural ventilation cooling, and 22.2% in contribution rate from passive solar direct gain low/high mass.

DISCUSSION OF FINDINGS

Through analyzing the effectiveness of the nine kinds of passive strategies used in 16 cities of the four provinces in the severe cold regions of Northeast China, it has explored suitable passive design strategies for rural residential buildings. The aim was to achieve the goal of rural residential buildings through a new standard of design, construction and operation that provides economy, comfort living and are ecologically energy saving.

In the first finding, there was 9% ~ 20% comprehensive effectiveness and potential of these passive design strategies in the severe cold regions and the order of the three sub climate zones was region C, region B and region A.

In the non-heating season, the comprehensive effectiveness and the passive design potential of nine kinds of passive design strategies was 40% ~ 80%, and its order was region A, region B and region C. At the same time, the comprehensive effectiveness and potential of passive design strategies in non-heating season had been found to be greater than that in the heating season. As such, the government needs to emphasize passive design when they formulate standards of building energy efficiency in the severe cold regions of Northeast China. The government should identify and give preferential policies to the buildings that using passive design strategies. The government should give incentives to encourage the development of passive design just like energy efficiency building honor, reducing taxes, giving the volume rate subsidies etc.

In terms of policies for passive building energy efficiency, government should set policies based on the economic and climatic situations of each province and considering habits of local residents, promoting the development of building energy efficiency in many aspects from settlements planning, building design, architectural detail, energy-saving equipment and energy habits.

In terms of recognizing the benefits of passive designed building, special awards could be established at the national and provincial levels. The government could select outstanding passive buildings in the country every year, and award stakeholders with due recognition in order to promote the development of energy-efficient building through passive building designs.

In terms of tax rate, the government could respond by giving passive building owners energy certification merits, and award different tax rates to them based on the assessment results. For instance, when a building saves more energy, the greater the tax benefits to encourage architects and developers to build passive energy efficiency buildings.

In terms of volume rate, the government could evaluate the passive buildings and award different volume rates to encourage developers to continue to build passive designed buildings.

In the second finding, whether in heating season or in non-heating season, passive solar direct gain had been found to be a very important passive design strategy. Its average contribution rate was 14.64% in the heating season, and 28.38% in the non-heating season. The government should vigorously promote the development and application of clean energy like solar energy because it can improve air quality and reduce rural residents' economic burden. The rural areas of China is a good place to use solar energy because it has vast terrain and plenty of sunshine. As such, the government should give sufficient economic subsidies to those rural residents who buy solar equipment in order to promote solar energy applications.

FUTURE DIRECTIONS OF THE RESEARCH

As future directions, the study will involve conducting in-depth research in the following areas:

1. The team will continue to research on passive design strategies and application of rural residential buildings in severe cold regions of northeast China, as well as conduct research on the potential and economical aspect of passive design strategies in the same latitude regions of the world.
2. The contribution of passive solar heat gain is large in the rural residential buildings to achieve energy efficiency. The team will study how to use passive solar heat gain more efficiently, and make buildings store and release solar energy better. Furthermore, it needs to reduce the cost of ancillary equipment for universal applicability.
3. Cold air infiltration acts against building energy efficiency. In the actual environment, cold air infiltration is almost inevitable and, therefore, in future studies, the team will explore how to increase air tightness of the building from the perspective of architectural design to better ensure that indoor air changes and air quality meet the national standards, maximizing the reduction of cold air infiltration which has adverse effects on building energy efficiency.
4. In rural areas of northeast China, burning coal to get warm is the main form of heat. But the problem is that coal combustion produces large amounts of gas like CO_2, SO_2, which can increase greenhouse effects that are harmful to

human health. Therefore, in future studies, the team will focus on exploring clean energy heating to replace coal in order to ease the greenhouse effect and protect human health.

5. The lack of building energy databases is an existing issue for accurate predictions of future trends of energy-efficient buildings. Hence, in order to better provide energy-saving building design recommendations, future studies of the team will establish a database of rural areas in northeast China to detect and predict building energy consumption trends.

CONCLUSION

Rural residential building design in severe cold regions of northeast China should fully consider climate features, rural economic levels and living habits. Using passive design strategies would reduce energy consumption of rural residential buildings and make rural residences more liveable, healthy and comfortable. Passive design strategies, from the consideration of cold adaptation, wind adaptation and light adaptation, put up with a proposal of energy-saving design of using passive solar direct gain contributes to 9% energy saving in heating season and 20% in non-heating season, and natural ventilation contributes to 18% energy saving in non-heating season. Generally speaking, the integrated use of passive design strategies can contribute to 15% energy savings in the heating season and 49% in the non-heating season. Hence, there is evidence to show that passive design strategies, adapted to climatic differences and natural environment will promote energy saving and sustainable development of rural residential buildings in severe cold regions of northeast China.

Building energy efficiency design can achieve significant development through the use of IT. The main benefits are in three aspects:

1. The use of IT ensures there is precision in the results obtained from the experiments carried out, improves the productivity and efficiency of researchers, and reduces research costs as direct benefits from using the EEB simulation software like Climate Consultant, EnergyPlus, DeST, and others.
2. Researchers can detect real-time air quality and global warming through the use of appropriate IT tools. The technology provided detailed data for accurate and efficient studying of global warming and air pollutions.
3. Architects can achieve integrated design and construction of multi-disciplinary domains like architecture, structures, water supply and drainage, HVAC and electrical services with an IT application such as BIM, and thereby able to improve on the quality of construction and at the same time reduce construction costs.

ACKNOWLEDGMENT

The research is wholly funded by the Chinese "twelfth five-year" national science and technology grant under the project name: Comprehensive Technology Demonstration of Green Village Construction in Severe Cold Regions of Northeast China (Number: 2013BAJ12B04).

REFERENCES

Barnes, C. S., Alexis, N. E., Bernstein, J. A., Cohn, J. R., Demain, J. G., Horner, E., & Phipatanakul, W. et al. (2013). Climate change and our environment: The effect on respiratory and allergic disease. *The Journal of Allergy and Clinical Immunology: In Practice*, *1*(2), 137–141. doi:10.1016/j.jaip.2012.07.002 PMID:23687635

Civil Building Thermal Design Code GB50176-93. (1993). Beijing: China Building Industry Press.

Cristóbal, J., Guillén-Gosálbez, G., Jiménez, L., & Irabien, A. (2012). Optimization of global and local pollution control in electricity production from coal burning. *Applied Energy*, *92*, 369–378. doi:10.1016/j.apenergy.2011.11.028

Design Standard for Energy Efficiency of Residential Buildings in Severe Cold and Cold Zones JGJ26-2010. (2010). Beijing: China Building Industry Press.

Gong, X. Z., Akashi, Y., & Sumiyoshi, D. (2012). Optimization of passive design measures for residential buildings in different Chinese areas. *Building and Environment*, *58*, 46–57. doi:10.1016/j.buildenv.2012.06.014

Jiang, Y., Lin, B. R., Zeng, J. L., & Zhu, Y. X. (2009). *Energy Efficiency of Residence* (Vol. 7). Beijing: China Building Industry Press.

Lam, J. C., Yang, L., & Liu, J. P. (2006). Development of passive design zones in China using bioclimatic approach. *Energy Conversion and Management*, *47*(6), 746–762. doi:10.1016/j.enconman.2005.05.025

Mao, Y. (2006). Study on Climate Adaptability of Human Beings to Thermal Comfort in China (Vol. 105). Xi'an: Xi'an University of Architecture and Technology.

Milne, M., Givoni, B., & Watson, D. (1979). *Energy Conservation through Building Design*. London: McGraw-Hill.

Morrissey, J., Moore, T., & Horne, R. E. (2011). Affordable passive solar design in a temperate climate: An experiment in residential building orientation. *Renewable Energy*, *36*(2), 568–577. doi:10.1016/j.renene.2010.08.013

Nema, P., Nema, S., & Roy, P. (2012). An overview of global climate changing in current scenario and mitigation action. *Renewable & Sustainable Energy Reviews*, *16*(4), 2329–2336. doi:10.1016/j.rser.2012.01.044

Qiu, J., & Liu, H. M. (2011). Northeast China rural residential passive solar design exploration. In *Proceedings of International Conference of Green Building Materials and Energy-saving Construction (GBMEC 2011)*. Harbin, People's Republic of China (pp. 200-203). doi:10.4028/www.scientific.net/AMR.280.200

Rodriguez-Ubinas, E., Montero, C., Porteros, M., Vega, S., Navarro, I., Castillo-Cagigal, M., & Gutiérrez, A. et al. (2014). Passive design strategies and performance of Net Energy Plus Houses. *Energy and Building*, *83*, 10–22. doi:10.1016/j.enbuild.2014.03.074

Yang, L. (2003). *Building Climate Analysis and Design Strategies. Xi'an*. Xi'an University of Architecture and Technology.

Zhang, Q., He, K., & Huo, H. (2012). Policy: Cleaning China's air. *Nature*, *484*(7393), 161–162. PMID:22498609

Chapter 3
Smart Homes and Sustainable Energy in Nigeria

Oluwasola Oni
Pan-Atlantic University, Nigeria

ABSTRACT

In many developing countries, there is limited access to electricity with the populace typically resorting to purchasing personal electricity generating units. This research explores the factors affecting the possibility of introducing smart homes that would conserve electricity, reduce a need for reliance on the national grid and lower energy costs. Advantages and disadvantages, as well as the impact on society and development in general are examined in addition to future perspectives on smart homes and sustainable energy.

INTRODUCTION

While sustainable energy has been researched and applied in developed countries with stable access to electricity, there has been limited research into such possibilities in the developing world in general and Africa in particular. A case can be made for the need of such innovation as access to power is limited and in many situations, unaffordable for the average person living in certain parts of the continent.

Many developing countries struggle with the generation and supply of power. Nigeria is one of such countries beleaguered with poor power supply and limited

DOI: 10.4018/978-1-5225-0302-6.ch003

access as currently, about 48% of Nigerians have access to electricity (World Bank, 2015). The country's population stands at approximately 160 million (United Nations, 2012) with a peak electricity output of over 4,000MW and per capita consumption just 3 per cent that of South Africa (Financial Times, 2012), one can conclude that this is insufficient to support the power needs of the growing and developing population. Subsequently, there are frequent power outages as a result of the limited supply. According to the Financial Times (2012), despite having large reserves of natural gas that can fire thermal plants, Nigeria's electricity supply and service is among the world's worst, with half of the population lacking access to the grid.

In addition to the limited supply available to many Nigerians, disposable income is also a problem that may hinder access to electricity. According to the Central Intelligence Agency (2009) in comparison to the rest of the world, 70% of Nigerians live below the poverty line. Higher income earners also have to deal with rising costs of energy costs for cooling and ventilation and an increasing amount of energy consuming appliances such as white goods and consumer electronic devices, which means that hidden saving potentials can be found in numerous areas of a modern smart home (Kolfer &Kastner, 2010). Furthermore, many sectors have responded to the need to reduce their levels of energy consumptions. However, in residential homes, sustainability is a relatively new phenomenon as there is a lack of sufficient technological advice for home owners (Reinisch, Kofler, Iglesias & Kastner, 2011).

As a result of the aforementioned challenges, this chapter aims to propose a framework that will identify factors affecting the adoption of smart home from the perspectives of different parties. This would allow for the exploration of the degree of openness towards adopting smart home technology enabling home owners to reduce their energy consumption, reduce a need for reliance on the national grid and lower energy costs. The chapter is structured as follows. This section introduces the research while the next session provides a background to the research area. The main focus of the chapter is discussed next, followed by the issues and recommendations arising. The future research directions are then discussed followed eventually by the conclusion.

SUSTAINABILITY AND SMART HOMES

Omer (2007) argues that the energy sector is encountering difficulties because increased production and consumption levels cause higher levels of pollution and in due course, climate change with possibly disastrous consequences. This implies that an alternative to excessive consumption must be sought to avoid such consequences. As a result, sustainable energy is a phenomenon that any country cannot afford to ignore. However, what constitutes sustainable energy may be subject to various interpretations.

Although sustainable energy may mean different things to different people, in this research, sustainable energy would be taken to mean an effective provision of energy that would meet the needs of the future without compromising the ability of future generations to meet their own needs (Obrecht & Denac, 2013). Sustainable energy comprises two key components which include efficient energy consumption and renewable energy sources (RES). It requires making a conscious effort to select appliances, the duration and modes of their use with the aim of decreasing overall consumption (Fischer, 2007)

Current research within Nigeria has also considered renewable and sustainable energy but mostly from the angle of the possibility of integrating smart grid technology into the Nigerian electricity grid system (Aniefiok, Igweonu, & Robert, 2013; Patrick, Tolulolope & Sunny, 2013; Dada, 2014; Vincent & Yusuf, 2014). According to Aniefiok et al (2013), the deployment of smart grid will provide a new set of tools for consumers to manage their usage and total energy bills, which indicates that users will start to become more aware of how much electricity they are consuming and possibly increase awareness about sustainability. Similarly, Vincent and Yusuf (2014) argue that the integration of smart grid technology and renewable energy into the Nigerian electricity grid system remains the only solution to the electricity crisis in the country. In addition to the limited access to electricity, pollution is also at an all-time high. As stated by Oyedepo (2012), Nigeria is one of the highest emitters of greenhouse gases in Africa. The practice of flaring gas by the oil companies operating in Nigeria has been a major means through which greenhouse gases are released into the atmosphere. While Aliyu, Dada and Adam (2015) argue that the unabated electricity crisis and the need to reduce the greenhouse gases should be the major drivers for the pursuing sustainable options in the Nigeria.

According to Lund (2007), strategies to support sustainable energy development typically involve three major technological changes: energy savings on the demand side, improved efficiency in the energy production, and replacing fossil fuels by various sources of renewable energy. As a result, large-scale renewable energy implementation plans should also include strategies for incorporating renewable sources in energy systems, which would take into consideration energy savings and efficiency measures (Lund, 2007). Issues surrounding sustainability have been of particular interest and have therefore been researched in a variety of research areas. However, according to Omer (2008), globally, buildings are responsible for approximately 40% of the total world annual energy consumption and while many sectors such as transport and manufacturing industry have made efforts to reduce their energy consumption, residential buildings have not enjoyed the same attention as research in this field is still relatively new (Reinisch et al, 2011). Additionally,

Scott (2007) argues that the environmental performance of homes is a growing concern and suggests that smart features in the home may contribute to improving the environmental standards of both new and existing homes. This suggests that on the demand side, smart appliances may prove to be beneficial for reducing energy consumption as smart home technology can facilitate sustainable ways of living by the use of efficient household appliances (Blumendorf, 2013). According to Paetz, Becker, Fichtner and Schmeck (2011), smart home technologies have the potential to help reduce energy consumption in private households. However, knowledge and awareness of this possibility appears to be limited.

Smart Homes

Although there are varying definitions of what constitutes a smart home, the UK Department of Trade and Industry (DTI, 2003) came up with the following definition. "A dwelling incorporating a communications network that connects the key electrical appliances and services, and allows them to be remotely controlled, monitored or accessed." They go on to identify four key aspects of a smart home, which include:

- An internal network through which devices talk to each other
- Intelligent controls for managing the system
- Sensors that collect information
- Smart features, such as intelligent heating systems, which respond to information from sensors or user instructions

Likewise, according to ITU (2010), a smart home allows for remote electronic control and management of smart appliances (heaters, air conditioners, washing machines etc) and represents the convergence of energy efficient appliances and real-time access to energy usage data, facilitated by a network of sensors and computers. While Blumendorf (2013) describes smart homes as acting autonomously and being proactive, based on artificial intelligence. Smart homes have several functions and as stated by Alam, Riaz and Ali (2012), can be useful for comfort, healthcare, safety, security, and energy conservation. Current research on smart homes places an emphasis on the possibility of health applications where the elderly or disabled people live in their own homes and assistance is provided electronically without necessarily moving such people into care homes (Bartram, Rodgers & Woodbury, 2011; Boyinbode & Akintola, 2009). Smart homes offer a better quality of life by introducing automated appliance control and assistive services. They optimize user comfort by using context awareness and pre-defined constraints based on the conditions of the home environment (Alam, Riaz & Ali, 2012). They also contribute towards sustainable energy. According to Peatz, Dütschke and Fichtner (2012) smart homes are a combination of variable electricity tariffs, smart metering, smart appliances and home automation. The question of how smart a home is or has to be, has been raised by Blumendorf (2013) highlighting the lack of measurements for the smartness of a system.

Smart Home Features

An example of what may be expected from a smart home is drawn from Cook et al. (2003), in their MavHome(Managing An Intelligent Versatile Home) project. The MavHome acts as an intelligent agent to maximize the comfort and productivity of its inhabitants, while minimising operational costs. As a result, the house is able to predict, reason about, and adapt to its inhabitants. Similarly, Blumendorf (2013) provides a compendium of the various aspects focused on when automating a house to the extent that it can be referred to as a smart home. These include temperature surveillance, air quality and indoor climate monitoring, heating, lighting, ventilation, and water heating control to minimize energy consumption. It predicts the actions of its inhabitants and automatically turns on and off heaters, lights and coffee machines in the morning, sprinkles the lawn, places grocery orders, and prepares hot tubs for inhabitants. All these features provide comfort though at the same time could be energy saving, contributing to sustainability (Table 1).

Advantages of Smart Homes

As highlighted in the table above, there are quite a number of benefits to adopting smart homes. Such benefits contribute to the growing interest in the adoption of both smart homes and smart appliances. As previously mentioned, smart homes help to increase the level of comfort. Such comforts could include adjusting blinds based on the sun's angle, control of multimedia and other devices based on the presence of the homeowner. It could also include planned activities from a digital calendar system (Blumendorf, 2013). One primary benefit that could be achieved from all these features is minimising operation cost of residential buildings (Cook *et al*, 2003).

Globally, smart homes and e-health have become common research issues in the past few decades (Chan et al, 2009). Health care is also a recurring theme in smart home research in Nigeria (Boyinbode & Akintola, 2009). However, the technology has been introduced into Nigeria for a variety of additional reasons. For example, to prevent the issue of carbon monoxide poisoning resulting from the improper use of alternative power supply such as fuel generators in densely populated environments (Mbonu, Kennedy, Godwin & Chinwe, 2013); instead of fragmented and unconnected subsystems which do not talk with each other, Oborkahle and Salatian (2012) proposed a smart home, which must consist of a single network of connected

Table 1. Several benefits that may be obtained from adopting smart homes

Technology	Benefit	Source
Assisted interactive dwelling house	A sensor system assesses vital signs and activity, and provides security monitoring and response. Environmental control technologies (doors, windows, and curtains)	Chan, Campo, Estève and Fourniols (2009).
	Monitoring elderly subjects with motor, visual, auditory or cognitive disabilities	Chan, Estève, Escriba & Campo (2008).
	Added functionality of the telecare system. E.g. Switch off the Energy source in a situation where a saucepan catches fire and automatically unlock doors for a safe escape and for entry by helpers	Tang and Venables (2000).
	Fosters independent living and enhanced quality of life	Demiris et al (2004).
	Can help to provide speedy responses to emergencies and need for treatment	
Home entertainment systems	Automated control of sound and video	Paetz, Becker, Fichtner and Schmeck (2011)
Energy-efficient building operation	Considerable savings of energy and optimising resource usage	Alam, Riaz and Ali (2012); Blumendorf (2013); Reinisch et al (2011)
Enhanced comfort	Context awareness, appliance control and assistive services	Alam, Riaz and Ali (2012)
Smart control over the environment	Temperature surveillance, air quality and indoor climate monitoring	Blumendorf (2013); Reinisch et al (2011)
Security	Monitor movement in the home, identify potential intruders, alerts about open doors and windows	Balta-Ozkan, Davidson, Bicket and Whitmarsh (2013)

subsystems facilitated by broadband and Oke et al (2013) developed a prototype based on a microcontroller device using SMS which would automatically control any electrical equipment remotely using a mobile phone. The link between smart homes and smart grids has also been highlighted (Vincent & Yusuf, 2014; Aniefiok et al, 2013) without necessarily linking it to energy optimisation and sustainability.

Smart homes allow a user to control home appliances and devices remotely and ambient intelligence systems, which monitor smart homes, sometimes optimise the household's electricity usage (Alam et al, 2012). This reason according to Paetz et al (2012) is likely to help home owners adopt smart home technologies in addition to environmental-friendliness, high levels of flexibility, greater transparency about electricity consumption and costs as well as enthusiasm for new technologies. In the same vein, features of smart homes such as smart electricity meters or intelligent heating controls have great potential to improve a home's environmental performance. They can run homes more efficiently and help to change behaviour by providing residents with a complete picture of their energy and resource use, thereby improving the sustainability of the building (Scott, 2007). Smart grids also contribute to the sustainability of residential homes, especially smart homes.

According to Tamburini, Rossi and Brunelli (2015), a smart grid is based on a concept of bi-directional electricity flow where buildings can also generate and store electric energy, whereas in the past, were simply supposed to be consumers. The authors go on to explain how this implies that households could become energy producers and such systems could potentially become a source of earnings and not just for saving money. In the same vein, smart homes can be transformed into intelligent networked nodes where a substantial portion of the energy is produced locally by renewables and the whole is intelligently managed (Gungor, et al, 2012). In such a smart home, the electrical appliances, control strategies, and network architectures will enhance the overall energy efficiency and improve the management of local and remote power grids with the aid of active demand-side energy management. Such a smart home would also integrate a large amount of low-power distributed renewable energy resources, which would as a result, reduce the use of traditional energy sources (Gungor, et al, 2012).

Challenges of using Smart Homes

The benefits and features of smart homes have been explored in the previous subsection. They are however fraught with several challenges as well. Chan et al (2008) discuss several issues that may impede the diffusion of smart homes. They include the need to study user habits and intentions in more detail and respect such whenever possible. The need for integration of construction, computing infrastructure, and service delivery aspects, has also been highlighted and there is the fact that new homes may be purpose built for smart homes but for existing homes, it may be a challenge as the homes may not be able to accommodate and integrate the technology (Chan et al, 2008; Edwards & Grinter, 2001). No single manufacturer will be able to supply all the components of a complete smart home system therefore interoperability is essential in order to succeed on the market (Eckl & MacWilliams, 2009).

On the one hand, Eckl and MacWilliams (2009) argue that smart homes tend to be too simplistic and limited in their applications and on the other hand can be too complex, too hard to build, too hard to maintain. Existing homes may become too complex for the occupants if the smart features are added on where users may have to become system administrators dealing with complex systems (Edwards & Grinter, 2001). Legal and ethical problems, user and provider acceptance, requirements and satisfaction have also been identified as potential deterrents (Chan et al, 2008).

In the context of this study, several issues have been raised as challenges facing smart home adoption in Nigeria. No common standards are available for equipment service providers and supplier, the current networks were not built for multimedia and, in particular, for applications that involve video communications, multimedia collaboration, and/or interactive-rich media although this is rapidly improving. Other issues include maintenance and power supply (Oborkhale & Salatian, 2011).

DIFFUSION AND ADOPTION OF SMART HOMES

According to Bartram, Rodgers and Woodbury (2011), a sustainable home is more than an efficient building: it is also a living experience that encourages occupants to use fewer resources more effectively. An energy efficient building may be impossible to achieve without the use of building automation systems which are also necessary conditions for a sustainable home or building (Kastener et al, 2005). Smart homes can fall into this category if all the components work together effectively providing a situation where the home is not only more comfortable but energy efficient and environmentally friendly. Considering the rapidly increasing energy costs, reduced energy consumption has economic benefits. However, in spite of this benefit and others mentioned in the previous section, there is little research on the use of smart homes in Africa. The literature review in the second section indicates that there is limited research on consumer perceptions regarding smart home technology in general (Paetz et al, 2012). Consequently, this study has identified various factors that may be worth examining further in order to achieve the benefits of using smart home technology particularly to ensure sustainability.

Possible Deterrents to the Adoption of Smart Homes

One deterrent could be the high upfront costs involved in the construction of smart homes (Eckl &MacWilliams, 2009). Although investment in building automation systems will result in higher construction costs, their use is mostly economically feasible as soon as the entire building life cycle is considered (Kastener et al, 2005). Arguably, adopting smart features into the home will not come cheap but at the same time, would not necessarily be prohibitive particularly if government intervention in housing takes the above mentioned issues into consideration and provides smart features in new developments. In paetz et al (2012), the researchers found that price, convenience, ecology, transparency and technical equipment were five key reasons why users were likely to adopt smart home technology even though they also agree that there would be a need for significant investments. Furthermore, smart home projects are far more feasible and cost-effective as a result of sustained progress and cost reduction in electronics, information and communication technology (Chan et al, 2008). Reinisch et al (2011) also argue that an overall reduction in operational cost is one of the reasons why home owners are motivated to include automation technology in their homes.

Another deterrent could be the complexity of the features of smart homes. The features and appliances appear to be provided independently by various vendors as opposed to being complete solutions (Fischer, 2007). This could contribute to making it difficult for the home user to adapt to varying appliances running on different platforms from diverse manufacturers. If homeowners are having to change their habits to suit the automation systems in their homes, they may be less than enthusiastic to buy in to such technologies but if their experience is not too complex they may be more likely to adopt smart homes (Edwards & Grinter, 2001). However, home occupants may need to change their habits if the goal of adopting smart homes is for sustainable energy (Paetz et al, 2011).

Privacy is another concern that could impede the diffusion of smart homes and meters (McKenna, Richardson & Thomson, 2012). The authors provide a list of some privacy concerns that could arise when data is collected from smart meters. These concerns include: illegal uses, commercial uses, uses by law enforcement agencies, other legal purposes and use by other co-inhabitants. There is the risk that adopting these technologies would make it easy to spy on home occupants and store data that can be easily accessible for any purpose.

Smart homes have been researched in a variety of fields such as healthcare, smart grids and accident prevention. However, few studies have examined the social aspects of smart home technology adoption and diffusion and a limited attempt has been made to understand how these technologies/services and challenges may interact with values, behaviour and society (Balta-Ozkan et al, 2013). This research therefore would consider not only the consumers' point of view of factors affecting the adoption of smart home but also the technology vendors and all other parties that may be involved in the diffusion process. This is particularly important because currently, although there are a number of smart home/appliance providers, they seem to provide services without common standards or regulatory framework within which they should operate indicating that interoperability could potentially be a challenge (Oborkhale & Salatian, 2011).

Innovation Diffusion Theory

According to King et al (1994), diffusion is "the spread of the capacity to produce and/or use innovation, and its use in practice". Rogers (2003) defines innovation, as 'an idea, practice or object perceived as new by an individual or other unit of adoption' and diffusion as 'the process by which an innovation is communicated through certain channels over time among members of a social system'. Ajzen and Fishbein (1987) argue that Roger's work focuses more on the perception of the innovation itself as opposed to actually using the innovation. Moore and Benbasat (1991) also examined innovation diffusion and extracted several constructs which were adequate predictors of innovation. However, their study focussed on individual adopters within organisations as opposed to individual use altogether. While Fichman (1992) agrees that diffusion theory provides a useful perspective on how to improve technology assessment and implementation and that the theory proved most useful when researchers have examined individual adoption, and/or independent-use technologies.

The perceived attributes of an innovation that are important in explaining the rate of its adoption include: Relative advantage, compatibility, complexity, trialability and observability (Rogers, 2003).

Perceived Innovation Attributes

Relative advantage in the theory of diffusion of innovations is described as the degree to which an innovation is perceived as being better than the idea it supersedes. The degree of relative advantage is usually expressed in terms of economic profitability, social prestige or other benefits. The nature of the relative advantage would be dependent on the innovation in question (Rogers, 2003). According to Rogers, diffusion scholars found relative advantage to be one of the best predictors of an innovation's rate of adoption.

Compatibility is described as the degree to which an innovation is seen as consistent with existing values, past experiences and needs of the potential adopters.

Complexity is described as the degree to which an innovation is perceived as relatively difficult to understand and use. Some innovations are quite clear in their meaning while others are not.

Trialability is described as the degree to which an innovation may be experimented with on a limited basis.

Observability is described the degree to which the results of an innovation are visible to others. This means that the results of some ideas are easily observed and described while other innovations are difficult to observe and describe.

Perceived Innovation Attributes and Smart Homes

Some of the aforementioned attributes feature in current literature, among other attributes which this paper will examine subsequently, and would therefore be applied in a framework to examine the issues surrounding the diffusion of smart homes in Nigeria. Although there are five of these attributes, not all of them have proved to be as effective in explaining the rate of diffusion of an innovation. From the literature examined, complexity, relative advantage and compatibility are the three attributes that have arisen as issues in previous research. Trialability and Observability are innovation attributes that less explain the rate of diffusion (Agarwal & Prasad, 1998; Cooper & Zmud, 1990; Crum, Premkumar & Ramamurthy, 1996).

Compatibility is an issue that could encourage or deter the diffusion of smart homes in the country (Edwards & Grinter, 2001; Fischer, 2007; Paetz et al, 2011). Smart homes may or may not be consistent with existing values, past experiences and needs of the potential adopters. As is usually the case with technology, it could take a bit of getting used to especially when people have lived in homes for years and suddenly introduce features they have to adapt to. The features in the smart home may not be easy to adapt to as they may be enforcing the users to behave in a way they are not accustomed to particularly where the smart homes are adopted for the purpose of sustainable energy and efficiency.

Relative advantage is another innovation attribute that features in existing research on smart home diffusion. This, within the context of this research, is the degree to which a smart home is perceived as being better than one without smart features. In terms of sustainability, environmental-friendliness, high levels of flexibility, greater transparency about electricity consumption; where these features were present, users were more likely to adopt smart home technology (Paetz et al, 2011). This suggests that if potential customers thought that living in a smart home would work out cheaper than a regular house, then they would be motivated to adopt one.

Complexity was an issue that was discussed earlier as one that can encourage or impede the adoption of technology. Smart homes can be perceived as relatively difficult to understand and use particularly by those who are not already technology enthusiasts. As previously mentioned, interoperability is currently a problem that leads to systems in smart homes becoming too complex for users. This is especially the case when smart features are added on making home owners system administrators (Edwards & Grinter, 2001).

A Socio-Technical Approach to Examining the Perceived Innovation Attributes

The diffusion variance model shows that compatibility, complexity and relative advantage are the most important attributes where Information Technology (IT) innovations are concerned (Tornatzky and Klein, 1982). The literature reviewed so far indicates that the last two perceived attributes of innovations are not pertinent to this research. As a result, they will not be examined any further. However, there are some other aspects that may also aid or hinder smart home diffusion as observed in existing literature. The cost of the innovation was frequently observed a concern in the literature examined. Rogers' work also agrees that the cost of an innovation may impede adoption as the less expensive such an innovation is, the more likely it is to be adopted. This study will therefore examine the first three innovation attributes (Tornatzky and Klein, 1982) in addition to cost (Premkumar & Potter, 1995) and privacy. Whilst the first four attributes describe the innovation itself, privacy concerns are more related to the users' perception as opposed to being a feature of smart homes. This implies that the features of the innovation on its own are insufficient for explaining the rate of diffusion (Papazafeiropoulou, Gandecha &Stergioulas, 2005) and as such, this study would have to take into cognisance the other aspects that could help to explain the diffusion process of smart homes in Nigeria.

The research area under consideration has to examine the effect of suppliers and competitors because several of the issues that have been mentioned earlier such as interoperability are centred on suppliers and competitors therefore, in addition to examining the influence of the features of smart homes, external factors will be included in the framework guiding this research. From existing literature, the external bodies that have a part to play in the diffusion process include home owners, the government and suppliers. As previously mentioned in the first section, this research aims to propose a framework that identifies factors affecting the diffusion of smart homes from the perspectives of different parties. Both the demand and supply side of smart home diffusion have an impact on the process, therefore it would be necessary to seek their views in detail and compare with what has already emerged in the existing literature reviewed in this study. Differing opinions and agendas of various groups involved in the process of the adoption of an innovation are not adequately represented in the innovation diffusion theory (Papazafeiropoulou et al, 2005) implying that it would be useful to provide an additional lens through which the situation could be better understood. Theories embracing socio-technical approaches are useful in such situations where there are different views and opinions to be taken into account. Moreover, technical and social factors are closely interconnected (Woolgar, 1996).

Using a socio-technical alignment approach, Kinder (2000) argued that innovating complex networking technology risks technological failure and social rejection unless attention is paid to (internally and externally) aligning the organisational and technological dimensions of the innovated technology. Accordingly, this research argues that diffusion theories are one-sided trying to identify characteristics of the innovation that would make the users adopt the innovation while the perceptions of other influencing parties are not taken into consideration and therefore would propose the combination of perceived innovation attributes with a socio-technical view as a way of examining the views of the different parties involved in the smart home diffusion process. This is represented in the framework below which combines innovation attributes with external views and not just focusing on the features of the innovation alone.

Smart Home Diffusion Framework

Figure 1 above illustrates the proposed framework that will guide this research. As previously explained, according to Rogers, innovation diffusion is the process by which an innovation is communicated through certain channels over time among members of a social system. Within the theory, some attributes have been identified as key predictors of an innovation's rate of adoption. These are known as the perceived attributes of an innovation, which include relative advantage, compatibility and complexity (Agarwal & Prasad, 1998; Cooper & Zmud, 1990; Crum, Premkumar & Ramamurthy, 1996).

Relative advantage is a particularly useful predictor of adoption as this has shown to be one of the attributes that consistently explains the rate of adoption. Relative advantage, compatibility and complexity are the attributes that have been found to be consistently related to adoption (Tornasky & Klein, 1982). Relative advantage infers that innovations are more likely to be adopted if they can demonstrate that using the new product or service provides clear advantages over what it replaces. Complexity and compatibility are also included in the framework as they have also been found to be adequate predictors of adoption. This is because innovations are more likely to be diffused quickly if they are compatible with existing practices,

Figure 1. A framework for examining different perspectives on the smart home adoption

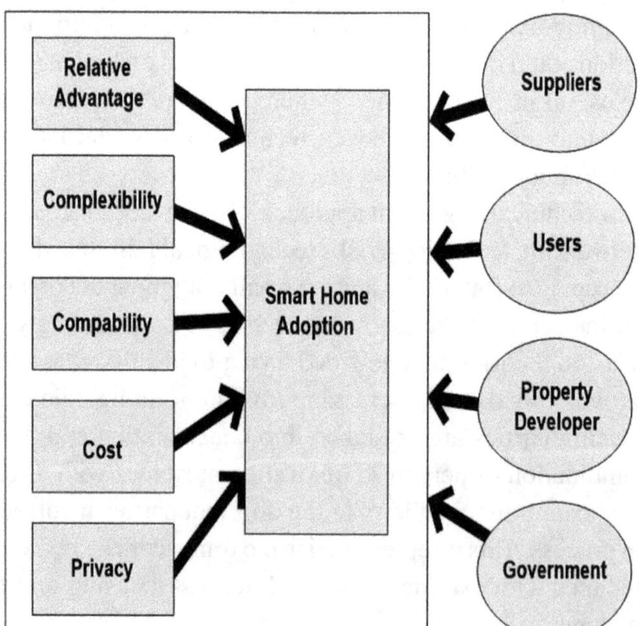

values, and skills of potential adopters, and are reasonably easy to understand and use. In this framework, the views of potential adopters as well as existing adopters would be elicited and examined to have a view of what exactly adopters have experienced or hope to experience, that makes it better to live in smart homes. Other features have been included in the framework including cost (Premkumar & Potter, 1995) and privacy (McKenna, Richardson & Thomson, 2012) as they have been found to be influential in existing literature and therefore may also be applicable to this research.

However, as a result of criticisms that Rogers' work focused more on the innovation itself as opposed to the influence of the members of the social system on the innovation (Papazafeiropoulou et al, 2005), the framework proposed in this study extends the innovation diffusion and diffusion variance models by including the views of various groups that may have an influence on the smart home diffusion process. This ensures that the influences of such groups are not missed, as a focus on the attributes alone may be insufficient to explain smart home adoption. Similarly, Heiskanen et al (2007) argue that consumers are not isolated individuals interacting with clearly defined technologies. Instead, new technologies enter a context of existing networks of artefacts and practices. It is therefore expedient to consider the social networks and effects on the new technology. Introducing a socio-technical approach helps when there are different views and opinions to be taken into account since technical and social factors are interconnected (Woolgar, 1996). Both perspectives are useful when looking at adoption in a smart home context. While the theory of diffusion of innovations perspective focuses on the characteristics of the innovation, the socio technical view examines the effect of various groups and their impact on how the innovation is adopted. Together, they provide a more comprehensive view that would help to investigate the diffusion of smart homes.

The various groups that have been identified in existing literature which may be influential in the diffusion process include users, private developers, government and suppliers. They have as a result, been included as part of the framework. These groups may be likened to social groups within the social construction of technology (SCOT) theory (Pinch & Bijker, 1984). They describe social groups as relevant groups related to the development of a technology that have shared meanings about the technology.

SOLUTIONS AND RECOMMENDATIONS

The energy situation in Nigeria has been examined in the first section of this chapter with the recommendation that energy efficiency would have to become a watch word if the populace has to make do with limited energy resources while at the same time ensuring sustainability. Energy efficiency is pertinent today across the world and one of the ways this can be achieved is with the use of smart homes. While this link has been made in literature, in reality, diffusion of smart homes has been quite slow and this study has attempted to identify factors surrounding this phenomenon. There are several benefits to adopting but there are challenges as well. Many of the challenges and factors affecting the adoption of this technology have been discussed in the preceding section.

A number of smart home vendors have begun operating in Nigeria (Smart home group Nigeria, 2014; Hausba, 2014) though adoption is still in its infancy. Advantages that may be experienced by users adopting smart home technology have been discussed and indicate without a doubt, that there are very good reasons to adopt the technology. However, there are some challenges that have been identified indicating that there is a need for studies of domestic settings to inform design (Edwards & Grinter, 2001). Providers need to ensure that users do not have to alter the way they live their lives completely when choosing to adopt smart homes. This way, potential adopters would not be put off by the possibility of having to develop new skills.

The technology would be replacing an existing system and therefore the evidence should be clear that smart homes supersede the regular home particularly in the area of sustainability and energy efficiency. If the difference is clearly visible, there is likely to be more of an interest in the adoption of the technology (Paetz et al, 2012). The problem of a lack of integration and interoperability was also an issue therefore it is important to offer market solutions that would take this issues into cognisance and provide solutions that would ensure that the features in the home are not operating as islands but are truly integrated into a common platform. These needs are currently not being met by any single technology (Paetz et al, 2011).

This research studies the complexities of the smart home diffusion process, examining a new phenomenon while the linkages among the organisations involved in the process are highly unstructured indicating a need to examine not just the attributes of smart homes but the external influences as well. Qualitative research has been selected for the initial data collection process because it is a useful approach to achieving a better understanding of the phenomena this research is studying. It enables researchers understand people and the social and cultural contexts within which they live (Denzin & Lincoln, 1994).

Key Challenges That May Have an Effect on Achieving the Solutions and Recommendations

Various benefits have been identified as good reasons for adopting smart homes ranging from individual comfort to sustainable energy. However, the diffusion process may not be an easy or straightforward one.

One of the major benefits of adopting a smart home as previously mentioned is the possibility of living sustainably and minimising waste as far as electricity and energy is concerned. However, as identified in existing literature, this may not be a sufficient enough motivator to get people interested in smart homes. Instead, the most common motivator is the cost saving potential. This implies that the adoption process may likely be hampered by the high initial costs that would be encountered in the construction of smart homes. Individuals involved in building projects may be unwilling to take on these additional costs and this is also likely to be the case for property developers as well. To ensure that the diffusion process is not slow, it may be necessary for vendors to be transparent about the kind of savings that adopters would possibly experience. Whether or not potential adopters would trust the figures advertised by smart home vendors is another issue altogether.

The changes to habits of would be adopters may also act as a deterrent and may hamper the path of the diffusion of smart homes. Living in homes that have been adjusted to sustain the environment would be a novelty in the region and may be resisted as an attempt to force expenditure when they are already unwilling to incur additional costs. A lack of awareness may also be a factor that may stand in the way of the adoption. The phenomenon is still in its infancy and as such, it may be difficult to conceptualise living in a different way to what the populace is already accustomed to. Awareness of sustainability and green living in the country under consideration is also in its infancy. As a result, this may also stand in the way of adopting smart homes considering the fact that a major benefit and a driver for ignoring significant start up costs, is the contribution of smart homes to sustainable energy. If the populace are not concerned about reducing waste, then there may be a lack of interest in adoption.

One challenge may perhaps be the supporting network infrastructure that would be necessary to ensure that smart homes can exhibit features that would render such a home sustainable. As previously mentioned in the literature, the current networks in the country were not built for multimedia applications. However, there are rapid improvements which may nullify this challenge. The environment in a developing country would most likely not be similar to countries where adoption of smart homes is more vibrant. Therefore there might be even further issues that have not already been identified in existing literature. In the same vein, current research focuses on western countries with very limited insight into developing countries so it might be the case that the issues that have already been identified would apply only loosely to the situation in Nigeria. Therefore, it would be expedient to ensure a degree of openness during the data collection process. This challenge would hopefully be mitigated by applying open ended questions for the qualitative aspect of the research. Although this method of data collection is particularly useful when studying relatively new phenomena, it is not without its challenges as well. Access may not be readily granted as interviews may sometimes be perceived to be intrusive.

Another challenge that may face the diffusion process is the fact that policy intervention is a factor that has contributed to the adoption of smart homes in western countries (Kinder, 2000). This may not be the case in the region under consideration. Current research has indicated interests from a health care angle and private initiatives with no mention of any contribution or interest from the government.

FUTURE RESEARCH DIRECTIONS

Smart cities have been described as cities where investments in human and social capital and traditional and modern communication infrastructure fuel sustainable economic growth and a high quality of life, with a wise management of natural resources, through participatory governance (Caragliu, Del Bo, & Nijkamp, 2011). Smart grids and smart homes fit into this overall plan as they can help to achieve energy efficiency and better management of diminishing resources

Whilst current research focuses on the themes discussed within the context of developed countries, this paper has considered a developing country with ongoing sustainability challenges. A model has been proposed with the aim of exploring issues that may stand in the way of the diffusion of smart features in the home and has identified various groups that may contribute to or impede the diffusion process in the country.

Future research would focus on semi-structured interviews with participants that have been identified as having a part to play in the diffusion of smart homes particularly for sustainability beginning with suppliers. Subsequently, questionnaires would be distributed to current adopters to obtain their views concerning the innovation attributes as discussed in this paper and any effect on their decision to adopt smart homes

CONCLUSION

Sustainability is a very important issue facing cities, countries and the entire world as the earth's resources are being depleted at an alarming rate. One-way sustainability can be achieved is through the use of smart grids and smart homes. Improving energy efficiency in residential buildings has not been a focus until recently with limited literature available on the issue.

This chapter has examined smart homes and the contributions of adopting such technology and impact on sustainable energy. Being able to control and manage the consumption of energy is essential in general but in particular in Nigeria where the current daily output is insufficient for the populace. The concerns of most of the influencing bodies have been examined from existing literature. Evaluating the views and agendas of all relevant parties involved in the smart home diffusion process. However, the resulting framework will be applied using primary data from Nigeria. An initial interview with suppliers has been planned, the results of which will be analysed with the view to get a better understanding of the current situation. This would in turn aid the further data collection process for potential and non-adopters. The initial findings indicate that the diffusion of smart homes although still in its infancy especially in Nigeria, is beset by challenges that could stand in the way of adoption.

It is hoped that the results from this research will inform policy and industry as the smart home industry in Nigeria is growing and results from this research may contribute to avoiding the pitfalls that have been experienced in countries far ahead in the process of smart home diffusion.

REFERENCES

Agarwal, R., & Prasad, J. (1998). A conceptual and operational definition of personal innovativeness in the domain of information technology. *Information Systems Research*, *9*(2), 204–215. doi:10.1287/isre.9.2.204

Ajzen, I., & Fishbein, M. (1980). *Understanding Attitudes and Predicting Behavior*. Englewood Cliffs, NJ: Prentice-Hall Inc.

Alam, M. R., Reaz, M. B. I., & Ali, M. A. M. (2012). A review of smart homes—Past, present, and future. *IEEE Transactions on* Systems, Man, and Cybernetics, Part C: Applications and Reviews, *42*(6), 1190–1203.

Aliyu, A. S., Dada, J. O., & Adam, I. K. (2015). Current status and future prospects of renewable energy in Nigeria. *Renewable & Sustainable Energy Reviews*, *48*, 336–346. doi:10.1016/j.rser.2015.03.098

Aniefiok, E., Igweonu, I. E., Eguzo, C. V., & Robert, B. J. (2013). Integrating Smart Grid Model in Nigeria Power Network. *International Journal of Advances in Engineering & Technology*, *6*(4), 1760.

Balta-Ozkan, N., Davidson, R., Bicket, M., & Whitmarsh, L. (2013). Social barriers to the adoption of smart homes. *Energy Policy*, *63*, 363–374. doi:10.1016/j.enpol.2013.08.043

Bartram, L., Rodgers, J., & Woodbury, R. (2011). Smart homes or smart occupants? supporting aware living in the home. In *Human-Computer Interaction–INTERACT 2011* (pp. 52–64). Springer Berlin Heidelberg. doi:10.1007/978-3-642-23771 3_5

Blumendorf, M. (2013). Building sustainable smart homes.*Proceedings of the 1st International Conference on Information and Communication Technologies for Sustainability* (pp. 151-158).

Caragliu, A., Del Bo, C., & Nijkamp, P. (2011). Smart cities in Europe. *Journal of Urban Technology*, *18*(2), 65–82. doi:10.1080/10630732.2011.601117

Chan, M., Campo, E., Estève, D., & Fourniols, J. Y. (2009). Smart homes—current features and future perspectives. *Maturitas*, *64*(2), 90–97. doi:10.1016/j.maturitas.2009.07.014 PMID:19729255

Chan, M., Estève, D., Escriba, C., & Campo, E. (2008). A review of smart homes—Present state and future challenges. *Computer Methods and Programs in Biomedicine*, *91*(1), 55–81. doi:10.1016/j.cmpb.2008.02.001 PMID:18367286

Cook, D. J., Youngblood, M., Heierman, E. O. III, Gopalratnam, K., Rao, S., Litvin, A., & Khawaja, F. (2003). Mavhome: an agent-based smart home.*Proceedings of the First IEEE International Conference on Pervasive Computing and Communications PerCom '03* (pp. 521–524) doi:10.1109/PERCOM.2003.1192783

Cooper, R. B., & Zmud, R. W. (1990). Information technology implementation research: A technological diffusion approach. *Management Science*, *36*(2), 123–139. doi:10.1287/mnsc.36.2.123

Crum, M. R., Premkumar, G., & Ramamurthy, K. (1996). An assessment of motor carrier adoption, use, and satisfaction with EDI. *Transportation Journal*, *35*(4), 44.

Dada, J. O. (2014). Towards understanding the benefits and challenges of Smart/Micro-Grid for electricity supply system in Nigeria. *Renewable & Sustainable Energy Reviews*, *38*, 1003–1014. doi:10.1016/j.rser.2014.07.077

Demiris, G., Rantz, M. J., Aud, M. A., Marek, K. D., Tyrer, H. W., Skubic, M., & Hussam, A. A. (2004). Older adults' attitudes towards and perceptions of'smart home'technologies: A pilot study. *Informatics for Health & Social Care*, *29*(2), 87–94. PMID:15370989

Denzin, N., & Lincoln, Y. (1994). *Handbook of Qualitative Research*. Thousand Oaks: Sage.

Eckl, R., & MacWilliams, A. (2009). Smart home challenges and approaches to solve them: A practical industrial perspective. In *Intelligent Interactive Assistance and Mobile Multimedia Computing* (pp. 119–130). Springer Berlin Heidelberg. doi:10.1007/978-3-642-10263-9_11

Edwards, W. K., & Grinter, R. E. (2001). At home with ubiquitous computing: Seven challenges. In Ubicomp 2001: Ubiquitous Computing (pp. 256-272). Springer Berlin Heidelberg.

Edwards, K., & Grinter, R. (2001, 30 September–2 October). At home with ubiquitous computing: Seven challenges. *Proceedings of the 3rd International Conference on Ubiquitous Computing*, Atlanta, GA, USA (pp. 256–272).

Fichman, R. G. (1992, December). Information technology diffusion: a review of empirical research. *Proceedings of theInternational Conference on Information Systems* (pp. 195-206).

Fichman, R. G. (1992, December) Information technology diffusion: a review of empirical research. In J. De Gross, J. Becker, & J. Elam (Eds.), *Proceedings of the 13th International Conference on Information Systems*, Dallas (pp. 195-206).

Fischer, C. (2007). Influencing electricity consumption via consumer feedback: a review of experience.*Proceedings of the European Council for an Energy Efficient Economy (ECEEE)* (pp. 1873-1884).

Gungor, V. C., Sahin, D., Kocak, T., Ergüt, S., Buccella, C., Cecati, C., & Hancke, G. P. (2012). Smart grid and smart homes: Key players and pilot projects. *Industrial Electronics Magazine, IEEE, 6*(4), 18–34. doi:10.1109/MIE.2012.2207489

Hausba (2014) *Hausba Smarthomes* Retrieved fromhttp://www.hausbassmarthomes.com

Heiskanen, E., Hyvönen, K., Niva, M., Pantzar, M., Timonen, P., & Varjonen, J. (2007). User involvement in radical innovation: Are consumers conservative? *European Journal of Innovation Management, 10*(4), 489–509. doi:10.1108/14601060710828790

International Telecommunication Union. (2010). Applications of ITU-T G.9960, In: Paper, I.T.U.T. (Ed.), ITU-T G.9961 Transceivers for Smart Grid Applications: Advanced Metering Infrastructure, Energy Management in the Home and Electric Vehicles (ITU-T, 06/2010). International Telecommunication Union Technical Paper.

Kastner, W., Neugschwandtner, G., Soucek, S., & Newman, H. M. (2005). Communication systems for building automation and control. *Proceedings of the IEEE, 93*(6), 1178–1203. doi:10.1109/JPROC.2005.849726

Kinder, T. (2000). A sociotechnical approach to the innovation of a network technology in the public sector – the introduction of smart homes in West Lothian. *European Journal of Innovation Management*, *3*(2), 72–90. doi:10.1108/14601060010322284

King, J. L., Gurbaxani, V., Kraemer, K. L., McFarlan, F. W., Raman, K. S., & Yap, C. S. (1994). Institutional factors in information technology innovation. *Information Systems Research*, *5*(2), 139–169. doi:10.1287/isre.5.2.139

Lund, H. (2007). Renewable energy strategies for sustainable development. *Energy*, *32*(6), 912–919. doi:10.1016/j.energy.2006.10.017

Mbonu, E. S., Kennedy, C. O., Godwin, O. U., & Chinwe, G. E. (2013). E-MART: A novel Smart Home Governance System for a densely populated power challenged environment. *Proceedings of theIEEE International Conference on Emerging & Sustainable Technologies for Power & ICT in a Developing Society (NIGERCON)* (pp. 100-105). doi:10.1109/NIGERCON.2013.6715644

McKenna, E., Richardson, I., & Thomson, M. (2012). Smart meter data: Balancing consumer privacy concerns with legitimate applications. *Energy Policy*, *41*, 807–814. doi:10.1016/j.enpol.2011.11.049

Moore, G. C., & Benbasat, I. (1991). Development of an instrument to measure the perceptions of adopting an information technology innovation. *Information Systems Research*, *2*(3), 192–222. doi:10.1287/isre.2.3.192

Nguyen, T. A., & Aiello, M. (2013). Energy intelligent buildings based on user activity: A survey. *Energy and Building*, *56*, 244–25. doi:10.1016/j.enbuild.2012.09.005

Obrecht, M., & Denac, M. (2013). A sustainable energy policy for Slovenia: Considering the potential of renewables and investment costs. *Journal of Renewable and Sustainable Energy*, *5*(3), 032301. doi:10.1063/1.4811283

Oke, A. O., Emuoyibofarhe, J. O., & Adetunji, A. B. (2013). Development of a GSM based control system for electrical appliances. *IACSIT International Journal of Engineering and Technology*, *3*(4), 443–448.

Omer, A. M. (2008). Energy environment and sustainable development. *Renewable & Sustainable Energy Reviews*, *12*(9), 2265–2300. doi:10.1016/j.rser.2007.05.001

Oyedepo, S. O. (2012). Energy and sustainable development in Nigeria: The way forward. *Energy. Sustainability Science*, *2*(1), 1–17.

Paetz, A. G., Becker, B., Fichtner, W., & Schmeck, H. (2011). Shifting electricity demand with smart home technologies–an experimental study on user acceptance. *Proceedings of the30th USAEE/IAEE North American conference online proceedings* (pp. 19-20).

Paetz, A. G., Dütschke, E., & Fichtner, W. (2012). Smart homes as a means to sustainable energy consumption: A study of consumer perceptions. *Journal of Consumer Policy*, *35*(1), 23–41. doi:10.1007/s10603-011-9177-2

Papazafeiropoulou, A., Gandecha, R., & Stergioulas, L. (2005) Interpretive flexibility along the innovation decision process of the UK NHS care records service (NCRS). Insights from a local implementation case study.*Proceedings of the Thirteenth European Conference on Information Systems*Regensburg, Germany.

Patrick, O., Tolulolope, O., & Sunny, O. (2013). Smart Grid Technology and Its Possible Applications to the Nigeria 330 kV Power System. *Smart Grid and Renewable Energy*, *4*(05), 391–397. doi:10.4236/sgre.2013.45045

Pinch, T. J., & Bijker, W. E. (1984). The social construction of facts and artefacts: Or how the sociology of science and the sociology of technology might benefit each other. *Social Studies ofScience*, *14*(3), 399–441. doi:10.1177/030631284014003004

Premkumar, G., & Potter, M. (1995). Adoption of computer aided software engineering (CASE) technology: An innovation adoption perspective. *The Data Base for Advances in Information Systems*, *26*(2-3), 105–124. doi:10.1145/217278.217291

Premkumar, G., Ramamurthy, K., & Crum, M. (1997). Determinants of EDI adoption in the transportation industry. *European Journal of Information Systems*, *6*(2), 107–121. doi:10.1057/palgrave.ejis.3000260

Reinisch, C., Kofler, M. J., Iglesias, F., & Kastner, W. (2011). ThinkHome: Energy efficiency in future smart homes. *EURASIP Journal on Embedded Systems*, 2011, 18.

Rogers, E. M. (2003). *Diffusion of innovations*. New York: Free Press.

Sawyer, S., & Tapia, A. (2005). The sociotechnical nature of mobile computing work: Evidence from a study of policing in the United States. *International Journal of Technology and Human Interaction*, *1*(3), 1–14. doi:10.4018/jthi.2005070101

Scott, F. (2007) *Teaching Homes to be Green*, London: Green Alliance Smart Home Group Nigeria (2014) *iSmart Home Technology Integrators Limited.* Retrieved from http://www.smarthomes.com.ng

Tamburini, L., Rossi, M., & Brunelli, D. (2015). Electronic and ICT Solutions for Smart Buildings and Urban Areas. In A. Vesco & F. Ferrero (Eds.), *Handbook of Research on Social, Economic, and Environmental Sustainability in the Development of Smart Cities* (pp. 163–192). Hershey, PA, USA: IGI Global. doi:10.4018/978-1-4666-8282-5.ch009

Tang, P., & Venables, T. (2000). 'Smart' homes and telecare for independent living. *Journal of Telemedicine and Telecare*, *6*(1), 8–14. doi:10.1258/1357633001933871 PMID:10824384

The World Bank. (2015). Access to electricity (% of population). Retrieved fromhttp://data.worldbank.org/indicator/EG.ELC.ACCS.ZS?order=wbapi_data_value_2010+wbapi_data_value+wbapi_data_value-first&sort=desc

Tornatzky, L. G., & Klein, K. J. (1982). Innovation Characteristics and Innovation Adoption-Implementation: A Meta Analysis of Findings. *IEEE Transactions on Engineering Management*, *29*(11), 28–45. doi:10.1109/TEM.1982.6447463

United Nations. (2012) *World Population Prospects: The 2012 Revision.* Retrieved from http://esa.un.org/wpp/

Vincent, E. N., & Yusuf, S. D. (2014). Integrating Renewable Energy and Smart Grid Technology into the Nigerian Electricity Grid System. *Smart Grid and Renewable Energy*, *5*(09), 220–238. doi:10.4236/sgre.2014.59021

Woolgar, S. (1996). Technologies as cultural artifacts. In W. H. Dutton (Ed.), *Information and Communication Technologies: visions and realities* (pp. 87–102). Oxford: Oxford University Press.

KEY TERMS AND DEFINITIONS

Adoption: Take up of new technology.

Developing Country: Countries with low human development index (HDI).

Energy Efficiency: Reducing the amount of energy needed to supply products and services.

Innovation Attributes: Attributes of an innovation that help to predict its rate of adoption.

Innovation Diffusion: An innovation is an idea, practice or object perceived as new by an individual or other unit of adoption and diffusion is the process by which it is communicated through certain channels over time among members of a social system.

Smart Grids: A smart grid is an evolved grid system that manages electricity demand in a sustainable, reliable and economic manner, built on advanced infrastructure and tuned to facilitate the integration of all involved.

Sustainable Energy: An effective provision of energy that would meet the needs of the future without compromising the ability of future generations to meet their own needs.

Section 3
Smart Innovations for Infrastructures

The section concerns smart solutions for creating and operating a city's infrastructures for the provision of public transportation, as well as all the basic utilities and amenities. Chapter 4 deals with the efficiency improvements in the transportation sector, with the aim of reducing pollution from engine fuel emissions, and concluded that it would necessitate behavior and urban planning changes as well rather than financial input alone. In chapter 5, a study of how smart transportation system can be adopted in Harare, the capital city of Zimbabwe, has found that besides the need to design and implement air pollution control measures in the urban areas, the integration of transportation, land use and decision making is also critical in the achievement of smart transport.

Chapter 4
Increase of Transportation Efficiencies and Emission Reduction within a City

Christian Jenne
University of Duisburg-Essen, Germany

ABSTRACT

On a worldwide scale population is increasing, necessitating efficiencies in transportation essential for a modern and healthy environment. Most cities have already reached their limitations on population and vehicle movement. Street layouts and cycling lanes, which were initially planned for a certain number of inhabitants, are now not able to cope with increasing demand. The same applies for fresh water supply and effluent treatment which have limitations also. With an increasing population in urban areas, restriction levels in these regions can be reached very quickly, creating a problem which needs to be addressed. A smarter city can only be developed when there is pre-planning on all of these issues, when there is an understanding of newly developed human behaviours, and when new inventions and technologies in urban areas can adapt to these changes.

INTRODUCTION

In order to develop a good eco-society, there are a number of elements that need to be in place before a city can be labelled as a "smart city". Refuse collection, recycling infrastructure and recycling amenities, low smog concentration, traffic flow of

DOI: 10.4018/978-1-5225-0302-6.ch004

vehicles and people, substantial investment in renewables, and energy efficiencies will all have a major influence in achieving a smart city status.

Industrially developed cities are fighting against pollution by increased decarbonisation programmes. Supporting energy efficient systems with the help of renewable energy technologies, while at the same time carefully targeting the spending of those investments, is a key part of this strategy. This chapter will focus mainly on methods and technologies in the transportation sector which until now have been ignored but which play a major role in the pollution of cities. Transportation emissions are a major contributor to smog levels and therefore need to be addressed in any assessment of this problem.

In order to have a modern eco-society, planners and engineers have to reassess their city in every area, evaluate required steps and take critical action against those problems identified. This chapter will highlight potential pitfalls and will demonstrate state of the art developments across the globe, enhancing a city's lifestyle and associated environmental management system accordingly.

BACKGROUND

Urban population is increasing at an alarming rate which, over time, will result in traffic flow problems, increased pollution density, crowded public transport, exhaustion of land availability for future property developments, and consequently a reduction of available social parks and woodlands. Reduced green zones in built-up areas result in a subsequent increase in CO_2 emissions, particles measured in parts per million (PPM), and noise levels.

According to the "Bevölkerungsfonds der Vereinten Nationen", the world population is predicted to reach 9.6 billion people by 2050, by which time more than two thirds of this population (approximately 6.4 billion inhabitants) will be residing within urban areas and cities (Malte Laub, 2015). This shocking prediction highlights the urgent need for more efficiency in cities, smarter traffic planning and innovative living accommodation.

In India, 30 people currently move from a rural area to a city every minute. This growth trend is expected to continue for the next 20 years. Experts are predicting that there will be a development of approximately 500 new cities within India in the next two decades (Heckel, 2015). This movement of people from rural to urban will result in enormous lifestyle changes, with a much greater proportion of people's lifestyle requirements having to be met in a smaller living space.

- **Issues:** With the constant growth of cities on a global scale, ownership of properties is becoming more limited and requires more land area. This demand on properties will increase rent and limit affordability, which will trigger higher living expenses in these cities. A higher population density for a city which has not been designed initially for this explosion of people puts pressure on waste disposal, the resources of treatment plants, and parking spaces for vehicles, and also leads to a surge in noise and emissions. This has a huge impact on current layout designs, beginning with traffic flow problems, public transportation systems, postal delivery, nearby parks and forestry, fresh water supply, and available grocery stores.

- **Controversies:** Living accommodation can be very limited and expensive in a growing city which consequently requires more housing solutions. This can apply when looking for new property developments while reducing social green field spaces like parks, amenities and small forests or nearby developments. Reducing social spaces will not please everyone, but doing nothing against a growing number of inhabitants won't solve the problem either.
- **Problems:** Some cities are not designed for rising population numbers and need to reinvent in order to adapt. In some cases not every city and urban area can deal with these until-now unknown problems and they will require professional help. Focus groups, advisors and other specialists are needed to deal with the above mentioned problems like waste treatment, water supply, the parking situation, limited public service transport, required rental properties, and other related pollution factors.
- **Solutions:** Pre-planning in response to forecasts of increased population is essential for a stable city design. Every city should have a planning and design team focusing on a specific area with the help of a focus group. These groups need to analyse current system setups, address current and future issues, and evaluate possible options. The forecast data, which will include expected population numbers, is essential for a pre-planning procedure. Available land area and lead time to build properties must be considered when pre-planning, as it can take up to three years from filing a planning application to initial completion.

Transportation efficiencies and emission reduction are very important topics, which must and will be explored by many cities. Growing urban areas have come under pressure, as these problems have only recently been taken seriously.

Research Statement

The expression "smart city" is in very common usage nowadays to cover a very broad range of areas stretching from sustainable living, eco-friendly dwellings, smart integrated solutions via mobile phone applications, modern waste water treatment facilities using the off-gas for power and electricity production, and smartly controlled traffic flow guided junctions. All of this integrated technology has already changed behaviours and will certainly shape the coming years and even decades. Computer assisted guiding systems for disabled and elderly people, and robots serving food and accomplishing easy tasks like cutting lawns when required, are already a reality. This evolution is all part of becoming a more connected environment, helping in decision making processes, which will subsequently lead to higher efficiency. Televisions integrated into mirrors in bathrooms, informing people of breaking news, is not a future idea any more. This all is part of a smarter city structure, embedded in various fragments of this system.

This section however deals mainly with transportation related topics ranging from traffic flow design, energy saving innovations, new technology developments, and efficient system integration using public transport services. A comprehensive framework of innovative strategies has been introduced. Those technologies will highlight what deficiencies exist in the process of transforming a smart city to the next level. Best practice and examples from around the world which demonstrate the potential of each city have been researched.

Small Solutions Can Make a Huge Difference

There is not just one solution to reducing urban waste and pollution. Several small changes have the potential to have a major impact on these problems. There is enormous room for efficiency and improvements and, when all enhancement programmes are running concurrently, there could be a noticeable change. Below is a list of suggestions using different technologies and inventions, some of which are still on a pilot scheme basis.

Action Against CO_2 Reduction Requires Several Steps

Traffic systems and public transport need to be constantly re-evaluated for future action which may be required. The newest available technology has to be used, keeping traffic flow moving despite the fact that the population in urban areas is growing. "Smart traffic light systems", higher efficiency with parking spaces, more public transport incentives, and an increase in cycling lanes, car sharing, and usage of electric vehicles are all things that can make a city smarter. All these improvements will have the effect of decreasing carbon dioxide and other emissions due to higher effectiveness. This will also positively impact on the health and quality of life of people living in those areas mentioned.

Renewable Systems Will Play a Major Part of an Eco-City

Hot water solar panels that can produce free hot water in combination with photovoltaic panels can generate localised electricity as has been done widely in mainland Europe. This is one of the key elements that can contribute to reducing emissions in cities. Energy efficiency, architectural design of south facing buildings in the northern hemisphere and north facing buildings in the southern hemisphere, and draught and heat loss reductions supported by thermal coated triple-glazed windows will enhance sustainable living. These improvements will also reduce heat load and electricity bills, which can then lead to reduced emissions due to lower energy requirements from public, domestic and industrial premises. These combined with highly efficient electrical devices will conserve electricity consumption which is traditionally produced by nuclear power or peat, coal, oil and gas power plants. Therefore, electricity reduction is one of the best methods of saving by reducing the emissions and environmental pollution of energy production.

Smarter Building Design Integrated with Innovative Renewable Technology has been Enforced by Tougher Building Regulations

Most counties in Europe have a localised standard of building regulations. In order to get planning permission, buildings need to meet those standards, and technology and some type of renewable energy contribution need to be installed to comply with these standards. In order to compare different system types, energy performance of dwellings are measured in Maximum Permitted Energy Performance Coefficient (MPEPC). This performance certificate indicates the emissions produced and the quality of insulation in the house. For each m^2 of living space there is a certain heat and energy requirement. The better a house is insulated, the lower the heat demand; therefore, efficiency is increased. Nowadays, these certificate values are indicated in kWh/m^2 per annum. With this system the buyer and property developer know exactly what energy costs will be expected after purchase. In Germany, these certificates have to be renewed every 10 years in order to include new technologies and modernisation improvements which were carried out in the intervening years.

Energy performance certificates for houses used for letting and sales are mandatory in Germany since 2008. This energy and house performance certificate (Energieverbrauchsausweis) indicates the energy requirement as a value but also has, for the less technical person, a scale in letters ranging from A+ to H (dena, 2009). A house with an energy rating of "H" would be the worst performing house with literally no insulation and poor window glazing. A house with the "A+" accreditation would be classified as a passive house which does not require any import of heat and electricity. In another words, an A+ house produces its own required heat and more electricity that it consumes. Most passive houses have solar thermal panels for hot water production and photovoltaic panels for electricity production. If the annual electricity production is greater than their need, surplus electricity will be exported. Electricity could be used for geothermal heat pumps and, therefore, no fossil fuel is consumed. It could also be used for wood pellet boilers, igniting the fire, powering the wood pellet auger and circulating the hot water for domestic heating purposes. Passive house designs are always costlier than normal dwellings, but eventual running costs are less over the life cycle of the building and it can be seen as an emission free house all year around (Figure 1).

Start and Stop Function in Trucks and Vehicles

Newer cars have a start/stop function which will stop the engine when the car is not moving, the gears are engaged and the clutch is not activated. This technology can reduce emissions in a traffic jam and will have the added bonus of saving fuel. All these systems work similarly and are ubiquitous on hybrids. When the car comes to a stop, the engine computer cuts spark and fuel. When the brake is dis-engaged, or the clutch engages, the engine fires back up. The engine starter has to be stronger than in normal engines to withstand the markedly increased on-off cycling. A more

Figure 1. Photovoltaic panel mounted on roof top for electricity production
Source: Own picture.03.2015

powerful battery is also required to satisfy the car's entire electrical needs with its more frequent engine off cycles. Otherwise, stop-start is about as easy a fuel-saving technology as driving downhill (Colwell, 2015).

This technology will help during traffic jams, reducing delay times at traffic lights and other situations where longer idling time is encountered before proceeding in driving off. Depending on the vehicle model, fuel type and engine size, the rule of thumb is that if the engine is switched off for more than 5 seconds a fuel saving will occur. This applied technology in cars is known as "blue motion technology", which has been developed for exactly these kind of situations. Most of the newer mid-range specification vehicles manufactured after 2009 will have this feature as standard.

Increased Road Tax Fees for Cars with High Emissions

It is generally becoming more common throughout Europe that road tax systems are based on emission levels only. The higher the emissions of a vehicle/truck, the higher the running costs and associated road tax. This will incentivise vehicle owners to invest in more modern engine technology which will lead subsequently to less road tax expenditure. Road tax is based on emissions of various fuel types like petrol, diesel, LPG, electricity, hydrogen, LNG and CNG due to various fuel densities and associated emission values. The higher the tax for a vehicle the more likely the owner will change it. Cars older than 10 years must also undergo an annual roadworthiness inspection which incentivises the vehicle owner to drive a newer automotive. This will help in reducing emissions and will likewise lead to less fuel consumption.

Travelling Ban for Amber and Red Emission Labels in City Centres

Particulate emission bands for vehicles in cities in Germany are classified in three categories: green, amber and red. Vehicles, therefore, have been classified into different emission bands, which must be denoted by a sticker displayed on the front windscreen. The vehicle registration office will issue those stickers according to the emissions of the vehicle. More than 50 cities in Germany have introduced areas where only low emission vehicles are permitted. Other locations are less restricted and allow amber coloured labels. A green labelled vehicle can drive through the city without any limitations, amber labelled vehicles are limited to certain areas and red labelled vehicles are prohibited from entering the city centre (ÖAMTC, 2014). This system has been in operation since January 2013 and will keep emissions lower in busy cities. Compliance is strictly controlled and any violation is prosecuted with a hefty fine.

Traffic Information for Vehicles and Public Service via Mobile App and Satellite Navigation

Commuters can be notified via mobile phone app, radio data systems or satellite navigation about current traffic status, enabling them to pre-plan required journeys and routes in a more efficient way. This has the potential to reduce traffic jams, making traffic flow more efficient, with the aim of reducing travelling time and emissions. Some companies are very efficient in this area, developing their own software and networks. One of the company developments is based in Zaragoza in Spain (Figure 2).

The Vehicle Traffic Monitoring Platform, part of the Libelium Smart Cities solution is a platform capable of sensing the flow of Bluetooth devices in a given street, roadway or passageway differentiating hands-free car kits from pedestrian phones. Sensor data is then transferred by a multi-hop ZigBee radio, via an internet gateway, to a server. The traffic measurements can then be analysed to address congestion of either vehicle or pedestrian traffic (Gascón, 2015).

Figure 2. Vehicle traffic monitoring platform with Bluetooth Sensors over ZigBee, 15.05.2015

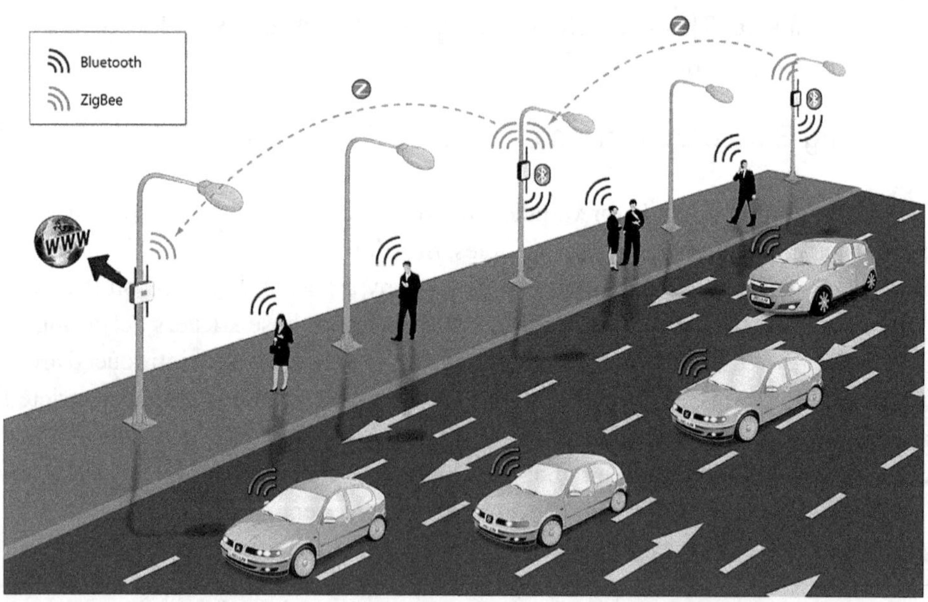

Intelligent Streets Lights and Public Places Illumination

Most street lights in cities are switched on at sunset and switched off at sunrise. This technology is dated and not conducive to energy savings. There are several ways of developing smarter street lights whereby the first step should be changing the light bulb to a LED lighting unit. The second step should be that streets lights are switched on and off automatically, measuring lux rate, and when light is required should be dimmed to a programmed light emittance. This will prevent lighting up when not required or during daylight. In addition to this, light sources should be upgraded with a presence sensor, detecting moving objects, so that the light can be dimmed when full light output is not required (Leccese, 2014).

Smart Junction Crossings and Traffic Light Systems

The first traffic light was erected in Berlin in Germany in 1914. More than 100 years ago it was a unique innovation. Nowadays a city without traffic light systems is not imaginable or practicable. In some situations, waiting for a green light can take up to several cycles depending on the traffic situation and road layout. Sometimes despite having no other traffic in the vicinity a driver has to wait on a red light. This is a typical situation where a smart traffic light would be very beneficial.

Smart traffic lights work in conjunction with small sensors that have been integrated into the street, measuring the traffic flow, and can therefore calculate the travelling speed of oncoming vehicles. These systems can also predict the time of arrival at a particular junction and as a result have the ability to switch the traffic light to green if optimal conditions prevail. In addition, autonomous cameras are connected to this system, recognising emergency vehicles like ambulances, fire brigade vehicles, police vehicles, etc. and regulating the traffic lights for priority. This makes journeys for emergency staff safer and has the potential to decrease travel time in reaching the emergency situation, when seconds can make a difference between life and death. This technology will automatically reduce passing-through times at traffic junctions, reduce "stop and go" behaviours, increase throughput of vehicles and eventually reduce particle emissions, thereby reducing the emission levels within cities. All sets of traffic lights are connected with sensors and smart technology, enabling communication between the various sets of traffic junctions.

According to Carlos Gershenson, a researcher from Universidad Nacional Autónoma de México in Mexico, the estimated upgrading cost for a full functional junction is in the region of only 25,000 US dollars which is approx. 19,000 Euro. Politicians in Mexico City are in talks with a researcher group in upgrading major traffic junctions and therefore reducing costs due to mass production down to 2,000 to 4,000 US dollars (Gershenson, 2015). This will enable smaller councils to afford this smart technology, reducing emissions and noise levels in their area.

Available Parking Space Display and Mobile App Functions

A mobile app should be designed for individual locations fed with live data, identifying empty parking spaces and locations, and notifying customers when a parking space becomes available. This will reduce time and effort for the vehicle driver looking for a free parking space, make parking space searches more efficient and reduce vehicle numbers on the road. Bigger car parks should have a display board outside their premises informing customers about the current number of car spaces available. The app could also inform electric vehicle (EV) owners about free parking spaces to charge up their vehicles. In the future these apps could also be used for charging point pre-bookings. This technology would secure an available parking space for an EV driver knowing the exact location and time of arrival in advance.

Photovoltaic (PV) Solar Panel Cycling Lane for Smarter Emission Free Electricity Production

The Dutch government has designed and installed the world's first photovoltaic solar panel cycling lane which was opened to the public in December 2014. This cycling lane has the potential to produce electricity during daylight while at the same time functioning as a cycling lane. The electricity production of this trial solar cycling lane has already excelled during the winter months, according to a recent announcement by the solar road spokesperson (Wit, 2015). The 70 metre test track has inserted solar panel crystalline cells into a cast concrete base instead of tarmacadam. This is protected by a tempered 10 mm safety glass. This surface is also proven to be slip resistant. The surface was tested with various protection coatings to compare their chemical properties but no complaints were submitted and reaction to this new innovation is very positive. Further investment and installations are expected by the end of 2015 with expansion to major towns in the Netherlands.

Horizontal PV on cycling lanes has the potential to produce emission free electricity, necessitating less need for oil rich tarmacadam and with no required uptake of available land space with conventional PV solar farms. This form of renewable energy production has been gaining acceptance by the Dutch community with no complaints filed so far (Figure 3).

Figure 3. Horizontal PV renewable energy
Source: http://www.solaroad.nl/en, 19.05.2015

Freephone Traffic Engineering Hotline

Since 2009, the City of Cologne in Germany has a Freephone hotline connected to the traffic engineering office where commuters can make suggestions on any traffic improvements they feel would be advantageous for the city. On average, the engineering office receives approximately 900 calls per annum with some very valuable ideas coming from people who are dealing with these traffic problems on a daily basis. Since this traffic efficiency initiative has begun, commuting in the city of Cologne has improved noticeably (Verkehrsdienst, 2015). There was one particular traffic junction about which motorists complained regularly. At one particular time during heavy traffic in the morning, waiting time was more than 8 minutes. Trams and Metro have priority before cars, followed by cyclists and finally pedestrians. After the traffic light setup was assessed and adjusted in conjunction with other surrounding traffic lights, the waiting time was adjusted to an average of 110 seconds, which was archived with an impressive waiting reduction time of more than 400%. These improvements show clearly the need for an efficient system like this. Initially, this hotline was only introduced as a trial phase for 6 months but, after a successful take up by the citizens, the service is still running.

Tax Incentives for Low Emissions Cars

In the USA there are fast lanes in place enabling cars and car pooling commuters to drive in the fast lanes once the car has more than three or more occupants. This incentivises car sharing in driving to work, for example, which correspondingly reduces the number of cars on the road. The same principle is being applied in Germany for hybrid and electrically propelled cars. Any violation of these rules can be prosecuted, with hefty fines imposed.

High-Occupancy Vehicle Lane (HVO) on Motorways

High-occupancy vehicle lanes are also known as HVO lanes, transit lanes, 2+ lanes and diamond lanes, and are designed to increase the enthusiasm for sharing vehicles on the way to work, for example. These lanes are normally less busy, vehicles can drive faster and, because they carry less traffic, they create a more relaxed driving experience. They are designed for peak times and can only be used for carpooling with at least two occupants in the vehicle. Depending on country and state there are other vehicle exemptions imposed for school buses, emergency and law enforcement vehicles, motorcycles, and low emission vehicles. These lanes are created with the intention of reducing the numbers of cars on the road and increasing average occupancy numbers in cars while reducing congestion on the roads (Figures 4 and 5).

Figure 4. High-occupancy vehicle lanes
Source: (Christopher K. Leman, 1994)

Figure 5. High-occupancy vehicle lanes
Source: (Transportation, 2015)

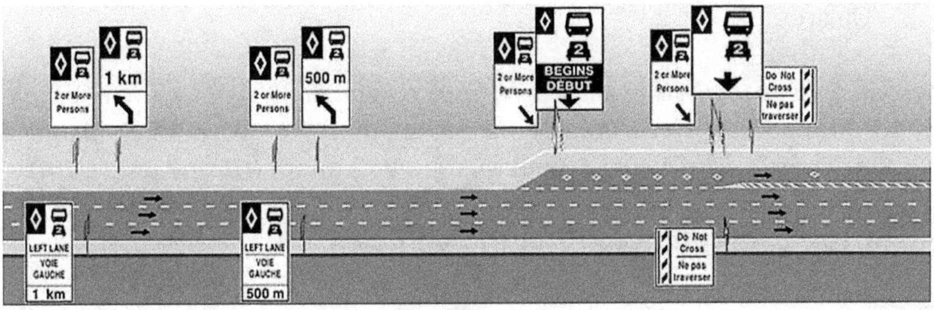

BENEFITS IN USING HVO LANES

- **Save Time:** It can avoid congestion and arrive at destination more quickly than those who don't carpool
- **More Reliable Commute:** Avoiding congestion means a quicker and more consistent commute time
- **Save Money:** It costs less to ride a bus or to share a ride than to drive alone every day. Regular carpooling could cut fuel costs by 50%
- **Conserve Fuel:** less fuel waste sitting in traffic
- **Less Stress:** Letting someone else drive while relaxing on the drive to work

BENEFITS FOR CITIES

- **Manage Congestion:** An HOV lane can handle a lot of growth in demand. Once a general traffic lane reaches capacity, it becomes congested and moves fewer vehicles
- **Make Better Use of Infrastructure:** One highway lane can carry 1,500 to 2,200 vehicles each hour. A lane full of buses and carpools moves many more people than a standard traffic lane.
- **Transit Priority:** Buses and transit occupants have priority. A single transit bus can replace 57 single-occupant cars!
- **Provide Choices:** HOV lanes make carpooling and public transit more effective and reliable choices for commuters
- **Support Mobility:** Taxis and airport limousines that use HOV lanes can return to duty faster after dropping off a fare or arrive sooner to pick up a fare
- **Support Electric Vehicles:** Electric vehicles with green plates can use HOV lanes until June 30, 2016

- **Re-Thinking HOV:** High Occupancy Vehicle Facilities and the Public Interest
(Ontario, 2013)

According to an analysis by the U.S. Census Bureau in 2004, the share of people carpooling was 12.2 percent in 2000, which accounts for the second largest share: 77 percent of people drive alone; 4.7 percent use public transit (Hideo Konishi, 2010).

Road Tax Exemption for Electric Vehicles

The German government has also extended road tax exemptions for any electrically propelled vehicles, initially agreed for five years, which has recently been extended to 10 years until 2025. This will also help to get more electric vehicles (EVs) registered, with the aim in having 1,000,000 EVs on the road by 2020.

Other countries in Europe have similar incentives to stimulate the EV market. Due to very high investment costs for EVs, mainly due to the high purchase cost of the battery, it is evident that market growth cannot be achieved without some form of incentive. For that reason, many countries have schemes to stimulate the EV sales market. In the Netherlands for example they have a low private motor vehicle and motorcycle tax called "BPM" for EVs, which is divided into various categories related back to the CO_2 emission band. The highest tax for emission > 110 gram/km is set with 25% BPM and pure EVs are only charged a tax of 4% if < 50 gram/km (TACA, 2015). Therefore, not all EVs have a low tax band, but the lower the emissions the less tax charged, and consequently the greater the incentive for introducing environmentally friendly vehicles (ACEA, 2015).

Tax incentives can be introduced in different formats, like a tax rebate when purchasing an EV, the introduction of road tax exemption, tax reduction for companies reducing their liabilities/annual circulation tax, and top-up grants. Some countries also differentiate CO_2 bands, which are then linked to associated support levels, while some countries have no incentive scheme in place. This highlights the fact that certain countries have no interest in supporting EVs, while others are very inventive with a broad range of incentive schemes introduced.

Most EVs are classified as emission free or CO_2 neutral but this is not a comprehensive reflection of the situation. This does not apply to EVs if they are using regular sockets where the electricity is only supplied from the national grid to charge vehicles. Countries have various emission levels depending on how electricity is produced, i.e. coal, turf, oil, gas, waste, etc. In reality, only an emission free EV is CO_2 neutral when charged by photovoltaic, wind, hydro, geothermal, and biogas-produced electricity which produce literally no emissions. It could be argued that some fossil fuel energy was required to build those systems but this equally applies to traditional fossil fuelled power stations. Many companies in Austria and Germany are charging their EVs on site with electricity produced from their own installed photovoltaic roof panels, which represents a totally CO_2 emission free charge.

Park and Ride Systems

Park and Ride (P&R) systems are a very effective way of reducing the numbers of vehicles in a city. This commuting system, whereby vehicle owners can park their vehicle at a car park outside the city free of charge, is very common in Germany. These car parks can accommodate a couple of hundred vehicles. The intention is that urban and rural commuters will travel to a P&R location, park their vehicle on that site and travel the rest of their journey by public transport to their intended destination. This system minimises the number of vehicles driving directly into the centre of the city, thereby reducing traffic flow, noise, pollution, and fuel emissions. This makes the centre of the city a more pleasant environment for tourists and shoppers and also for residents.

Gigaliner Trucks for Long Haul Logistics

Depending on the availability and cost, cargo can be transported by airplane, ship, truck and/or train. A combination of multiple shipping solutions is also possible but this depends very much on the timeframe for goods. If goods are transported via trucks, there might be a clever way of dealing with this delivery by road. Long haul goods which have to be shipped more than 250 miles can be carried in so called "super liner", "gigaliner" or "giant" trucks. These trucks are longer and heavier, and therefore specifically designed for long distances. The maximum total weight of these giant trucks, which are 25.25 metres long, and is 60,000 kg. Two of these trucks can replace three ordinary long distance trucks due to a higher storage capacity. This makes the running costs of fuel, insurance, the number of qualified drivers required, and maintenance cheaper and therefore more competitive. Currently, there is only one section of the German motorway approved by the government which has a special license for testing these trucks on the motorway.

These gigaliner trucks are longer than regular trucks by as much as 7.5 metres and therefore more dangerous for overtaking, as they create a longer obstacle for any overtaking vehicle. One hundred trucks have been granted special permission in Germany for a trial period beginning in 2012 and ending in 2016 (Figure 6).

The aim of these trucks is to transport goods from the collection point to a central collection location close to the destination. These huge trucks are prohibited

Figure 6. Gigaliner trucks
Source: (Leduc, 2009)

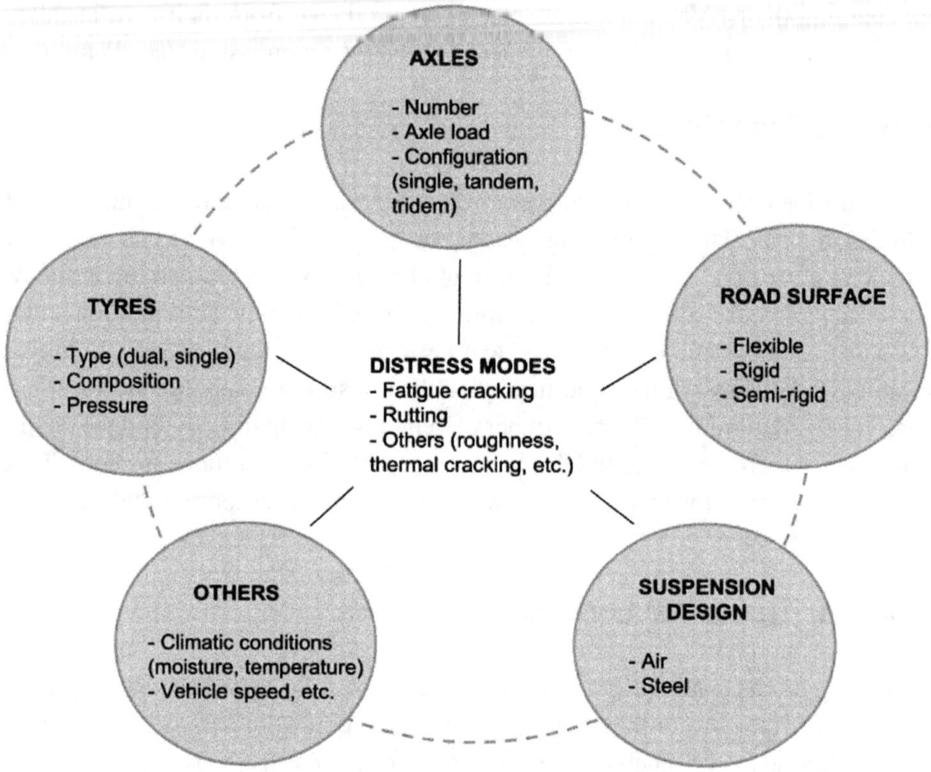

from driving inside the city and only designed for long haul journeys. At this central point, or "Umschlagepunkt", which is naturally located outside bigger cities, delivery goods can be reloaded to smaller vans and delivery trucks. Delivery to nearby places can be arranged with electric vans with few emissions coupled with shorter journeys.

Within the frame of the Logistics Action Plan, the European Commission is analysing the impacts of a potential introduction across the EU of Longer and Heavier Vehicles (LHVs), i.e. vehicles measuring up to 25.25 m and weighting up to 60 tons. Such vehicles are expected to improve the efficiency of the road freight sector but their impacts on the environment, infrastructure and safety need to be analysed in more depth. The present study aims at providing an overview of the technical aspects associated with LHVs regarding energy efficiency, infrastructure (road wear and bridges) and safety issues. This is carried out at vehicle level based on a comparison with standard 40t heavy duty vehicles (HDVs). Some technologies that can potentially improve the performance and reduce the damage caused by LHVs are described and issues that require further and detailed research are also identified (Leduc, 2009)

Tax Saver Commuting Ticket Scheme

Since 2000, Ireland has introduced a commuting tax saving scheme incentive for public transport, which can have a total tax saving for employees of up to 51%. This incentive reduces tax on the total gross income while travelling to work using a multi ticket valid for bus, train, Luas or ferry pass for one year from purchase.

Cycle to Work Scheme

The Irish government has also introduced a second incentive aimed at health conscious professionals and has found a unique advertising strategy for this incentive. The scheme is officially called "Burn Calories Instead of Emissions" which has proven popular over the last number of years. Any regular/electrified bicycle or Pedelec can be purchased up to a total value of €1,000 including safety equipment, with tax savings of up to 51% similar to the tax saver commuting ticket.

Bicycle Rental from Multiple Points within a City

In various cities across Europe bicycle rental schemes are available. These schemes are very effective for tourist cyclists, enabling them to discover the city emission free. Cities usually have several bicycle hubs spread around the city centre, facilitating customers in collecting and dropping off bicycles whenever suits them. Any person can hire a bicycle for a day, three days or a full week depending on the user's requirements. In order to take a bicycle, the customer has to register at a hub, using a booking terminal while giving their credit card details as a deposit for security. This registration process normally takes less than 5 minutes and rental fees are in the region of €10 to €20 a day which is equivalent to $12 to $24. Those central hubs usually have more than 15 bicycles available and most cities offer a free app for mobile phones as well. These apps can show the nearest hub location for picking up or returning bicycles. This is a very attractive, environmentally friendly and healthy option for discovering a city with a great degree of flexibility. This has a big advantage in that cyclists can be flexible in their cycling trips with the option of not cycling back to their starting point if unable to do so.

Eco-Label Introduction in Pubs, Hotels and Restaurants

Pubs, bars and restaurants could introduce an eco-label showing customers their commitment to producing and sourcing sustainable food. Eastern located countries have already an embedded tradition of buying food and groceries from the local market. This reduces transportation emissions while avoiding unnecessary imports of food and also impacts positively on local employment. At the same time, it strengthens the local community and keeps money in the local economy. The idea is to reduce waste and packaging to a minimum, and this can then be displayed in the form of a certificate which will be issued annually.

All restaurant customers can see the benefits of using locally produced food for the same price and most customers would opt to frequent these particular restaurants. This in not only a marketing incentive, but also a sustainable way of sourcing local food with low packaging requirements, electricity coming from renewables, 80% of light bulbs fitted with energy Ratings of A, etc. (Commission, 2009). Achieving an eco-label is not easy and will take some time and effort but customers will certainly appreciate it. Spot checks and annual inspections are just one part of ensuring standards are met. There are several criteria that have to be achieved in getting this certification with a 34 point questionnaire. Depending on the size of the business, some smaller hotels need to fulfil 5 fewer points if they do not have a leisure centre/spa on their property.

Efficiency of Public Transport

Public transport also has to be affordable, efficient and convenient, making it easy for people to choose this form of commuting. Only an efficient and reliable transportation system can be attractive for customers, which means that expectations are set high. Affordability is also one of the major factors in decision making when choosing public transport. It goes without saying that reliability would be a number one priority. Affordability would be the second consideration in attracting customers followed by frequency of trains, buses, metros and so on. Comfort and modern layout would also be an optimum feature when choosing public transport systems. Modern technology like free Wi-Fi broadband on board will certainly attract business people on their way to work or on business trips.

Incentivise Compressed Natural Gas and Biogas Powered Vehicles and Trucks

Compressed Natural Gas (CNG) powered vehicles have much fewer emissions when compared to petrol and diesel powered vehicles. The savings on emissions have been scientifically proven ranging from 12% up to 99% on carbon monoxide, carbon dioxide, soot, nitrogen oxide, hydrocarbon and sulphur. Biogas, when measured in kWh in comparison to fossil fuels, is also cheaper in almost all countries, is cleaner, and creates less engine noise and, in Germany, for example, road tax is lower due to lower emissions. Also quieter noise levels for early morning delivery vehicles is becoming increasingly popular in cities.

Low CO_2 Footprints for Packaging and Wrapping of Food/Goods

Low emissions with reduced oil rich packaging content for wrapping of goods is becoming important for many people. Therefore, companies using environmental eco wrapping need to focus on promoting the environmental benefits when advertising. Reducing emissions is even more important for multinational companies which need to show their efforts on their websites and inform people of how environmentally responsible they are (Figure 7).

Customers will naturally prefer a certain product if there is little or no price difference. This eco driven behaviour is good for the environment and certainly reflects an attitude change in Europe which has been noticed over the last number of years.

One of these multinational companies could be, for example, DHL. This company has a separate ordering section for environmentally connected customers who

Figure 7. Green logistics
Source: (Logistic cluster, 2015)

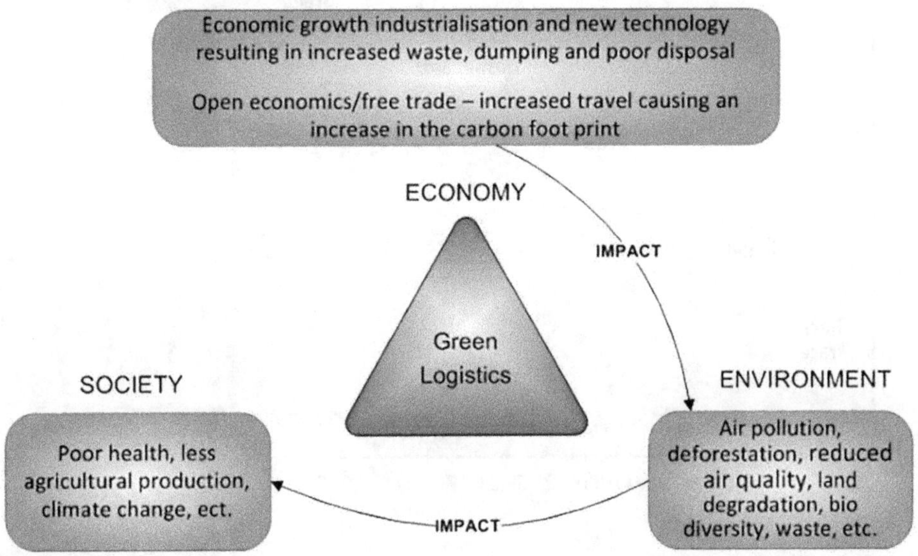

want to send a parcel emission free. This is good for the environment and is also good public relations. A CO_2 emission free parcel order has the advantage of a "feel good factor" for the consumer and helps the company in reducing their generated emissions. These "green logistic shippings" are a little bit costlier but in return the company is doing something to offset emissions like planting a tree, supporting a project in a third world country, etc. (Figure 8).

Automatic Waste Collection Notification

Waste bins should be fitted with weight sensors whilst transmitting a collection notification to the nearest waste provider when required. This technology would reduce unnecessary collection journeys picking up partially filled waste bins. This system has the potential to increase efficiency and subsequently reduce fuel consumption when applied to the waste collection fleet, including all waste, recycling, paper, and food waste collection bins. These intelligent systems are costlier than a regular waste bin but, with fewer collection intervals and greater fuel savings, this will have the advantage of recouping the higher investment cost very quickly. With the implementation of this technology, waste collection noise due to less frequent collections can reduce noise levels, which will also improve the quality of life for residents.

Figure 8. Maturity of Green Logistics Program
Source: (Department, 2014)

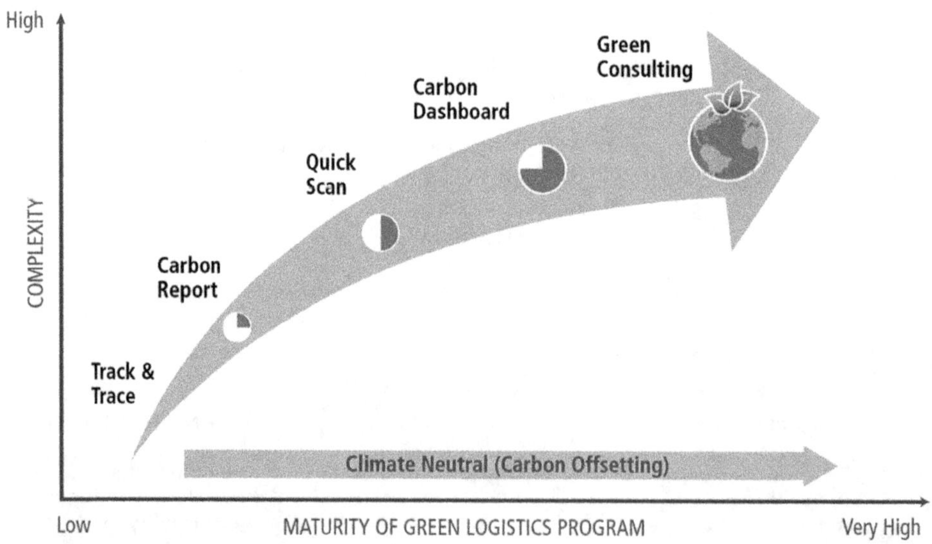

Soot Filters and Emission Limitations for Industrial Buildings in Germany

Air pollution in urban areas can cause health issues, and other associated problems. For that reason the German government has introduced a legislative law enforcement introducing tighter restrictions on particulate filters on vehicles and for industry. The automotive industry has developed particulate filters, first for petrol and later for diesel powered engines, and these have become much more effective in keeping these emissions trapped inside the particulate filter (Melkonyan D. A., 2012). The industry has also had to comply with legislation, which has improved the air quality in urban areas significantly. There are three major reasons for the improved air quality in industrialised areas. Firstly, the quality improvement was linked to very strict law enforcement. Secondly, the filter technology improved during research, with newer materials and better practical applications. Thirdly, an allowed "grace period" for industries concluded in the early 1980s, followed by lower emissions on a year by year basis (Figure 9) (Erste Verordnung zur Durchführung des BundesImmissionsschutzgesetzes - Verordnung über kleine und mittlere Feuerungsanlagen - 1.BImSchV, 2010).

Figure 9. Steel industry in Qian City in North China
(Source: Spiegel online, 01.07.2015)

Since the early 80s, soot filters installed at commercial premises have reduced smog levels significantly in Germany. The impact on coal and oil fired businesses for steel processing, aluminium smelting and major energy consumers was especially noticeable. This was due to cleaner air, fewer smog emissions and brighter days in cities, which also improved the health and quality of life of citizens.

District heating coming from power plants is also a very efficient way in using exhaust heat from incineration plants, for example. This heat has already been produced in burning waste which can then be used further in supplying heat via heat exchange to buildings. This heat is also known as by-product, which means that heat was produced during incineration which was subsequently consumed further via heat transfer. These are smart ways in using a by-product for something useful, making heating systems affordable.

Reduction of Emissions Coming from Vehicles are a Major Step Forward in Having a Cleaner City

Car manufacturers are also obliged to reduce their emissions, which has required more research on the engine efficiency of their cars over the last few decades. The European Economic Community as it was known at the time was initially formed by six industrial countries after the Second World War, namely Belgium, Germany, France, Italy, Luxembourg and the Netherlands, with more countries joining over the years (Commission, Erst sechs, dann 28 - Europäische Nachbarschaftspolitik und Erweiterungsverhandlungen, 2015). At the moment there are 28 Member state partners in the EU (European Union). The EU members' medium term plan is that the proposed legislation will be agreed, with a universal regulation for all EU member states eventually in place. Since January 2015, heavy goods vehicles cannot be registered at the licencing office if they do not comply with the newest Euro particulate 6 norm. The same rule applies for vehicles below 2,610 kg in total weight but the deadline for those vehicles to comply was September 1st 2015 (Wolkenstein, 2014). The Euro 6 norm for diesel powered engines, known as the 715/2007/EC document, requires lower emissions of mg of NOx /km. Flexi-fuel vehicles with a bioethanol mix of 75% and more have different emission ratios as well as petrol powered vehicles (Industry, 2006). Emission ratios are measured in seven different categories, based on engine fuel requirements and vehicle type – for example, passenger cars, light duty vehicles, heavy duty vehicles, etc.

The impact of these stricter requirements is evident and can be seen in the chart below. Emission measurements from three different locations in the state of North Rhine Westphalia in Germany were taken over a time period of 26 years and accumulated accordingly. The three different colours and locations indicated in red (busy motorway), blue (industrial area) and green (local forest) suggest a very similar indication of emission reductions (Figure 10).

Emission levels within urban areas can fluctuate very much depending on time of day and day of week. The graph below illustrates that there is a noticeable NOx increase during mid-week which takes place between 5.00 and 8.00 a.m. This incline from low to peak can reach up to 250% from early morning emission levels. This emission chart represents the traffic frequency of vehicles on the road. The traffic participation is more distributed and, even on a Saturday and Sunday, traffic increased only slightly after 4.00 p.m. It can be concluded that weekend traffic behaviour is totally different to weekdays. Mid-week traffic is therefore mostly related to working commuters travelling to work in the morning and returning back home in the evening (Figures 11 and 12).

It needs to be highlighted that over the last 30 years high emission reductions have been possible even though vehicle registration numbers are increasing.

Figure 10. Yearly emissions emittance in Duisburg of NO/NO_2
Source: (Melkonyan D. H., 2015)

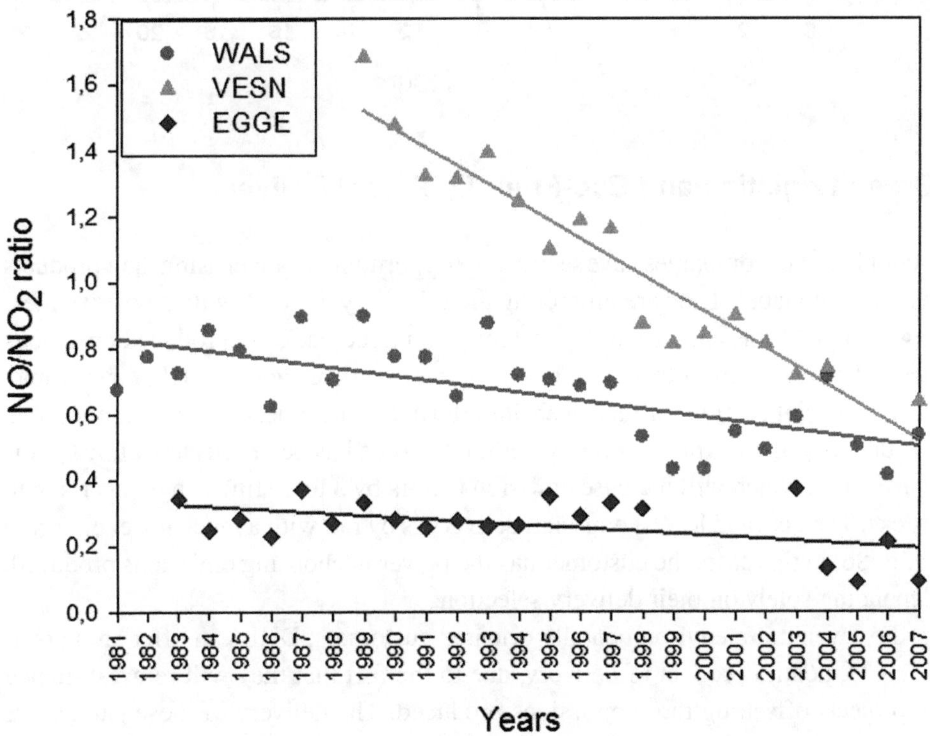

Figure 11. Hourly emissions of NOx in urban areas within one week
Source: (Melkonyan D. A., 2012)

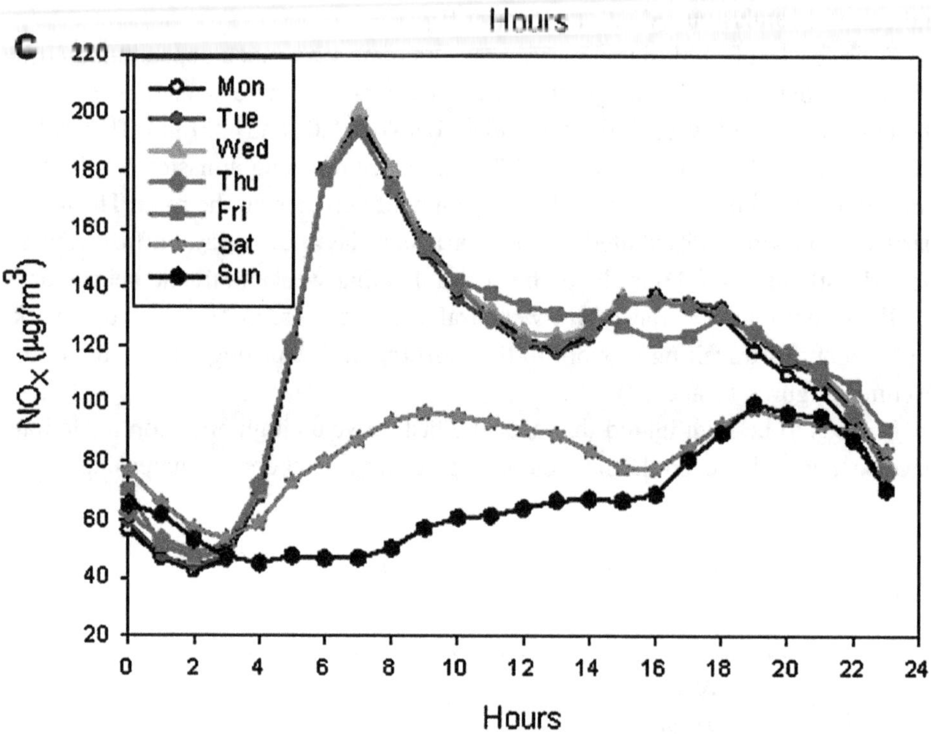

Green Logistics and Eco-Friendly Parcel Delivery

Major logistics companies have seen a huge opportunity in advertising their products with a green label. They are promoting their delivery of goods with a positive message, while at the same time the customer can trace back the produced emissions depending on the importance of the arrival time of ordered goods. DHL, for example, have a special section on their website informing customers on emission loading per delivery. If an express delivery within 24 hours has been selected, air freight is considered, which will increase emission factors by 3 to 10 times in comparison to trucks. The best and lowest emission delivery is by rail with a much lower emission ratio. So, in this case, the customer has the power in choosing emissions produced, depending solely on their delivery selection.

For even more environmentally oriented customers, DHL can offer an innovative CO_2 neutral product of delivery, due to the fact that they will reinvest money in projects offsetting those emissions produced. The delivery of these parcels are

Figure 12. Hourly emissions of CO in urban areas within one week
Source: (Melkonyan D. A., 2012)

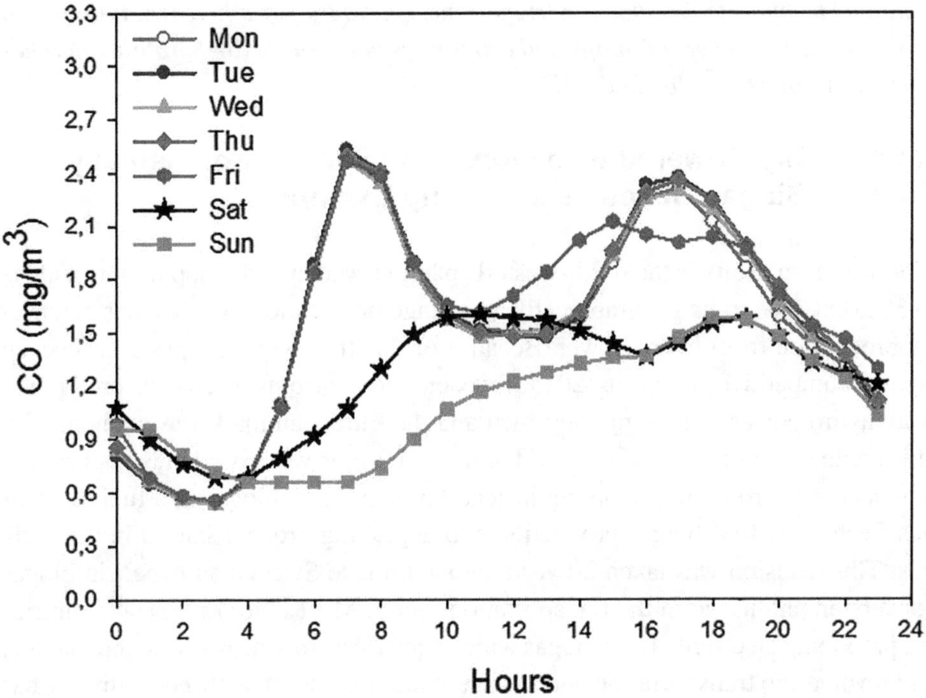

approx. 5% to 10% more expensive, but money will be reinvested in sustainable projects like wind turbine farms, tree planting, electric delivery vans, etc. Since its introduction, this idea looks as if it will have huge potential on a long term basis. If companies are using this type of service, they can also highlight their contribution in reducing emissions and therefore they can show this commitment to their own customers.

Green logistics are seen to be reducing emission impacts and action on the environment. The main activity in green logistics is coordination of activities within a supply chain in such a way that emissions are kept as low as possible. This all comes down to economics, which has an impact on low emission goods delivery, which can only be achieved in pre-planning routes in a very efficient way. In the past the haulage industry focused on having low running costs while making high profits. This business model has changed direction dramatically. Nowadays, customers are much more selective in picking delivery companies. The cost for air pollution, dumping waste and packaging costs, climate change contribution, soil degradation, noise, emissions and selective transportation methods now appears in the calculated cost.

Logistics need ICT - The use of the most modern information and communication technologies (ICT) plays a significant role in logistics: Information and communications technology became the trigger and guarantor for efficient solutions and processes. The internet of things and cyber physical systems are heralding another new era in logistics (Hompel, 2015).

A Small City Powered with More Than 950 Environmentally Friendly Biogas Buses as a Leading Example

Malmö is a small city in the south-westerly part of Sweden with a population of close to 300,000. Due to its geographically advantageous location, Sweden has electricity production from hydro power. Because of this the electricity price in Sweden is very competitive. In spring 2015, Sweden had a surplus electricity production from hydro power, supplying their own and the European market as a whole with affordable electricity. This is one of the major reasons why Sweden is not producing electricity from biogas, opting instead for converting biogas as a fuel for their bus fleets. The first biogas production and upgrading process started in the early 90s. This decision was taken 20 years ago and made Sweden an expert in biogas conversion and usage in the transportation sector. Most other European countries are producing electricity from biogas which needs less treatment, is easier to handle, and involves no transportation and storage issues but, because Sweden already has ample production from hydro power, it is not financially worthwhile for them to convert biogas to electricity.

Skånetrafiken is the name of a district in Sweden which regulates the public transportation market, a semi-state organisation funded by the taxpayer in Skåne County, which is one of 25 counties in Sweden. This organisation does not have an active role in owning, maintaining or running these transport vehicles; its role is of a supervisory nature. At the moment there are 10 contractors operating approximately 950 buses of which 98% are biogas/natural gas and biodiesel buses, 5 electric buses and 20 ordinary diesel buses (Sörensson, 2015). The aim of this particular region is to have 100% environmentally friendly buses replacing natural gas and diesel buses with biogas and electric buses by 2020.

A new innovation has also been introduced – a hybrid bus which can carry more people using only one instead of two buses for popular routes and peak time compensation which is powered by biogas and electricity. This is a 24-meter hybrid CNG-electric bus, using regenerative braking power charging a battery which, when accelerated, produces electricity which can then be used to power the bus, which subsequently reflects positively on fuel consumption (Cody, 2009). This bus is able to carry up to 151 passengers and uses smart technology, saving electricity where available. The doors, for example, open only if passengers are pushing the door opening button and sophisticated software shows the bus driver's daily performance, fuel consumption ratios between recouping electricity by braking and biogas. The company is also able to trace back each individual driver's performance which can be easily measured and improved with training (Figure 13).

Delivery Restriction for Trucks within Malmö Which has a Huge Impact on Their Future Businesses

Malmö has even more ambitious plans for reducing traffic journeys from heavy goods vehicles by introducing a very strict enforcement for companies driving through the city. Several politicians and environmental groups are supportive of this idea. Currently they are working on how this can be implemented without impacting on business negatively. The premise of this proposal is that trucks will only be allowed to enter the city if their load storage capacity is greater than 90%. This would mean that several businesses would need to do a joined-up thinking in sharing one truck for different deliveries. This will reduce transportation movements which will also improve profitability due to there being fewer trucks in the fleet and more efficient delivery systems. Plans and proposals have been discussed and the system is currently under review.

Figure 13. Van Hool ExquiCity 24 hybrid CNG-electric
Source: Own picture, 2015

Nowadays the Choice of an Environmentally Friendly Taxi is in the Customers' Power

Since 2010, Malmö central train station has a highly sophisticated taxi rank system in place. Customers can select their taxi of choice on a touch screen terminal, choosing the size – normal or multi-seater – and fuel type individually. On the display it will be shown if there are any taxies available and if not how long the estimated waiting time will be. Initially when this eco-friendly taxi initiative (Miljöbil) was introduced, it was more for promotional purposes and seen as experimental, but is now being readily accepted. It started off with a couple of eco powered taxies; which can be hybrid, gas, biodiesel, bioethanol or electric vehicles, but nowadays these taxies are the preferred choice for the majority of customers. This shows that consumer behaviour is changing, leading towards a more environmentally friendly focused future. Taxi fares for all choices are the same but the customer can clearly dictate their support for those requested services. Since the introduction of this system, customer choice has dictated that eco vehicles can, if offered for the same fare, be the preferred choice.

CO_2 Neutral Produced Electricity from Renewable Energy Sources

Electricity suppliers also have a large variety of environmentally supportive products that the consumer can select. In Germany, for example, there are more than 1,100 electricity providers competing with each other in offering the best possible deal for customers. Again, the consumer can select the production source of electricity coming from wind power generation, solar photovoltaic farms, hydro, geothermal, up to biogas produced electricity. Over the last decade, electricity production costs from renewable sources have gone down and are currently very competitive. In some cases, renewable energy production can be even more cost effective than nuclear and coal fired power plant production with the benefit of producing few or no emissions.

Some electricity providers can also indicate the location where electricity is produced. A mix of various renewable energy systems are also possible for those who do not care where the electricity comes from as long as it comes from a sustainably produced source. This enables ordinary customers who do not have the knowledge about this technology to select a good product from a climate friendly source.

From this it is obvious that the customer has a large selection of choices and the power to make an active decision on emission production. According to an existing European law, all electricity providers need to show the CO_2 emissions produced per unit which is given in kWh. If electricity is produced from nuclear power this also has to be displayed as waste production measured in weight per produced kWh.

Customers with a sustainable mindset can see on their monthly bill an indication of zero emissions produced. This positive message is a good motivator in supporting this type of industry for a similar price with the benefit of getting locally produced, clean, emission free electricity.

Energy Cities in Europe

Energy Cities, which is a trade mark, is operated by a European semi-state organisation based in Brussels. It has the aim of developing an existing or new city in a more sustainable, motivated environment. This presents the advantage of living in a more environmentally friendly area, with less pollution, more cycling lanes and electricity which is generated by a greater share of renewables. Water harvesting systems and waste water treatment plants are designed to use a limited amount of chemicals and work with natural plants like Norfolk water reed.

Since the unification of Europe more cities are representing themselves as eco cities offering a new way of sustainable living. “Energy Cities” was an initiative which was established more than 25 years ago, and has achieved great success, with astonishing participation rates. Since 2009 more than 6,000 city mayors have signed the “covenant of Mayors” and committed to go beyond the EU’s CO_2 reduction target (Roumet, 2015). The principle behind this competition is to bring people together, sharing ideas, improving their living situations in the city, and learning from each other. There are several categories that any individual city can apply for, and nowadays more than 1,000 cities and local authorities from over 30 countries participate in this annual competition, highlighting their best show cases. Cities can be found on a “virtual” map, and it is possible to do searches for particular projects with associated contact details (Cities, 2015).

There is also promotional literature assisting cities and policy makers in preparing for these competitions. One particular document, “The big five”, explains the five most important steps which are:

1. Include cities in energy policies and in international climate negotiations
2. Set new investment priorities to increase energy security
3. Use local resources to tap the full efficiency potential
4. Involve all actors to foster social innovation
5. Design future-proof cities with radical reduced energy needs

These guidelines are assisting cities – even with no previous knowledge of this innovation – to enter the competitions with the aim of fostering smarter living in their particular city.

FUTURE RESEARCH DIRECTION

Taking into account all possible technologies and lifestyle choices that can make a difference, more research is this area is required. Future research should focus on the green economy and sustainability. Transportation of goods is also a very crucial factor when emission reduction is concerned. Every situation and circumstance is different; consequently, technology needs to be adapted to individual environments.

Discussion Section

This chapter predominantly deals with transportation systems and technologies. Every individual citizen can make a lifestyle choice even without actively doing something other than making positive selections. Decisions need to be made every day, but the crucial question is whether someone would sometimes like to go for a sustainable option for the same price as one that does not contribute positively to the environment.

Fuel and emissions pollution from transportation vehicles is unhealthy, noxious and not good for the environment. Therefore, every effort needs to be made in reducing those levels caused by traffic. Car-pooling, travelling by train instead of taking a short distance flight, cycling to the next bakery instead of taking the car, and electricity purchase from a local sustainable source are all choices we can make without huge sacrifices.

Older diesel engines are heavy polluters - Exhaust fumes from older diesel vehicles are the main source of fine particulate matter in cities. Some 1.7 million diesel cars in Austria are not equipped with particulate filters as standard, and older Lorries emit vast amounts of fine particulate matter. If the pollution levels in Vienna stayed below the WHO threshold for fine particulate matter (PM 2.5), people who are 30 years old today would live nine months longer. Living near busy roads causes a number of chronic diseases: One in three cases of coronary heart disease and one in four cases of chronic obstructive pulmonary disease in over 65-year-olds in Vienna can be attributed to traffic pollution (Gansterer, 2013).

CONCLUSION

All incentives need to be assessed carefully from a social, technical, political, and practical point of view with the objective of contributing towards a more environmentally friendly society as a whole. In order to complete these suggestions positively, a special project group should be in charge of coordination, implementation and project revision. It is also advisable to engage with communities during open days, theme days and workshops to include citizens in this project as much as possible. Support of those proposed improvements will only be visible if a high level of people engagement is present. These schemes should also have a deadline towards which all participants can work.

Education of young pupils and students in sustainable lifestyle choices is also vital. As mentioned in this section, there are feasible options without compromising lifestyle in any way. It is not a question of what we should do; what is more important is how we can do this without harming the environment.

Engineers and planners have the power to make changes in the city due to legislation and enforcement. There are so many showcases on a worldwide scale that some of these technologies can be introduced on a small trial basis and, when successfully completed, expanded on a national scale. A comprehensive framework for innovative strategies was emphasised to give the reader a more in-depth knowledge about various technology implementations across the world. Some improvements could be achieved within a couple of months while others need to be planned for a longer period.

This chapter as a whole has dealt with efficiency improvements in the transportation sector with the aim of reducing pollution from engine fuel emissions. It can be seen that, in order for this change to come about, behaviour and urban planning changes rather than financial input are required. Sometimes it is easier to achieve a solution if the solution required is financial rather than one requiring a radical change in behaviours and attitudes. However, in order to see progress in this area, urban planners and city councils will have to be open minded and proactive in relation to a different way of urban planning. It is probable that the expected increase in migration towards cities will force planners to consider alternative solutions for reducing urban waste and pollution.

The aim for this chapter was to cover all important areas of environmental reduction impacts, ranging from modern transportation systems, emission reduction of industries, residential heat and electricity production coming from renewables, tax incentives for eco vehicles, car sharing options and eco-label introductions for restaurants to smart building designs. These efficiency proposals cover a vast range but they have already been introduced by all participating cities successfully. It has been observed that, once transition has taken place gradually, moving away from petrochemical fuels towards more sustainable alternative solutions like hydrogen, electricity and biogas powered vehicles, subsequent consumer behaviour can have a positive impact in emission reductions. This has the effect of enriching air quality. With only limited fuel and energy resources available, every positive impact can help improve environmental quality and help conserve precious natural resources on this planet.

ACKNOWLEDGMENT

Conversion factors can be found in the book "Measure for measure: the story of imperial, metrics and other units" ISBN: 0-8018-7072-0, Maryland, The John Hopkins University Press, 2003.

REFERENCES

ACEA. (2015). *Overview of purchase and tax incentives for electric vehicles in the EU in 2015*. Brussels: ACEA - European Automobile Manufacturers Association.

Christopher, K., & Leman, P. L. (1994). *Re-Thinking HOV - High Occupancy Vehicle Facilities and the Public Interest*. California: California Department of Transportation.

Cities, E. (2015, September). Retrieved from Energy Cities in Europe. Retrieved from http://www.energy-cities.eu/-Members-

Cody, J. (2009). *Regenerative breaking in an electric vehicle*. Maszyny Elektryczne: Zeszyty Problemowe.

Colwell, K. (2015, May 25). *Car and Driver*. Retrieved from Engine Stop-Start Systems on Nonhybrid Vehicles. Retrieved from http://www.caranddriver.com/features/engine-stop-start-systems-explained-tech-dept

European Commission. (2009). The EU Ecolabel for Tourist Accommodation - 2009/578/EC. Brussels.

European Commission. (2015). Erst sechs, dann 28 - Europäische Nachbarschaftspolitik und Erweiterungsverhandlungen. Brussels: European Commission.

dena, D. E.-A. (2009). *Leitfaden Energieausweis Teil 3 – Energieverbrauchsausweise, IBSN: 978-3-9812787-4-3*. Berlin: dena.

DHL - green services, our solution for you. (2014). Berlin: DHL Headquarters.

2010. *Erste Verordnung zur Durchführung des BundesImmissionsschutzgesetzes - Verordnung über kleine und mittlere Feuerungsanlagen - 1.BImSchV*. Saarbrücken: Bundesministeriums der Justiz und für Verbraucherschutz in Zusammenarbeit mit der juris GmbH.

Gansterer, M. (2013). *Managing traffic and transport in urban areas. Sustainable solutions for traffic problems*. Vienna, Austria: VCÖ - Mobilität mit Zukunft.

Gascón, D. (2015, May 15). Vehicle Traffic Monitoring Platform with Bluetooth Sensors over ZigBee. *Libellium World*. Retrieved from http://www.libelium.com/vehicle_traffic_monitoring_bluetooth_sensors_over_zigbee/#!prettyPhoto

Gershenson, C. (2015, May). *Smart traffic light systems as a pilot project at Universidad Nacional Autónoma de México*. Retrieved from http://green.wiwo.de/ampeln-mit-kopfchen-forscher-entwickeln-das-schlaue-verkehrslicht/

Heckel, M. (May 2015). *Exportchancen, Wachstum Ohne Grenzen*. Von Smart cities entstehen in aller Welt - auch mit deutscher Hilfe. Retrieved from http://www.der-betrieb.de/content/hbnews/ abgerufen

Hideo Konishi, S.-M. (2010). *Carpooling and Congestion Pricing: HOV and HOT lanes*. Boston: Department of Economics Boston College.

Hompel, M. t. (2015). *Cyber physical systems in logistics*. Mülheim an der Ruhr: EffizienzCluster Management GmbH.

Industry, C. d. (2006). *Impact Assessment for Euro 6 emission limits for light duty vehicles*. Brussels: Commission of the European Communities.

Leccese, Z. L. (2014). *Intelligent wireless street lighting system*. Poland: Street lighting paper.

Leduc, G. (2009). *Longer and Heavier Vehicles - An overview of technical aspects*. Luxembourg: JRC Scientific and Technical Reports.

Logistic cluster. (2015, May). Retrieved from http://log.logcluster.org/operational-environment/green-logistics/index.html

Malte Laub, T. M. (2015, May). *Gebäude gehen auf Sendung*. Von Genios - Fachmedien und Mittelstand. Retrieved from http://www.genios.de/presse-archiv/artikel/HB/20150515/gebaeude-gehen-auf-sendung/F75850E1-EF6D-44D2-9A4B-B2E490F7BEB7.html

Melkonyan, D. A. (2012). *Long-term analysis of NO, NO2 and O3 concentrations in North Rhine-Westphalia, Germany*. Essen: Elsevier.

Melkonyan, D. H. (2015, May). Statistical analysis of long term air pollution in North Rhine Westphalia. Duisburg, NRW, Germany.

Ministry of Transportation, Toronto, Ontario. (2013). *Using High Occupancy Vehicle (HOV) Lanes*.

ÖAMTC. (2014). Umweltzonen in Deutschland. Augsburg: Österreichischer Automobil-, Motorrad- und Touring Club.

Roumet, C. (2015). *25 years of Energy Cities*. Brussels: The European association of local authorities in energy transition.

Sörensson, K. (2015, June 26). Transportation situation in Malmö, County Skåne coordinated by Skånetrafiken. (C. Jenne, Interviewer).

TACA. (2015). *Belastingdienst*. Heerlen, The Netherlands: Tax and Customs Administration.

Ministry of Transportation Ontario. (2015, June). *High Occupancy Vehicle (HOV) Lanes*. Retrieved from http://www.mto.gov.on.ca/english/ontario-511/hov-lanes.shtml

Verkehrsdienst, O. u. (2015, February). Freephone traffic engineering hotline since 2009. City Cologne.

Wit, S. d. (2015, May). *Sonnenseite - Oekologische Kommunikation*. Retrieved from http://www.sonnenseite.com/de/energie/erster-solar-radweg-effizienter-als-erwartet.html

Wolkenstein, P. (2014, September). Euro-6-Neuzulassungen - Saubere Sache. Stuttgart, Baden Wuerttemberg, Germany. Retrieved from http://www.auto-motor-und-sport.de/news/euro-6-neuzulassungen-saubere-sache-6053193.html

KEY TERMS AND DEFINITIONS

Eco-Society: A cultural driven society, focusing on environmental, efficient and health orientated civilisation.

Gigaliner: A haulage truck that is specially designed for long haul journeys with a total length of 25.25 metres and weighing up to 60 tons. These trucks are only allowed on specialised licensed motorway sections for a trial period until 2016.

Green Economy: Buildings and infrastructure are designed to be environmentally friendly and economically sustainable, made up from local sources and alternative biofuels.

Logistical Urban Central Hubs: Can be used for parcel delivery for distances of less than 5 miles, whereby the delivery vehicle will be electrically powered.

Semi-Central Urban Hubs: Logistics centres designed for short haul deliveries within a 20 mile20-mile radius of their customers.

Smart City: Sophisticated, well designed and efficient living environment within a highly densely populated city.

Smart Traffic Light Systems: This is a new technology suitable for infrequent traffic, whereby traffic lights are controlled by sensors switching lights, depending on traffic flow.

Chapter 5
Towards Smart Urban Transportation System in Harare, Zimbabwe

Elmond Bandauko
University of Zimbabwe, Zimbabwe

Tinashe Bobo
University of Zimbabwe, Zimbabwe

Gladys Mandisvika
University of Zimbabwe, Zimbabwe

ABSTRACT

The concept of smart transportation systems is increasingly becoming critical in addressing the challenges posed by an increasing number of mega cities in both developed and developing regions in maintaining safety, smooth traffic flow, and an environmentally friendly and sustainable urban environment. The Government of Zimbabwe deregulated its transport sector in the early 1990s. This development ushered in the informal public transport operators, locally referred to as 'kombis'. Major cities such as Harare are characterised by a disjointed and chaotic urban public transport system. The major problems are the impacts this has on the quality of the environment. Currently, the urban public transport system is contributing greatly to both air and noise pollution within the confines of the city, especially in the Central Business Districts. The problem is further exacerbated by the massive importation of used vehicles from outside the country. These developments have resulted in high emission rates of major air pollutants resulting in a deterioration of the ambient air quality especially in the major cities such as Harare. Transportation is a major source of air pollutants. Vehicles are probably the largest single source

DOI: 10.4018/978-1-5225-0302-6.ch005

of pollutants such as hydrocarbons, nitrogen dioxide and carbon monoxide. Other harmful emissions include as lead, benzene, arsenic, aldehydes, sulphates, particulate matter and the secondary creation of ozone. In Harare the number of registered vehicles increased from 192 901 in 1994 to 292 862 and by August 1999 showing that the increase in the number of vehicles is mostly in the cities. This chapter seeks to explore how smart transportation system can be adopted in Harare, the capital city of Zimbabwe. Harare was purposefully selected as it is experiencing rapid urbanization and motorization in the country. Using documentary analysis, discourse analysis and textual analysis, the chapter also describes and examines the challenges, constraints and opportunities of adopting smart urban transportation system in Harare. From this chapter the major conclusions are that the main problem associated with this rapid growth in vehicle population in the major cities is increase concentration of line and area emission sources due to traffic congestion at peak times. The stock of vehicles is quite old and they lack emission control equipment. Major constraints and limitations are observed in the current pieces of legislation. For example, the current Atmospheric Pollution Prevention Act (1971) does not require vehicles to be fitted with emission control equipment. Most of the vehicles use leaded fuel resulting in emission of the dangerous lead particulate matter in the urban areas. There is therefore an urgent need to design and implement air pollution control measures in the urban areas of the country. It is also critical to develop smart and eco-friendly transportation infrastructures so as to achieve sustainable urban communities. The integration of transportation, land use and decision making is also critical in the achievement of smart transport.

INTRODUCTION

Urbanisation is making smart cities more important. Globally, 54% of the world's population lives in cities, a figure the United Nations predicts will rise to 66% by 2050. For cities to manage an ever-increasing influx of people, they must adopt new smart technologies and processes to remain attractive places to live and work. Over the last few years, the rapid development of Information Technologies (IT) has created new opportunities for the sustainable growth of cities. Nowadays, cities are being approached as complex systems which increasingly attract people who wish to work and live in them. For this reason, cities have to deal with various problems, such as traffic congestion, noise and air pollution, energy efficiency, high densities, the lack of green space and an increasing demand for services. The concept of smart transportation systems is increasingly becoming critical in addressing the challenges posed by an increasing number of mega cities in both developed and developing regions in maintaining safety, smooth traffic flow, and an environmentally friendly and sustainable urban environment. The term "Smart City" is used to capture the overall vision of an urban environment with well-managed processes such as traffic control and energy distribution; safer and more efficient services such as parking or emergency response; and a new generation of innovative services yet to be developed. From a technological point of view, at the heart of a Smart City is a cyber-physical infrastructure with physical elements (e.g. roads, vehicles, power lines) which are continuously monitored through various sensors to observe, for instance, air/water quality, traffic conditions, occupancy of parking spaces, the structural health of bridges, roads and buildings, as well as the location and status of city resources including transportation vehicles, police cars, police officers, and municipal workers. The data collected need to be securely communicated (mostly wirelessly) to information processing and control points. These data may be shared and the control points can cooperate to generate good (ideally, optimal) decisions regarding the safe operation of these physical elements (e.g. vehicles guided through the city). It is important to emphasize that what ultimately makes the city "smart" is not simply the availability of data but the process of "closing the loop", consisting of sensing, communicating, decision making, and actuating. This chapter seeks to explore how smart transportation system can be adopted in Harare, the capital city of Zimbabwe. Harare was purposefully selected as it is experiencing rapid urbanization and motorization in the country. Using documentary analysis, discourse analysis and textual analysis, the chapter also describes and examines the challenges, constraints and opportunities of adopting smart urban transportation system in Harare.

RESEARCH STATEMENT

Developing countries have over the years experienced a rapid growth of cities and improvements in car ownership ratios. This phenomena has enhanced a challenge in transport systems of major cities such as Harare in Zimbabwe. The development of the increase in the number of mega cities in the developing part of the world has caused a steaming challenge in the quest to manage safety, smooth traffic flow and an environmentally friendly and sustainable urban environment in terms of transportation. Transportation challenges in some countries such as Harare have been caused largely by policy interventions in the transport sector such as the Zimbabwe Deregulation of the transport sector in the early 1990s mainly after the failure of the Economic Structural Adjustment Policy (ESAP). The deregulation and other factors in the developing world facilitates the informal public transport system which is mainly associated with urban evils such as congestion and chaotic urban public transport system which impacts on the quality of the urban environment. The public transport system also causes a great deal of air and noise pollution within the confines of cities especially in the densely populated and developed city centres. Due to these adverse impacts that results from the use of the kombis and taxes as forms of the informal transport systems which are mostly outmoded planners and city strategists concerned with the transport ecosystem have devised a smarter transportation system as a means to mitigate the impacts of informal transport system. This paper tries to discuss the scope and plight of adopting and improving the concept of the smarter urban transport system in Harare, Zimbabwe looking at the opportunities, challenges, cases, issues and some relevant information that is necessary.

RESEARCH METHODOLOGY

The research is based on a desktop study which consists of literature review, document analysis, discourse analysis and textual analysis. Literature review means the use of published researches from accredited scholars and researchers and various scholarly work has been studied for the establishment and development of this paper. Literature form accredited scholars provides the necessary knowledge and ideas used to develop this paper. Document analysis is a social research method which focuses on reading materials such as public records, the media and visual documents, (Bryman, 2001). The outlined research methods have been used to develop this paper looking at the challenges, constraints and opportunities of adopting smart urban transportation system in Harare.

BACKGROUND TO THE CITY OF HARARE

Harare is the capital and industrial and commercial city of Zimbabwe. It was declared a municipality in 1897 and became a city in 1935. Formerly dubbed 'the sunshine city', Harare has in the last few years been affected by the country's debilitating macroeconomic situation. The once glamorous streets are now characterised by potholes and litter. Harare once prided itself with a reasonably good 4,000 km of surfaced road network. The road network system is failing to cope with increases in the number of vehicles. Notwithstanding the adverse economic environment, the people have found a reasonably cheap source of motor vehicles from Japan and Asia. Vehicles arrive in the country daily in great quantities and the majority of them are destined for Harare. According to the Central Vehicle Registry (CVR, 2012), the number of vehicles in the country increased by approximately 6% from 522 682 in 1999 to 973188 by 2009. Albeit the non-availability of figures by city, it is estimated that about 70% of these vehicles are in Harare. It is evident that there is a clear mismatch between increases in the number of vehicles plying the streets of Harare and provision of road space resulting in severe congestion. The city experiences high levels of congestion which can be protracted for several hours. Residents in Harare at times casually tell a joke that the number of vehicles plying the streets outnumbers the population. Clearly, as this is an exaggeration, it however highlights the magnitude of congestion in the city. Congestion is compounded by small vehicles which are used as modes of public transport. Up to 1983, public transport in Harare was solely provided by conventional buses. Progressively, the increasing public supply/demand gap compelled the proliferation of pirate taxis (informal sector) which started to surface on high density corridors. The growth of the informal sector was given impetus by the Economic Structural Adjustment Programme (ESAP), which was introduced in 1990 (Government of Zimbabwe 1993). In essence, the main objective of ESAP was to liberalize the economy. This culminated in the deregulation of urban passenger transport in 1993 when privately operated public transport vehicles were for the first time legally permitted to operate in urban areas. Privately operated vehicles in the form of minibuses (officially known as commuter omnibuses in Zimbabwe) have mushroomed and now dominate the entire public transport market in Harare resulting in the dearth of the conventional bus operator. In summary, the city of Harare is facing serious challenges which inter alia include an increase in population and the number of motor vehicles, a deteriorating transport infrastructure, severe congestion, an inefficient public transport and a high rate of accidents. All these snags have implications on achieving a sustainable transport and hence the need to raise the question on the requirements of achieving sustainable transport for the city.

Poor Urban Transport System in Harare

The transport system in the city of Harare has deteriorated over the past 25 years after the formal transport system shrunk mainly due to the impacts of the ESAP which has led to the expansion of the informal public transport sector. This experience is explained by Gibson and Kelly (1994) who provides that when the formal sector shrinks, the informal sector expands. Transport woes in Harare range from congestion, pollution and shortages which have caused a hitch hiking of any mode of transport such as pick-ups, Lorries and private cars (Dube & Chirisa, 2012). A decline of the road network in the city of Harare has been largely due to the lack of funding for routine and periodic maintenance in the 1990's up to the present time. According to African Development Bank (2011) the period 2005-08 was characterised by a sharp decline in the state of an already ailing economy and in general little was done in terms of road rehabilitation and maintenance. The poor conditions of the road network as shown above has negative implications to the urban transport system as it affects the transport service industry. For the past decade the Zimbabwean public transport services in urban areas such as Harare have been declining and this has caused high transport costs as well as an inadequate and unreliable public transport system. The advent of in formalism in the transport sector after the year 1993 facilitated the massive importation of transport vehicles by private individuals and companies and these have by now served their time and have reached the end of their economic life and needed to be replaced. However in Harare it can be noted that old vehicles are still operating and causes stress on roads and environmental problems which affects the urban fabric. Due to the problems associated with the current public transport system it can be noted that there are times of long queues of passengers and long waiting times for certain routes. From this informed position of the city of the city of Harare it can be agreed upon that there is need for a sustainable smart urban public transport system in Harare and Zimbabwe as a whole.

THEORETICAL FRAMEWORK AND LITERATURE REVIEW

The problems of traffic congestion, air pollution and unaffordable bus fares have justified the need for smart urban transport systems. The current international discourse of sustainable development has also challenged governments to implement strategies that do not affect the natural ecosystem. Smart transportation systems are very crucial in achieving smart cities- cities that have advanced infrastructure which help to achieve sustainable development (Clarke, 2010). This section explores the different theoretical perspectives on smart cities and smart urban transport systems. The section also highlights the nature of informal transport system and how it affects the overall transport system. This will help to justify the need for the adoption of smart urban transport systems.

Smart Cities

A 'smart city' is an urban region that is highly advanced in terms of overall infrastructure, sustainable real estate, communications and market viability. It is a city where information technology is the principal infrastructure and the basis for providing essential services to residents. There are many technological platforms involved, including but not limited to automated sensor networks and data centres. Though this may sound futuristic, it is now likely to become a reality as the 'smart cities' movement unfolds in India (Clarke, 2010).

In a smart city, economic development and activity is sustainable and rationally incremental by virtue of being based on success-oriented market drivers such as supply and demand. They benefit everybody, including citizens, businesses, the government and the environment. As world urbanization continues to grow and the total population expected to double by 2050, there exists an increased demand for intelligent, sustainable environments that reduce environmental impact and offer citizens a high quality life. A smart city brings together technology, government and society to enable the following characteristics:

- **A Smart Economy:** Smart economy has various factors all linked to economic competitiveness such as innovation, entrepreneurship, trademarks, productivity and flexibility of the labour market and also integration in the national and international market.
- **Smart Mobility:** Smart mobility entails the local and international accessibility through various strategies and modes of communication and transport systems. The concept of smart mobility also includes the availability of information and communication technologies and modern and sustainable transport systems.
- **A Smart Environment:** Smart environment is the attractiveness of the natural conditions (climate, green space), pollution, resource management and also efforts towards environmental protection.
- **Smart People:** Smart people are described by the level of education of the citizens and also the quality of social interactions regarding integration and public life and the openness towards the outer world.
- **Smart Living:** Smart living includes various aspects of quality of life as culture, health, safety, housing, tourism. Other components of smart living are cultural facilities, health conditions, individual safety, housing quality, education facilities, touristic attractivity and social cohesion.
- **Smart Governance:** Smart governance consists of aspects of participation in internal politics, services for citizens and the functioning of the administration.

Smart Urban Transport System

A smart urban transportation system is a transportation system that creates walkable communities, provides a range of transport choices and provides access to a wide range of origins and destinations, (Santore et al, 2003). Some of the elements of a smart urban transportation system include a dense, frequent public transit service; a variety of street types that provides both access and mobility and sidewalks and bicycle facilities that provides direct, safe travel routes. A smart urban transportation system should be closely related to land uses such as residential and commercial in a specific environment to support quality transit services and to reduce trip distances. According to the Smart Transport Guide Book from the New Jersey Department of Transport and Pennsylvania Department of Transport (2008), smart transportation proposes to manage capacity by better integrating land use and transportation planning. This shows a strong relationship between transport planning and land use.

The transportation guidebook from the New Jersey and Pennsylvania Departments of Transport provides some principles of smart transportation which include the design of roadways which should respect the character of communities and the current planned land uses by transitions from one area to another, transportation projects should be planned through ongoing partnerships with local communities, considerations of the needs of pedestrians, bicyclists and transit users in designing all roadway projects and side walk networks that should be well connected with opportunities for regular, safe street crossings and finding transportation solutions that can be executed in a sensible phase.

Many developing countries require smart transportation systems due to the limitations and problems being caused by the established disjointed informal urban transportation system, (Fadare, 1990). To build a smart urban transportation system there is need for a number of strategies and innovations in the transport sector such as increasing the capacity of the existing transport networks by accurately envisaging demand and aligning capacity for both passenger and freight transporters, develop advanced traffic control systems and designing a traffic interchange working as an integrated network. There is also much need in endowing traffic users with information about their transportation choices and reducing harmful emission and reducing energy use while at the same time assuring passenger safety.

Approaches that can be used in trying to achieve smart urban transportation systems include predicting demand and optimizing capacity, assets and infrastructure. Many cities in the world encounter the glitches of traffic congestion and underutilization of transportation infrastructure due to privation of demand prediction or inapt prediction of demand and the capacity of the infrastructure is poorly managed. Transport operators need knowledge and insights into the demand in the transport networks so there is need to analyse historic and real-time data to see patterns based on recurring factors such as location and time.

In a smarter urban transportation system there is need to empower travellers and freight customers with the necessary information and tools that enables them to determine for themselves the best way to travel from their origins to destination areas across all modes of transportation available and considerations such as cost, time, convenience and environmental impact. The tools outlined can work best in providing information about alternative route systems and schedules and highlighting information about connecting networks. Smart transportation systems also have the ability to reduce delays through advanced traffic management, improve on time performance through better planning of routes and schedules, and help make road and rail terminals more efficient and convenient through technological innovations.

Smart urban transport system promotes the notion of developing sustainable cities. It therefore hinges on the need to balance of economic, social and environmental goals, as illustrated in Figure 1. Smart urban transportation system can be achieved through increasing operational efficiency and reducing environmental impact. Today's transportation systems are inefficient and have caused deteriorating service provision, high energy use and environmental impacts. Frost and Sullivan (2008) argue that in urban areas we are reaching the limits of how land can be dedicated to new roads, rails and terminals, thus there is need for smart urban transportation systems. Smarter systems in the transportation sectors optimize the already existing assets and infrastructure, minimizing the need for increased investment. Operating efficiency and reducing environmental impacts requires a formal public transport planning process which is four staged. The stages of the formal public transport planning are network planning, time tabling, vehicle and crew scheduling. The above is supported by IMB Corporation (2010), who argue that energy use, carbon emissions and cost can be reduced by monitoring vehicles in real time and by better planning for supply chains and parts inventories. IBM further argues that crew planning tools ensures personnel with the necessary skills are available where and when they are needed.

Equally important, is the concept of achieving social goals in a transportation system (Frost and Sullivan, 2008). Security, fairness, equity, affordability and safety cannot be compromised in any smart transportation system (see Figure 1). A

Figure 1. Sustainable urban transport system goals
Source: Litman (2014)

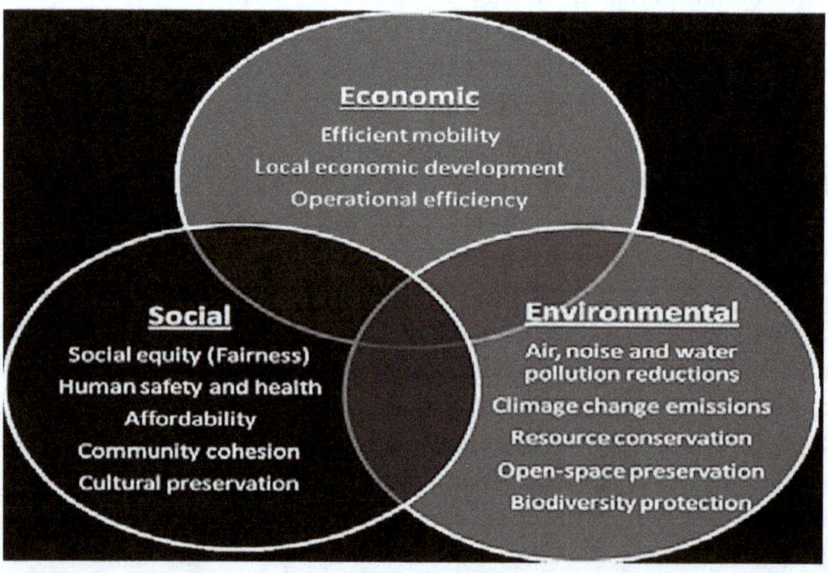

smarter transportation system can therefore improve safety and security while costs are reduced. Sensors can be used in smart transportation initiatives to monitor the condition of equipment and infrastructure in real time consequently avoiding imminent failures. Other tools include the use of video cameras which can constantly scan roadways and rail lines to identify any compromises and hazards in the transportation system. There are also advanced traffic management systems that can be used to reduce congestion and traffic accidents and improving security by detecting and evaluating threats through analysis of passenger information and electronic surveillance.

The 21st century transport system should be smart (Frost & Sullivan, 2008), it needs to be able to react quickly to the demands placed on it and antedate them. This means having the characteristics of satisfying travel demands of people and predicting the future trends in the transportation system. Frost and Sullivan gives four main challenges that are emerging from the current transportation system that have to be addressed by the new smart transportation system. The smarter transportation system should address the challenges outlined below to make transportation efficient, safe and eco-conscious (Figure 2).

Figure 2. Challenges for a smarter system
Source: Frost and Sullivan (2010)

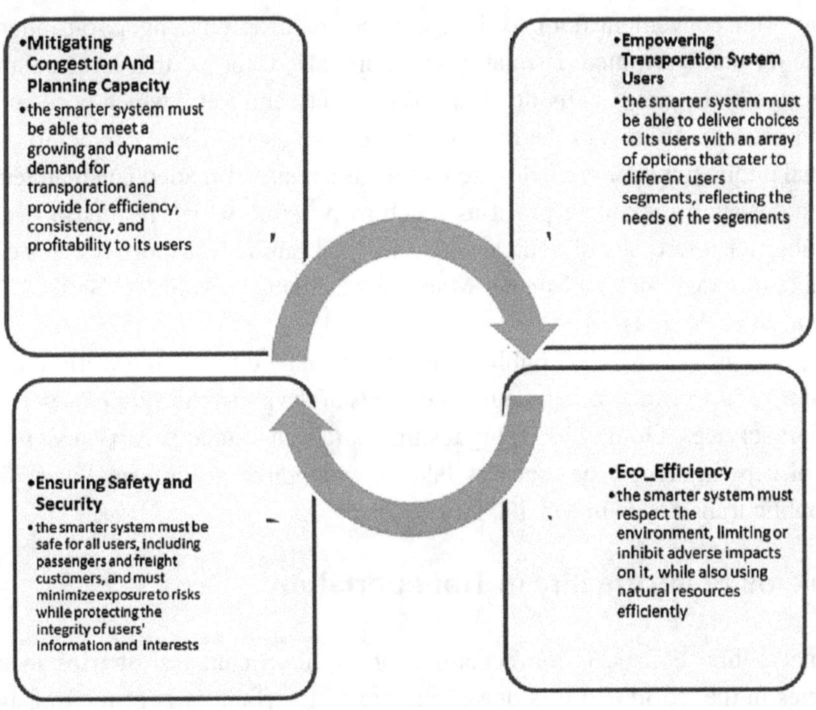

ADVANTAGES OF A SMARTER TRANSPORTATION SYSTEM

A smarter transportation system offers a variety of merits such as maximizing use and efficiency of existing infrastructure, this optimize the use of existing infrastructure and delay future heavy investment in new transportation facilities such as highways and rail lines; improve value for money of transport services; improves city attractiveness, this culminates from the reduced traffic congestion and improved public transport improves urban lifestyle; improve environmental sustainability by having better traffic management and development of public transport reduces air pollution and carbon emission,(Palmarini, 2012).

Informal Transport

Informal public transport also known as the intermediate public transport (IPT) or Para transit is an important component of the transportation services in many Asian, African and Latin American cities, (Golub, 2003). Informal transport involves vehicles such as mini-buses that lack a rigid schedule or route and sometimes even lack a preset fare, (Golub, 2003). Informal transport offers personalized services, particularly for passengers going to specific directions or places, (Aworemi et al, 2008). Another distinguishing feature of informal transport in urban areas is that informal public transport vehicles offer more flexible services than mass transit as they offer convenient door to door services or drop passengers on request at specific un-predetermined destinations. Fadare, (1990) argue that informal public transport vehicles are usually noisy, smoky, rickety and jolty, which pose specific environmental problems to the road and none road users. These characteristics of informal public transport vehicles are largely as a result of inadequate maintenance and the desire to maximize profit as much as possible which results in over-use of vehicles. These types of vehicles and informal public transport systems comes in different names such as Matatu, Marshrutka, Jeepney, Angkots, Bermos, Taxi, Minibus and Microbus.

The presence of informal public transport in many cities in the world indicates the failure of governments in meeting the needs of the people through formal public transport services. Golub (2003) argues that para transit modes carry a significant share of trips in many cities such as Jakarta where 50% are informal and 70% of total public transit trips in Manila, Philippines.

Problems of Informality in Transportation

Informal public transportation accounts for as significant transit trips in many countries in the world and it is one of the most important parts of the transportation system. Although it has an important role to play given the inefficiency in formal transportation systems it has a plethora of problems. Some of the problems as presented by Golub (2003)includes erratic scheduling and service, competition in the market whereby operators fight for waiting passengers at bus terminals and the driving becomes aggressive and dangerous, cream skimming where informal operators operate especially during peak hours because of perceived demand and costs, safety is compromised especially by lack of driver training and the use of inappropriate vehicles and poor vehicle maintenance, lack of accountability where lack of official regulation leaves passengers no avenue for complaints about the services, evasion of taxes, insurance, social security where the informal transport operators are undocumented so can therefore evade paying taxes; labour abuses where informal operators may disregard minimum salary laws and social security payments, lack of capacity, inefficient business practices and environmental problems such as air pollution due to use of old vehicles that even lack maintenance and proper servicing.

LEARNING FROM OTHER REGIONS OUTSIDE ZIMBABWE

This section gives a description of how Stockholm, Curitiba, Lima and Hong Kong have adopted smart transport systems to alleviate the problems of conventional transport systems and to achieve smart cities. The thrust of this section is to draw lessons from cities that have implemented different smart transport systems. Latin American and Asian cities adopted the Bus Rapid Transport system which is efficient in reducing congestion, choking pollution, and alarming accident rates that was as a result of the largely unregulated bus transit systems in the cities (ECLAC, 2014). The system uses high-capacity buses which are placed in committed bus lanes to avoid traffic jam with other vehicles. The dedicated lanes channel buses to a series of fixed stations. European cities adopted Intelligent Transport Systems which uses advanced technology to improve efficiency in availability of travel information, traffic and demand management, smart ticketing and urban logistics (Transport Research and Innovation Portal, 2014).

Asian Cities

Asia is experiencing rapid urbanisation like any other continent. Accompanied with rapid urbanisation, Asia is also experiencing a huge and unprecedented increase in economic productivity and prosperity. A research by OECD indicated that Asia's economic output recently exceeded that of all other world regions, and is projected to triple by 2032. The negative impact of this prosperity is that as more households attain middle-class incomes, vehicle ownership tends to increase. In affluent Asian countries, vehicle ownership peaks at 400 to 600 vehicles per 1,000 residents which pose the risks of air pollution and traffic congestion. With this background, Asian cities are more enthusiastic to adopt the concepts of sustainability, sustainable development, sustainable transportation planning, and liveability. Asian cities have gone a milestone towards the achievement of the smart transportation systems by adopting the Bus Rapid Transit (BRT) which is commonly understood as a system that emphasizes priority for and rapid movement of buses by securing segregated bus ways. The BRT system is a smart transportation system because it has social, environmental and economic benefits as has been experienced in Asian cities such as Hong Kong. Intelligent Transport Systems are also adopted in managing the public transport systems. A good picture of how Asian cities operate smart transport systems can be drawn from Hong Kong which is explained below.

Hong Kong

Hong Kong is one of the most crowded cities in the world with a dense population density, more than seven million people live in an area covering 1 100 square kilometres. The city of Hong Kong is dominated by high rise buildings in its built environment to maximize urban capacity. Despite its dense population in the city Hong Kong boasts of having one of the most efficient public transport networks in the world. (Audenhove et al, 2014), provides that public transport and walking make up to 92% of the modal split. The transport system of Hong Kong is multi-modal, that is, it is based on many modes of transport notable includes rail transport which is supported by the bus minibus, tram, ferries and taxis. Hong Kong's public transport system is well integrated and this can be supported by the use of the Octopus smart card which allows customers of the public transportation system to use all modes of transport and for paying parking services.

Hong Kong has been chosen as a case study for this paper due to its performance on the urban mobility index where it has been ranked top in the 2011 and 2013 studies of the urban mobility index. According to the urban mobility index the city of Hong Kong tops the index with regard to the financial attractiveness of the Public Transport, share of Public Transport in modal split, smart card penetration, number of vehicles per capita, traffic safety, climate impact of transport and public sector initiatives. The backbone of the transport system in Hong Kong is the rail transport which is suitable in the city due to the geography of the city and the volume of users of the transportation network. The rail network has been developed since the 1980's and the objective was to reach a synergy between rail public transport supply and construction. Investments in the Hong Kong transportation sector led to the reduction of car ownership ratio, 73 per every 1 000 citizens. The low car ownership ratio is largely a result of a dominant role for public transport, and the most notable investor in the Hong Kong's urban transportation system is the Mass Transport Railway Corporation Limited (MTR) which turned Hong Kong's high population density into an opportunity rather than a threat. MRT is driven by its policy of combining public transport development and urban development. The work of the MRT is supported or supplemented by the Octopus Card which is an IT facility in the Hong Kong's transportation system.

European Cities

The European Commission encourages all administrative authorities to adopt an integrated Intelligent Transport System approach which covers different transport modes and mobility services. The system should be tailored to local customers´ and market needs (European Commission, 2013). Intelligent Transport Systems (ITS) which also known as Transport Telematics refers to Information and Communications Technologies (ICT) applied in the transport sector to enable authorities, operators and travellers to make more informed and 'intelligent' decisions. ITS was adopted in European cities to improve efficiency in availability of travel information, traffic and demand management, smart ticketing and urban logistics (Transport Research and Innovation Portal, 2014). The implementation of ITS in Europe have increased mobility, reduced congestion and plays a critical role in cutting greenhouse gas emissions. ITS uses tools such as CCTV and ANPR technologies which help to improve transport management, and also to increase efficiency and safety while at the same time reducing energy consumption and environmental impacts, thus contributing to achieving a more sustainable transport system. The Transport White Paper highlighted that all transport users benefit from real time information on current network conditions and online information for journey planning and seamlessly integrated services (European Commission, 2011a). Stockholm offers a good example on how European cities have adopted ITS.

Stockholm

The city of Stockholm is located in Western Europe and it is ranked the second after Hong Kong in the urban mobility index, (Audenhove et al, 2014). Stockholm is built on different islands; these are interconnected by bridges. The geographical setup of Stockholm presents challenges to the transport system in terms of access to the city centre and control of traffic. Stockholm has a smart urban transportation system as supported by its performance in the Urban Mobility Index. The 2013 version of the Urban Mobility Index ranks Stockholm second from Hong Kong and it regards Stockholm as the best performing European city, (Audenhove et al, 2014). Some of its performance and achievements in the urban mobility spheres are a sustainable modal split with environmentally friendly modes, increase of shared cars per million citizens from 138 in 2011 to 400 in 2013, decrease in transport related fatalities from 21 per million in 2011 to 9 per million in 2013, decrease of carbon dioxide emissions from 1430kg per capita in 2011to 1348kg per capita in 2013, (Audenhove et al, 2014). Stockholm's cycle path network density is the third densest in the world with 4041km of lanes per 1000square km.

To gain recognition and status in the urban mobility index Stockholm implemented a number of initiatives in its urban transportation system. Those initiatives made its urban transportation system smart and they are reorganization of the bus network with high speed lines and feeder lines, improvement of quality and capacity of all modes of transport, provision of real time information on traffic and bus arrival times and the improvement of the cycle path network. These initiatives helped urban public transportation to achieve a comparative advantage. The introduction of a permanent congestion charge in 2007 led to a reduction in congestion and traffic numbers in the city. According to the IBM corporation (2010), traffic was reduced by 18% and carbon dioxide emissions by 14% in the city centre and increased public transportation ridership by 80 000 passengers per day.

Latin American Cities

Latin American cities used to depend on traditional mass transit systems but these had many problems which are similar to the ones being faced in African cities (Council on Hemispheric Affairs, 2009). The system was characterised by slow, overcrowded, expensive private bus lines, and small buses, which were poorly maintained and highly polluting. The traditional mass transit system was dysfunctional due to regulatory failures and ill-advised bus operation schemes. The bus transit system was dominantly for private operators and they aimed for profit maximisation which led them to charging very high prices. These higher fares promoted an influx of infeasible numbers of private companies in the hopes of reaping high profits. This resulted in a catastrophic oversupply of buses, high costs for impoverished commuters. This consequently increased congestion and produced chaotic routes as bus operators catered to increasingly niche markets. Cities like Quito and Santiago also suffered immensely from air pollution from bus emissions. For instance, in 1995 Quito spent over twenty-eight million dollars in addressing respiratory ailments caused by suspended particulates. Vehicular emissions accounted for over 80% of the suspended particulates (ECLAC, 2008). Informed with this background, cities throughout Latin America strived to achieve a significant revolution in urban design through the adoption and refinement of Bus Rapid Transit (BRT) systems (UN-Habitat, 2012). Despite the fact that BRT is a fast, reliable, and cost-effective means of transportation, private car ownership in the Latin American region keeps on increasing (ECLAC, 2008). This is largely because the higher social status is offered by owning a car. Private car ownership is as high as 97 per 1,000 population in Chile, 138 per 1,000 population in Mexico and in Colombia, car sales increased by approximately 50% between 2004 and 2005 (Becerra, *et al*, 2013). To restrict car use, cities across the region have implemented strategies such as *"Hoy no Circula"* in Mexico City and "*Pico y Placa*" in Bogota, that help to tackle the effects of traffic congestion on air quality by prohibiting use of cars during peak hours. Curitiba and Lima offers good lessons on how the BRT system can help to achieve smart cities and also to improve efficiencies in managing transport systems.

Curitiba, Brazil

Curitiba is the capital city of the state of Parana in southern Brazil. The city is located about 250 kilometres southwest of Sao Paulo near the coastal mountain range. The BRT system was pioneered in Curitiba in the 1970s (Council on Hemispheric Affairs, 2009). Car ownership has been reported to have been 295 per 1000 in 1997 and about 500 per 1000 in 1999. Curitiba has one of the best Bus Rapid Transit in the world and its bus ways are viewed as a model bus rapid transit system. The bus system includes about 60 kilometres of median bus ways and carries about 2 million people per day. The system carries up to 11 100 passengers one way on the busiest bus ways in the peak direction during the peak hour. Bus speeds average 20kmh along the bus way and about 30kph on the direct express routes. The Curitiba BRT system development has been estimated at $1.5 million (U.S dollars) per kilometre. The usage of the bus system is about 70%.

The Curitiba bus system comprises of a trunk and feeder bus system in which buses are routed through a series of terminals where passengers transfer between bus way vehicles, feeder routes and inter-district links with no further payment of fares. Buses which are privately owned and operated are under municipal supervision and they use a common colour coding system. some of the key feature of the Curitiba bus system include physically separated median bus lanes flanked by two local service streets, 26 mid-route and end-of line terminals for transfer among bus takers, an integrated bus fare structure, tube stations with off-vehicle fare collection and platform boarding of buses, direct express service in the parallel one-way arterials and distinctly colored bi-articulated buses along the bus way each with five doors and designed for level boarding at tube stations. These features and the performance of the Curitiba BRT system make the system a smart urban transportation system which developing countries such as Zimbabwe can learn from.

The Curitiba bus system has proven successful in many respects and has been cited as the model for the BRT development in other cities in the world. The popularity of the Curitiba bus system has attracted many motorists to use public transport despite a still high rate of car ownership relative to the rest of Brazil. A 1991 survey reported that about 28% of direct bus users have previously travelled by automobile. The BRT system has been praised for reducing automobile trips and reducing fuel consumption by about 27 million litres annually. Curitiba uses about 30% less fuel per capita because of its heavy transit usage and its ambient air pollution is one of the lowest in Brazil. These are some of the factors leading to Curitiba being labelled as having a smart urban transportation system.

Lima, Capital of Peru

Latin American cities provide a good example of smart transportation. The case study for smart transportation in the Latin American context for this section of the paper is the city of Lima in Peru. The city of Lima is situated along the Pacific Ocean and the Lima metropolitan area is a sprawling conurbation with a population of about 8.4 million inhabitants, (Audenhove et al, 2014). The city is ranked 5th in size in Latin America, (Audenhove et al, 2014).

The urban transportation system in Lima in the 1990s was deregulated and the decade saw the liquidation of the national urban transport company, (Audenhove et al, 2014). The implication of the two events that is deregulation and the liquidation of the national urban transport company was the development of a fragmented and chaotic urban transport environment. The city has 30 000 buses and 200 000 to 300 000 taxis that operate on more than 600 routes throughout the city. Buses and taxis are owned by private individuals and small operators and (Audenhove et al, 2014) provides that only a third of the minibuses are licensed. There is a stiff competition in the Lima transport system between the buses and the taxis. The Lima's transport system is marked with poor environmental and safety performance and lack of public transport integration, (Audenhove et al, 2014) and it is largely informal as supported by the fragmentation and poor performance of the transport system.

Efforts were made in the city of Lima to improve the inefficient and largely fragmented transport system in Lima that was posing environmental and safety threats to the inhabitants of the Lima metropolitan area. Improvements in the city of Lima transport system contributed immensely to the achievements of the transport system towards a smart urban transportation system. the city of lima adopted the Bus Rapid Transit (BRT) locally known as the EL Metropolitano and the opening of a metro line, known as Tren Electrico, (Audenhove et al, 2014). These outlined developments in Lima culminated into an attractive public transport, modal split, agglomeration density and the density of vehicles registered which reduced the number of vehicles in the city to only 139 per every 1000 citizens implicating a reduction in the occurrence of traffic jams. These policy strategies constitute a smart urban transportation system that can provide a good learning ground for the developing countries such as Zimbabwe.

The Lima BRT system was inaugurated in 2010 after it took some years of construction. Buses circulated on special bus lanes that are segregated from the rest of the traffic. Lima's BRT system links principal points in the Lima metropolitan area from the north of the city to Chorrillos in the south. (Audenhove et al, 2014), the headway varies between five and ten minutes and the system carries about 350 000 persons per day. The metro line was first of its kind in Lima and runs for 22km above ground and it has 16 stations. The metro line links Villa El Salvador to downtown Lima and the headway is around 15 minutes. Smart urban transportation system in Lima did not end with developments in the BRT system and the Metro Line, but it went further to the regulatory framework with which the new urban transportation system would work in. notable among the regulatory reforms is the provision of financial incentives to transport operators to take the oldest and most polluting and least safe buses and minibuses off the road to improve the efficiency of the remaining vehicles and the formalization of taxi services. These regulatory reforms had the objective of modernizing and improving the quality of the fleet based on age and road worthiness.

RESULTS AND DISCUSSION: THE HARARE CITY NARRATIVE

The paper has been informed by cases from various places or cities in different parts of the world where problems of transportation in cities similar to Harare have been experienced and measures to improve the transportation systems have also been devised owing to better transport systems in some cities. Cities such as Curitiba, Lima, Hong Kong and Stockholm have been looked at and these have provided various lessons which strategists and planners from cities of the developing world particularly Harare can learn from and help them in moving towards a smart transportation system. Hong Kong is widely known internationally due to its endeavour in attaining a smart transport system to alleviate problems of conventional transport systems such as congestion. Some of the strategies devised in a bid to attain smart urban transport systems are the BRT system and good examples of such cities are Lima in Peru and Curitiba in Brazil. These strategies help in reducing congestion, pollution and urban accidents which have and are still challenges being experienced in the developing parts of the world such as Harare.

Cases presented from various parts of the world showed that traffic related problems largely originate from the rate at which car ownership increases. The increase in car ownership in urban areas causes pressure on the road systems and transport corridors and thus affect the smooth flow of traffic and causes chaos. Knowledge about car ownership as experienced in already developed and developing transport systems of the world informs struggling cities on how they can devise strategies aimed at reducing car ownership through the importation of vehicles from other developed parts of the world. What is inspirational in examples provided in this paper is that cities such as Stockholm and Lima did not necessarily discourage car ownership by policy strategies such as banning importation of cars from other countries but through designing better transport networks and strategies such as the Bus Rapid Transit and modification of roads to enhance the smooth flow of buses. For instance, the development of the Lima Bus Rapid Transit reduced car ownership to 139 per 1000. The Bus Rapid Transport System (BRT) is a smart transport system because it has social, environmental and economic benefits as has been experienced in Asian cities such as Hong Kong.

From the discussion of cases presented in this paper it has been noted that smart urban transport system does not rests on automated traffic components of the transport system but also other modes of the transport system. Other modes of the transport system include rail, minibuses, tram, ferries and taxis. These modes of transport are important for the development of a smart urban transport system thus developing countries such as Zimbabwe and cities within such as Harare should emulate such developments as a strategy to alleviate problems of the outmoded transport components and the simmering challenges posed by informal public transport operation and in that way moving towards the smart urban transport system.

The transportation of people and goods marks the backbone of economic growth and sustainable development and hence the environmental effect of traffic is usually neglected as it is viewed as a threat to economic growth. Sustainable development is also based on the environment being it natural or man-made environment. Traffic movement is characterized by the demarcation of urban activities in terms of land use as people tend to live away from work, services and other urban facilities. This now prompts the movement of traffic and hence it influences the health, safety and attractiveness of the urban environment (Talvitie, 1997). Fuel which is mainly used for automobiles is usually liquid and this has effects to the environment after combustion. Air pollution is one of the major impacts of traffic to the environment. In Zimbabwe, the Environmental Management Authority (EMA) regulates the pollution levels by setting pollution standards but the legislation is not tallying with the huge increase in vehicles in the country. Motor vehicle emissions are no regulated yet they contribute immensely to the pollution of the environment. Exhaust fumes affects the composition of the air and they contribute to the destruction of the ozone layer hence leading to global warming. Another major concern of urban traffic is noise pollution. In Harare, the observations done reviewed that there is much vehicle smoke during the peak hours that is 0600 hours to 1000 hours as this period is characterised by people rushing to work, schools and colleges. The same scenario is experienced during the 1500 hours to 1800 hour period as most people will be driving or travelling back home. These peak hours also showed a high level of noise pollution, air pollution and traffic congestion. The increased air pollution from transport is a threat to sustainable urban transport systems.

CONCLUSION AND POLICY OPTIONS

This chapter has demonstrated that there is a growing realization that the present urban transport system in Harare which is dominated by minibuses is unsustainable as it is characterised by high levels of traffic jam, travel delays and unreliability as well as irregular bus fares. The introduction of mass transit may help to curb this problem and in creating smarter urban transportation systems. This is a reasonable suggestion because the present public transport system is characterised by small vehicles and the operations are fragmented and inefficient as they are run by individuals. Mass transit systems carry more efficient engines; they can save fuel better than small vehicles. Finally, their impacts (negative effects) on the external environment are lower compared to small vehicles. From the foregoing discussion, mass transit provides solutions to the economic, social, energy and environmental challenges in the city of Harare. Alternative fuels should be encouraged primarily by higher gasoline and diesel prices, particularly with carbon taxes. Some alternative fuels may deserve public support for basic development, but these should be evaluated critically to insure they are justified, taking into account all economic, social and environmental costs. Electric vehicle development should be encouraged but their production and use should not be subsidized since their overall benefits are modest.

REFERENCES

African Development Bank (2011). Infrastructure and Growth in Zimbabwe: An Action Plan for Sustained Strong Economic Growth.

Ang-Olson, J., Ecola, L., & Santore, M.K. (2003, July 28-30). Characteristics and Performance of Smart Growth Transportation Systems. Proceedings of the 2nd Urban street symposium, Anaheim, California, USA.

Aworemi, J., Salami, A., Adewoye, J., & Ilori, M. (2008). Impact of Socio-Economic Characteristics on Formal and Formal Public Transport Demands in Kwara State, Nigeria. *African Journal of Business Management*, 2(4), 72-76.

Bogota European Commission. (2011a). *Roadmap to a Single European Transport Area – Towards a competitive and resource efficient transport system* (White Paper). European Commission, Brussels.

Central Vehicle Registry. (2012). Vehicle statistics given verbally.

IBMCorporation, . (2010). *The Case for Smarter Transportation*. USA.

Council on Hemispheric Affairs. (2009). Bus Rapid Transit and the Latin American City: Successes to Date, But Miles to Go. Retrieved from http://www.coha.org/bus-rapid-transit-and-the-latin-american-city-successes-to-date-but-miles-to-go/

Dube, D., & Chirisa, I. (2012, May). The Informal City: Assessing its Scope, Variants and Direction in Harare, Zimbabwe. *Global Advanced Research Journal of Geography and Regional Planning*, 1(1), 16-25. Retrieved from http://garj.org/garjerr/index.htm

ECLAC. (2008). *Developments in public-transport systems in the cities of Latin America. Economic Commission for Latin America and the Caribbean*. ECLAC.

European Commission. (2013). Mobilising Intelligent Transport Systems for EU cities. Commission staff working document SWD527, Brussels.

Fadare, S.O., & Hay, A.M. (1990). Housing Densities, Life Style and Travel Behavior in A Third World City, The Example of Ibadan. *Nig. J. Econ And Soc. Stud.*, 32(2).

Frost and Sullivan. (2010). A Smarter Transportation System for the 21st Century.

Golub, A.D. (2003). Welfare Analysis Of Informal Transit Services. In *Brazil And The Effects Of Regulation*. Institute Of Transportation Studies, University Of California at Berkeley.

Government of Zimbabwe. (1993). *Economic Structural Adjustment Programme*. Harare: Government Printers.

Litman, T. (2014, November 19-21). Implementing Transport Policies and Programmes toward Realizing "Bali Vision Three Zeros - Zero Congestion, Zero Pollution, and Zero Accidents towards Next Generation Transport Systems in Asia." Paper presented at the Eighth Regional Environmentally Sustainable Transport (EST) Forum In Asia.

New Jersey Department of Transportation and Pennsylvania Department of Transport. (2008). *Smart Transportation Guidebook: Planning and Designing Highways and Streets That Support Sustainable and Liveable Communities.*

Palmarini, N. (2012). *Smart Transportation Systems. Human Centric Solution Centers Europe*. IBM Corporation.

Transport Research and Innovation Portal (TRIP). (2014). *Intelligent Transport Systems: Communicating Transport Research And Innovation*. European Commission.

Van Audenhove, F.-J., Korniichuk, O., Daubey, L., & Pourbaix, J. (2014). The Future on Urban Mobility 2.0: Imperatives to Shape Extended Mobility Ecosystems of Tomorrow.

Wagner, A. (2010). *Informal Public Transport; Transport Policy Advisory Services: Federal Ministry for Economic Cooperation and Development Bryman, A 2001, Social Research Methods*. Oxford University Press.

Section 4
Smart Innovations for Cities and Communities

The section covers innovative concepts and solutions developed on the macro-scale of cities and communities. Chapters 6, 7 and 8 present possible urban smart solutions to help reduce noise pollution, as well as minimize waste generation. In one study, accounting for noise pollution by incorporating appropriate mathematical optimization models for designing the layout of facilities during the planning stages of smart cities was found to be effective. The issue of managing and treating construction waste and organic fraction of municipal solid waste (OFMSW) in urban areas has been examined in another study, which included a case-study analysis of the related Dutch, German and Polish guidelines for the sustainable siting of waste processing infrastructure within an urban unit. Concerning urban medical waste, a study has introduced a statistical learning method to estimate the rate of medical waste generation to allow for a more effective way of storing, transporting and disposing the waste. The topic of building smart resilience in cities is next. In chapter 9, the proposed study aims to document the potential and need of developing smart cities in India, as well as understand smart city principles and various dimensions of smart city adopted in various parts of the world which will further help in deriving recommendations and lessons for India's future smart cities. In chapter 10, the building of smart resilience in cities that allows various stakeholders of urban infrastructures, including the local community, to adapt to changes and uncertainties, via a facility information platform for managing the needs of different user groups, was explained through a new concept of building resilience information modeling. Finally, chapter 11 presents an investigation of the role of total quality management (TQM) in smart city development which culminated in the design of an integrative framework with building information modeling and TQM serving as the information architecture and the ideological premise, respectively.

Chapter 6
Accounting for Noise Pollution in Planning of Smart Cities

A.W.A. Hammad
University of New South Wales, Australia

A. Akbarnezhad
University of New South Wales, Australia

D. Rey
University of New South Wales, Australia

ABSTRACT

The incorporation of sustainable design measures in urban planning and development has been steadily increasing in the recent years. Achieving a sustainable urban environment requires accounting for the economic, environmental and social impacts of the development involved. An important factor affecting the social and environmental sustainability of urbanised areas which is commonly overlooked in urban planning is the noise pollution level. Despite the proven impacts of noise pollution on the general wellbeing of individuals within an urban setting, there remains a lack of systematic methods to integrate the impact of noise within the design of urban areas. This chapter seeks to raise awareness of the issue of noise pollution in urban settings while proposing novel approaches for its incorporation as a design parameter in planning the layout of smart cities.

DOI: 10.4018/978-1-5225-0302-6.ch006

INTRODUCTION

The rapid growth in population of cities due to the increasing rate of urbanisation is among the major issues faced by many city planners around the world (Jenkins, Smith, & Wang, 2007). With the increase in population comes a massive surge in the demand for infrastructure and thus the need to undertake substantial construction work, the results of which are construction sites associated with a number of adverse social and environmental impacts. This is yet on top of the environmental and social impacts associated with the operation of the new infrastructure needed to cater for the needs of the increasing population. Therefore, this necessitates the need for adopting appropriate environmental protection strategies to be taken into account during the design and planning phases of urban cities (Ding, 2008; Neirotti, De Marco, Cagliano, Mangano, & Scorrano, 2014). Enhancing environmental and social sustainability by taking advantage of advances in technology and engineering has been emphasised as a core objective in planning of smart cities ((Anthopoulos & Tsoukalas, 2006))). This requires accounting for the adverse environmental and social impacts of various factors in urban planning (Tiwari, Cervero, & Schipper, 2011).

One of the critical environmental and economic impacts associated with construction and operation of new infrastructure is the noise pollution (Goines & Hagler, 2007). Due to its significant and immediate health and safety implications, accounting for noise pollution in planning of sustainable cities is crucial (Adams et al., 2006; Næss, 2001). The level of noise pollution in the cities can be affected significantly by a number of design and planning decisions including the location of different facilities in the city. Accounting for the noise pollution in planning the location of facilities, however, requires the availability of reliable methods for estimating the level of noise pollution generated at the source and the level of noise pollution at different noise-sensitive receivers as well as the availability of optimisation models that incorporate such estimation methods to select the optimal location of the facilities.

This chapter is centred around the issue of smart planning to tackle the environmental and social aspects of urban design, by placing emphasis on reducing urban noise pollution, through facility layout planning. The aim of this chapter is to therefore highlight the importance of accounting for noise pollution during the planning stages of smart cities, which has received little attention in the literature. In addition, this chapter aims to present a methodology for minimising the noise pollution at noise-sensitive facilities across a smart city by incorporating appropriate mathematical optimisation models in urban layout planning.

The chapter starts by reviewing some of the major negative impacts of noise pollution, where reference is made to the relevant literature. The chapter then continues by discussing various methods for quantification and minimisation of noise pollution in smart cities. Developments within a city that are known to produce high levels of noise are analysed, and guidelines for their noise assessments are reviewed. A mathematical modelling method for incorporating noise minimisation as an objective in planning of smart cities is presented. The application of the method is demonstrated by developing optimisation models for certain scenarios applicable to smart city planning and applying them to hypothetical case studies. Finally, a framework for planning the layout of facilities in a smart city setting under the consideration of noise is proposed.

DEFINING NOISE

There is a difference between what can constitute being labelled as noise and what is in general known as sound. Sound is physically defined as a longitudinal mechanical wave, characterised by the variation in pressure that it causes in the elastic medium through which it traverses (Berg & Stork, 2005). Noise is any sound that is considered to be a disruption. Throughout this chapter noise levels from different activities will be reported in the A-weighted decibel scale. This is a filter which considers the non-linear frequency response of the ear to noise at different frequencies (Fricke, 1985).

Distinguishing between different noise sources within a city is critical when assessing the overall impacts of noise pollution. Some of the characteristics present in the sound emitted from disturbing noise sources include intermittent sound levels that are irregular in nature, sound with conspicuous tonal components and sound with dominant low frequency waves (EPA, 2000; Rossing, Wheeler, & Moore, 2002). To clearly understand the level of disruption caused by various sound sources within a city, it is necessary to group noise sources based on the nature of activities producing the noise. In this chapter, four noise source types that are typical of any city are discussed; namely construction noise, traffic noise, railway noise and industrial noise. The assessment of these noise source types is carried out separately; the calculation of a single sound pressure value that encompasses all source levels is not possible as each source propagates noise in a defined manner that is different from any other (EPA Victoria, 2014).

EFFECTS OF NOISE POLLUTION

Figure 1 displays some of the common impacts of noise on individuals, on the environment and on the economy as a whole. Noise has been linked to various ailments and health hazards. Sørensen et al., 2014 studied the link between the cause of ischemic and fatal strokes amongst a cohort of 5700 people due to exposure to

Figure 1. Impacts of noise pollution

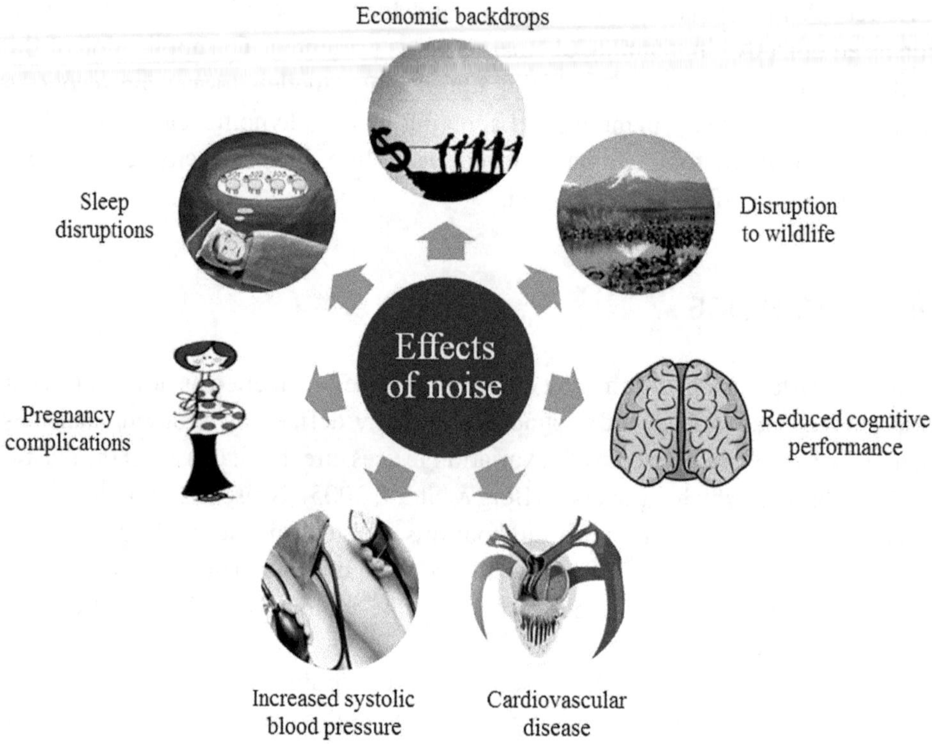

noise and air pollution from road traffic. Based on their results the susceptibility of the subjects to mild strokes was shown to be high, with exposure to noise at levels as low as 10 dB (A).

The risk of serious conditions such as cardiovascular disease in individuals has been linked to exposure to noise levels that exceeded 60 dB (A) (Babisch, 2008).

Noise that is a direct result of construction activities is a major cause of psychological disruptions to residents within neighbouring vicinity ((NG, 2000). By monitoring the blood pressure of workers at a metallurgical factory, Fogari, Zoppi, Vanasia, Marasi, & Villa, 1994 presented a correlation between exposure to noise levels greater than 80 dB (A) and susceptibility to hypertension and increased blood pressure (systolic blood pressure). Osada, 1988 emphasised the important physiological and emotional impacts of noise pollution and proposed a method for quantifying such impacts.

Exposure to high noise level has been also found to increase the risks of hearing loss in patients receiving aminoglycosidic antibiotics (Falk & Woods, 1973). Pregnant women have been observed to suffer from complications due to exposure to noise, with recorded effects on the birth weight of babies (Gehring, Tamburic, Sbihi, Davies, & Brauer, 2014; Nurminen, 1995). In a cross sectional survey conducted by Franssen, van Wiechen, Nagelkerke, & Lebret, 2004 to assess the impacts of noise on health status of residents living in close proximity to Schiphol airport, statistical results indicated that the prevalence of use of sleep and cardiovascular disease medication had an association with exposure to noise from aircraft. Cataract surgery patients needed to stay for longer durations at hospitals due to exposure to high noise levels (Fife & Rappaport, 1976).

Lercher, Evans, Meis, & Kofler, 2002 studied the impact of noise from traffic, on roads and railways, on the mental health of primary school children. A link was found to exist between a child's poor behaviour in classrooms and high levels of noise recorded within the class. Industrial noise has also been investigated in the literature where in one study it was concluded that people living close to industrial wind turbines were more prone to sleep disorder during the night, and reduced cognitive performances during the day (Nissenbaum, Aramini, & Hanning, 2012). Noise pollution induced by wind farms has been associated with disrupting sleeping patterns and causing negative emotions in people living in close proximity to wind farms (Pedersen & Waye, 2007).

Disruptions due to noise exposure are not only a major cause of health issues, but their impacts also span to cover various sectors of the industry. In manufacturing industries precision of equipment could be altered due to vibrations induced by high noise levels (Gilchrist, Allouche, & Cowan, 2003). Economical loss and drops in productivity can be expected to be incurred as a result of decreased cognitive performance by workers exposed to noise (Abo-Qudais & Abu-Qdais, 2005; Belojević, Öhrström, & Rylander, 1992). In areas where tourism play an integral role in the income generation of the region, high levels of noise may dispel tourists and can also lead to a reduction in the number of wildlife habituating the area (King & Davis, 2003). Noise pollution may also impact wildlife in various other ways, particularly in relation to the communication system in animals (Hanna, Blouin-Demers, Wilson, & Mennill, 2011; Penna, Pottstock, & Velasquez, 2005).

MEASURING NOISE LEVELS

For an area where its occupants are expected to be stationary, area monitoring sound devices, namely class Type 1 and 2, Figure 2(a) and (b), may be used to get an overall estimate of the noise level reaching a particular receiver, depending on the degree of precision required, as specified in AS 1259.1 (*Standards Australia*, 1990). On the other hand, the exposure level by a mobile person may be measured using noise dosemeters (Byrne & Reeves, 2008). In such cases since the worker

Figure 2. Sound level meters

(a) Type 1 meter (b) Type 2 meter (c) Noise Dosemeter

happens to be moving the dosemeter is worn by the person and the device records sound pressure levels over a specified duration. The device, Figure 2(c), comprises a microphone that is worn by the person, connected to a unit that records the data of sound levels. This data can then be cumulated to obtain the sound pressure over a specified period. The interval of recording the data can be defined according to the frequency of movement of the worker so that if the worker happens to be stationary for prolonged periods then interval recording period can be set to longer intervals, and vice versa (Crocker, 2007).

NOISE IN URBAN CITIES

The majority of noise in an urban area can be traced to two main sources: construction sites and road traffic (Ballesteros, Fernández, Quintana, Ballesteros, & González, 2010), though other sources of noise are also common. Figure 3 depicts an overview of the sound sources that are found in a city setting. Highways are sources of noise nuisance since their noise levels are known to disrupt sleep patterns (Vallet, Maurin, Page, Favre, & Pachiaudi, 1978). Airports are sources of disturbance to residents within close proximity (Black, Black, Issarayangyun, & Samuels, 2007). Another type of noise source found in cities is the operating industrial developments. Sources of noise in industrial environments include rotations of mechanical devices and working plants, electrical machines, hydraulic presses etc. A characteristic that differentiates industrial noise from construction or traffic noise is the frequency of the sound produced. While industrial operations tend to generate sound over the

Figure 3. Noise sources in cities

entire audible frequency range with sound being emitted at a steady level, the noise pollution associated with construction is generally intermittent (Razavi, Ramezanifar, & Bagherzadeh, 2014).

The geometric and topological features of cities are distinguishing elements contributing to the intensity of the sound emission within urban areas. Prominent noise amplification within cities is due to the presence of buildings that act as reflection surfaces (Egan, 2007). The predominance of linear, rather than point noise sources is also one of the common characteristics of noise in cities (Katti & Warren, 2004). For specific facilities, such as open space parks, noise level perception by park-goers can be directly related to the expectation of noise environment. Certain limits are imposed in hospitals for maximum noise levels that patients and healthcare workers within the facility can be exposed to. For instance, The Department of Environment and Climate Change NSW, 2009 specifies a maximum limit of 45 dB(A) for the equivalent continuous A-weighted sound pressure level, L_{Aeq}, threshold measured over a 15-minute period for internal hospital wards. Compliance with this limit is hardly observed however (Busch-Vishniac et al., 2005).

In the following sections, the characteristics of various aforementioned noise sources typical of urban cities are described in detail. In addition, the recommended procedure for assessing the overall impact of each noise category on noise sensitive receivers, as set by different regulations and standards, is discussed. A number of noise mitigation methods for reducing the noise levels that violate the specified requirements are presented.

Noise Sources on Construction Sites

Construction sites are well renowned for the intensity of the noise levels produced by their associated operations. Unlike noise that is produced in other trade settings, noise levels on a construction site tend to be highly variable in nature (Lusk, Kerr, & Kauffman, 1998). This means that workers and residents in the vicinity of the construction site are likely to be exposed to short intermittent periods of high intensity sounds. Studies reveal that exposure to intermittent levels of noise is perceived more disruptive to an individual than noise that carries on at a steady level (U.S. Department of Health and Human Services, 1988).

Many regulations are put in place to limit disruptions that can occur due to the activities taking place within a construction site. In Australia, and in many other countries, these regulations are set and controlled by the council overlooking the locality defining the area within which construction work is undertaken (Ballesteros et al., 2010; Standards Australia, 2010). The controls aim to limit the total noise that surrounding residents are subjected to by defining specific work hours. It is necessary to combine along with these regulations measures that aim to attenuate the overall noise levels during the operation of activities. A number of these measures have been outlined in AS 2436 together with their estimated overall impacts on dampening the sound levels (Standards Australia, 2010).

The majority, if not all, of noise emanating from construction sites can be attributed to the machinery and equipment utilised during construction activities. Noise levels within a construction site can reach levels of up to 114 dB (A) (Kenney & Ayer, 1975). Examples of delinquent noise sources contributing towards such high levels are ample. For instance, excavation/earth work is responsible for the highest levels of noise pollution emitted from a construction site. This can be attributed to the large equipment needed to handle material at this particular stage of a construction project. On the other hand, the finishing stage is known to produce the least level of noise during its execution on a construction site, due to the nature of the equipment involved in its implementation (Ballesteros et al., 2010).

Heavy mobile machinery that are not limited to one specific area on the construction site pose the most challenging encounter to noise control regulators, as these types of machinery tend to traverse over large sections, hence propagating sound in all directions and to many receiving ends located outside the construction site. Equipment size however is not a sole determinant of the intensity of sound produced as there are plenty of examples of small equipment that produce sound levels greater than 75 dB (A) (As a relative guide, an acceptable upper bound threshold for traffic noise in a neighbourhood is around 63 dB (A) (Department of Infrastructure, Energy and Resources, 2011)). Daily time weighted average sound level samples collected by Neitzel, Seixas, Camp, & Yost, 1999 on various construction activities undergone within a construction site indicate that the stage of construction producing the highest noise level is the structural work phase. Tools such as jackhammers, bulldozers, roto-hammers and chipping guns were found to be producing the highest mean sound levels.

Table 1 shows the threshold noise levels for different receivers around the construction sites, as specified by the Department of Environment and Climate Change NSW, 2009. Such noise level limits for sensitive areas may be compared with predicted or measured noise pollution around the construction site to identify potential disturbance.

Based on the standards for prediction of site noise levels, presented in AS 2436 and BS 5228, estimation of construction noise levels is performed according to the nature of the construction plant involved (BSI British Standards, 2009; Standards Australia, 2010). Depending on the type of plant operating during construction processes, several methods are available for initiating the estimates. Figure 4 depicts the procedure for predicting noise on construction sites. Commonly adopted methods are discussed in the forthcoming sections.

ACTIVITY EQUIVALENT CONTINUOUS A-WEIGHTED SOUND LEVEL METHOD FOR STATIONARY PLANT ON CONSTRUCTION SITE

As specified in standards AS 2436 and BS 5228, the preliminary step to obtaining the equivalent continuous A-weighted sound level, L_{Aeq}, for stationary equipment on a construction site is the identification of the locations of each of the concerned

Table 1. Noise level limits for different receiver categories

Receiver Type around Construction Site	Noise Level Limit $L_{Aeq(15\ min)}$ (dB (A))
Residents	75
Schools	45
Hospitals	45
Parks	65

Adapted from Department of Environment and Climate Change NSW, 2009

Figure 4. Noise prediction on construction sites

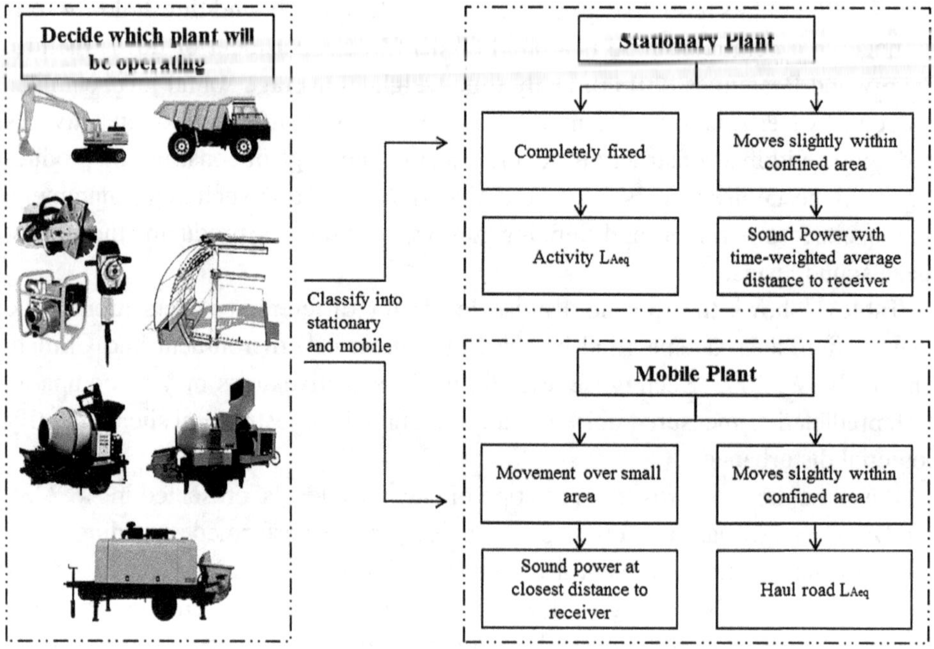

equipment with respect to the receiver positions (BSI British Standards, 2009; Standards Australia, 2010). This is mainly due to the impact that the location of equipment has on total noise level propagation with respect to receivers around the construction site. Facilities on a construction site tend to accommodate stationary equipment. The important effect of construction site facility location on noise propagation provides planning opportunities in terms of minimising the noise level at particular receivers, by optimising the location of different facilities on a construction site (A.W.A. Hammad, Rey, & Akbarnezhad, 2014). Identifying the optimal locations of facilities on a construction site to achieve particular objectives, including reducing material handling costs and reducing total noise levels directed towards noise sensitive recipients located outside the construction site boundary is commonly referred to as site layout planning. In the following sections, site layout planning with the objective of reducing the noise pollution in cities is discussed in detail as a planning strategy to improve sustainability in smart cities.

Once the positions of the equipment have been determined, sound level data table sheets, such as the one shown in Table 2, can be used to obtain estimates of the 10 m L_{Aeq} levels of the identified construction equipment. The 10m L_{Aeq} is then corrected if the distance between the source point and the receiver, D, is greater than 10m. The distance correction, K_h and K_s, is calculated to factor in the surface type

Table 2. Extract of a sound level data table

Equipment	Equivalent Continuous Sound Pressure L_{Aeq} at 10 m dB (A)	Sound Power L_{WA} dB (A)
Truck mixer	84	112
Tracked excavator	85	113
Poker vibrator	81	102
Pneumatic breaker	86	113
Disc cutter	84	112
Electric drill	78	105

Adapted from BS 5228 (BSI British Standards, 2009)

of the ground over which noise traverses. If the surface happens to be classified as hard then Equation 1 is deployed; else for soft ground Equation 2 is used.

$$K_h = 20\log_{10}\left(\frac{D}{10}\right) \tag{1}$$

$$K_s = \left(25\log_{10}\left(\frac{D}{10}\right)\right) - 2 \tag{2}$$

The calculated distance correction factor is then subtracted from the L_{Aeq} value (at 10 meter from source) specified in Table 1. The ensuing step is to account for reflections from buildings surrounding the receiving point, as well as any screening that is a result of noise barriers put in place between the path of direct sight of the sound source and the receiver. The reflection from the adjacent buildings is accounted for by considering an average value of about 3 dB (A); this intensifies the sound levels and is therefore added to the L_{Aeq}. Accounting for the screening effect due to noise barriers is, however, more challenging and an approximation is usually made based on the visibility of the sound source from the receiving end. If more than 50% of the sound emitting source is visible from the receiving end, then 5 dB (A) is deducted from the L_{Aeq}. Otherwise, if the sound source happens to be completely obscured, then a total of 10 dB (A) is subtracted from the L_{Aeq} (BSI British Standards, 2009).

Sound level summation is logarithmic in nature and hence noise from different equipment cannot be added linearly. Therefore, a scale indicating the effect of the disparity between the individual noise levels should be adopted for estimation of the total sound level.AS 2436 (Standards Australia, 2010) recommends two methods which may be selected depending on the nature of the operating hours of the plants. In cases where plants operate simultaneously for the whole duration of the assessment period, a logarithmic correction factor is used such that for a pair of given sound levels that are similar in magnitude, the highest correction factor applies to the maximum of the pair of assessed sound levels. As the disparity between the sound levels increases, the correction factor decreases, until it reaches zero, where no effect on the highest sound level is prompted. On the other hand, in cases where activities tend to take place for durations that are less than the assumed assessment period, the following equation may be used to specify the differences in the operating times of different equipment.

$$L_{Aeq(T)} = 10\log_{10}\left(\frac{1}{T}\right)\sum_{p} t_p 10^{0.1L_p} \quad \forall p \in P \tag{3}$$

In Equation 3, an assessment period, T, which is a measure of the average working hours for a typical day on a construction site, needs to be specified. For each equipment, p, the corresponding operating time over the assessment period, t_p, and the equivalent continuous sound level, L_p, are plugged into the equation to obtain the overall estimate of the L_{Aeq} at the receiver point. To achieve an acceptable noise level for the receivers surrounding the construction site during the construction phase in smart cities the aim is to minimise such a value.

NOISE PREDICTION METHOD FOR MOBILE PLANT OPERATING IN A DEFINED REGION OF THE CONSTRUCTION SITE

In order to identify the sound levels resulting from mobile plants, it is necessary to look up the sound pressure values, L_{WA}, for the respective equipment from sound level databases (Table 1). This measure differs from the L_{Aeq} descriptor discussed previously for stationary equipment as it represents the logarithm of the ratio of sound power emitted from a source to a reference level of 10-12 Watts. The subsequent step involves determining the resulting sound pressure, L_{pa}, using the closest distance between the mobile plant and the receiving point. This is done by examining the traversal path of the mobile plant and deriving the point on the path that is closest to the receiver. Once this distance has been determined, Equation (4) or Equation (5) can be used to obtain the L_{pa}, depending on whether the construction ground surface is hard or soft, respectively.

$$K_h^{'} = 20\log_{10}\left(D\right) + 8 \quad (4)$$

$$K_s^{'} = 25\log_{10}\left(D\right) + 1 \quad (5)$$

Sound attenuation and amplification due to sound barriers and building reflections, respectively should be also accounted for, as described for stationary plant. The ratio of the distance of the closest operation of the plant away from the receiver to the total length of the traversal path of the plant is calculated and used to obtain the equivalent on-time of the plant. This is a correction factor, which when combined with the estimated total operating time of the plant within a selected assessment period, gives the corrected on-time period of the corresponding plant (Equation (6)). In Equation (6), F refers to the on-time correction factor, T refers to the time the plant is expected to operate during a specified operating period and t reflects the corrected on-time period. The assessment period duration is chosen to reflect the typical operating hours of the plant during a single day of work. The final step involves calculating L_{Aeq} over the assessment period using the corrected on time period and the sound pressure level.

$$t = FT \quad (6)$$

Noise Due to Roads

Relevant regulatory authorities usually provide guidelines on common strategies for controlling the road traffic networks and their adverse environmental disturbances within a city. These strategies may have a significant impact on the planning and development of smart cities, since as the demographics of a particular city increase, car ownership and the use of other transportation means is also likely to increase, which may result in intensifying the noise-induced externalities associated with the transportation network. When compared to other sources of noise disturbance, traffic noise is one of the most negatively perceived nuisance to residents (Gan, Davies, Koehoorn, & Brauer, 2012; van Kempen & Babisch, 2012). A primary focus in the planning of smart cities should be the relevant use of land planning strategies to mitigate noise levels from traffic (Department of Environment, Climate Change and Water NSW, 2011).

A common irritant that is a direct result of traffic noise reported by residents is the associated sleep disturbance. According to World Health Organisation (WHO) an important point to note when assessing the relationship between sleep and traffic noise is the difference in the resulting noise level as measured with respect to the level of background noise (WHO, 2011).

Within a city, different categories of roads, including arterial roads, sub arterial roads, collector roads and local roads, are usually present (Roads and Maritime Services, 2009). Traffic noise should be assessed and accounted for in accordance with the category of road present at a specific location. The limits proposed in the EPA guidelines are based on the L_{Aeq} values of noise levels emitted from roads at different times throughout the day (Department of Environment, Climate Change and Water NSW, 2011). For day-time measurement, noise levels are given over an assessment period of 15 hours for assessing noise at receiver points corresponding to residential areas; during the night time, the period is specified to be between 10 pm and 7 am, and the equivalent sound levels are assessed over a 9-hour duration.

In Australia, the noise from airports is also taken into consideration in the local guidelines where annoyance of residents is measured using the 20 Australian Noise Exposure Forecast (ANEF) ("The Australian Noise Exposure Forecast (ANEF) System," 2014). This measure corresponds to the noise level point at which 10 percent of the residents in close proximity to the airport are highly annoyed due to the aircraft noise.

Figure 5 depicts a flow chart of a possible assessment process that may be conducted to assess the impact a road project may have on the neighbouring noise receivers within a specified zone in an urban setting (Roads and Maritime Services, 2015). The process starts by determining the function class of the road project. The second step involves identifying the type of the project to be undergone. Project type may be divided into two categories: 1). New road construction projects, where a road is placed in an undeveloped corridor or when an existing road alignment is completely altered; 2). Road redevelopment projects, where traffic capacity of existing road is increased.

In the third step, the presence of transitional zones is investigated. Transitional zones are characterised by junctions that are found between new and redeveloped roads, between road projects and between existing roads or between different road functional classes. The transition zones mark a gradual change in the noise classification criteria. The fourth step is to delineate the study area and the receivers to be considered. For residential receivers, an assessment criterion from Table 3 is applied. Table 4 is applicable for non-residential land noise exposure limits.

Noise Due to Railways

Along with their economic and environmental advantages, public transport systems also have a number of undesired negative environmental and social effects which

Figure 5. Road noise assessment for urban setting

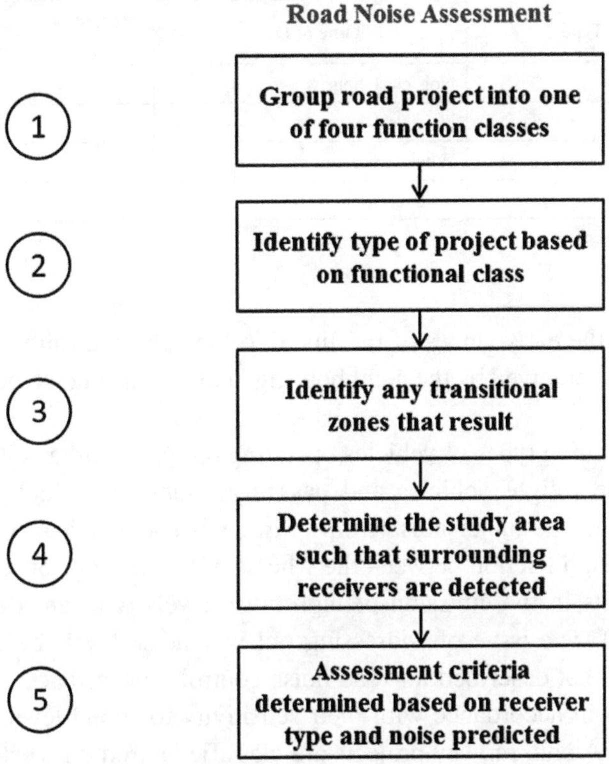

Table 3. Criteria for road assessments near residential areas

Road Category	Type of Project	Assessment Period for Measuring L_{Aeq} (in hours)		Assessment Criteria ($L_{Aeq,\ (period)}$ dB (A))	
		Day	Night	Day	Night
Freeway/arterial/ sub-arterial	New road	15	9	55	50
	Redevelopment & Existing roads affected by additional traffic	15	9	60	55
Local roads	New, redevelopments and existing roads affected by additional traffic	1	1	55	50

Adapted from (Roads and Maritime Services, 2015)

need to be considered in smart city/urban planning. One of such negative impacts is the noise pollution associated with their operation.

A guideline for assessing and mitigating noise disruption from rail services has been published by EPA, 2013. These guidelines essentially propose a series of

Table 4. Recommended noise levels

Receiver Type	Time of Day	Recommended L_{Aeq} Noise Level dB(A)
Schools internal	Noisiest 1-hour	35
Hospitals-internal	Noisiest 1-hour	35
Place of worship	When in use	40
Recreation Area	When in use	55

Adapted from (EPA, 2000)

regulations for the early stages of rail line development to ensure that acceptable noise levels are perceived by the neighbouring facilities during the operation of the rail track.

The two common types of vehicles operating on typical urban city railway line systems include freight vehicles and passenger transport vehicles (ENVIRON Australia, 2013). The major noise from railways is usually caused by the rolling stock, where rough friction between the wheels of the carriage or locomotive and the tracks results in the emission of high noise levels (Craven, Bewes, Fenech, & Jones, 2013). The issue of addressing rail way noise levels bears some close resemblance to that described for road noise control. For instance, receivers need to be classified in accordance with their sensitivity to sound levels from nearby rail operations. Again, railway projects are classified based on whether they are a new development, where the land on which the rail line is to be constructed has no operating services, or redevelopments, such as existing rail corridors which are expanded to increase capacity.

Quantifying Noise from Rail Lines

Equation (7) can be deployed to quantify the level of noise emitted from the rail tracks. In this equation, each event, v, which represents the train type traversing the rail tracks, is assessed separately. The estimated noise of events is then combined logarithmically to determine the total noise. $L^{v}_{Aeq(15hr/9hr/1hr)}$ measured over a given period, being day or night, for each event, is converted into an A-weighted sound exposure level, by the addition of a $10\log_{10}T^{v}$ term to the L_{Aeq} level over the assessed period duration. The individual time period for each event, T^{v}, and the overall time period, T, are measured in seconds (EPA, 2013).

$$L_{Aeq(T)} = 10\log_{10}\frac{1}{T}\sum_{v}\left(n_{v}10^{\frac{L^{v}_{Aeq(15hr/9hr/1hr)}+10\log_{10}T^{v}}{10}}\right) \tag{7}$$

As depicted in Figure 6, the assessment for railway noise takes place over five steps. In the first step, the study area surrounding the railway project is delineated to identify receivers affected by the noise levels. The trigger levels for the associated noise sensitive receivers considered can be derived from Tables 5 and 6. L_{Aeq} and L_{Amax} are used to refer to the average noise levels from events of individual train pass-by and the maximum noise level that is not surpassed by 95 per cent of train pass-by, respectively.

In the second step, following the identification of receivers, the rail way project is categorised as either a new rail line project or a redevelopment project for an existing line. Assessment levels are more rigorous for new rail lines as there is usually more flexibility in selection and implementation of noise mitigation measures (EPA, 2013). This is explicit in the trigger levels that are presented in Table 5. The third step involves classifying the rail way project as either being light rail or heavy rail line. This classification is required to determine the trigger levels associated (Table 5). Light rails have lower capacities compared to heavy rails and thus are used only for passenger transportation (Korve, Farran, & Mansel, 1995).Ground

Figure 6. Flow chart for railway noise assessment

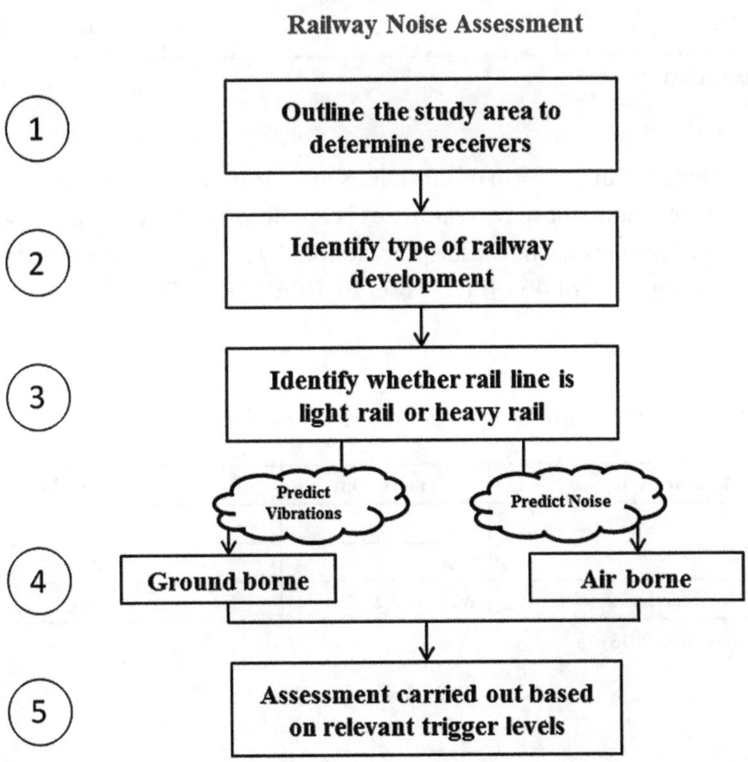

Table 5. Criteria for airborne road noise assessment near residential areas

Rail Development Category	Type of Development	Assessment Criteria 1 ($L_{Aeq, (period)}$ dB (A))		Assessment Criteria 2 (L_{Amax} dB (A))	
		Day (15h)	Night (9h)	Day	Night
Heavy Rail	New Rail	60	55	80	80
	Redevelopment	65	60	85	85
Light Rail	All development types	60	50	80	80

Adapted from (EPA, 2013)

Table 6. Recommended Air-borne trigger noise levels for sensitive receivers

Receiver Type	Time of Day	New Rail Recommended $L_{Aeq(period)}$ Noise level dB(A)	Redevelopment Rail Recommended $L_{Aeq(period)}$ Noise level dB(A)
Schools-internal	Noisiest 1-hour	40	45
Hospitals-internal	Noisiest 1-hour	35	40
Place of worship	Noisiest 1-hour	40	45
Recreation Area	Day time-15h	65	65

Adapted from (EPA, 2013)

borne noise emitted in the form of vibrations and air borne noise are estimated and the higher value among these two estimates is applied. For the case of ground borne vibration measurements, the descriptor adopted is the L_{Amax} measure where a maximum trigger level of 45 dB (A) is set by EPA, 2013 (Table 7).

Table 7. Recommended ground-borne trigger noise levels

Receiver Type	Time of Day	L_{Amax} Noise Level dB(A)
Residential	Day	40
	Night	35
Other sensitive receivers	When in use	40-45

Adopted from (EPA, 2013)

Industrial Noise

The industrial sites located within the city boundaries are among the major contributors to urban noise pollution, though with a small influence zone (Brown & Lam, 1987). Such sites encompass warehouses, commercial complexes and power source equipment like electrical generators etc.

Figure 7 outlines a possible procedure for assessment of noise pollution associated with existing and newly proposed industrial developments. In this flowchart, two sets of measurements are required to assess the impacts of noise from industrial sites on the sensitive receivers. The first set of measurements are used to assess the noise intrusiveness. This assessment is a short term one that accounts for the annoyance caused by the high noise levels, from industrial facilities, targeted at neighbouring communities, making use of background noise levels. Background noise level plays a key role in magnifying the level of disruptiveness of noise experienced by the residents (Department of Environment and Resource Management, 2010).

The second required set of measurements are used for the Amenity Assessment. The noise levels at specific noise-sensitive locations should satisfy certain sound level thresholds requirements, as determined by the type of users of the respective location. Table 4 can be adopted for use in the industrial development Amenity Assessment, with the addition of other criteria for residential threshold limits, including 60 dB (A) for the daytime period and 50 and 45 dB (A) for the evening and night time period durations, respectively.

The two aforementioned assessments are used to determine the limiting criteria; i.e. the appropriate noise level below which sound due to the proposed industrial

Figure 7. Noise assessment for industrial developments

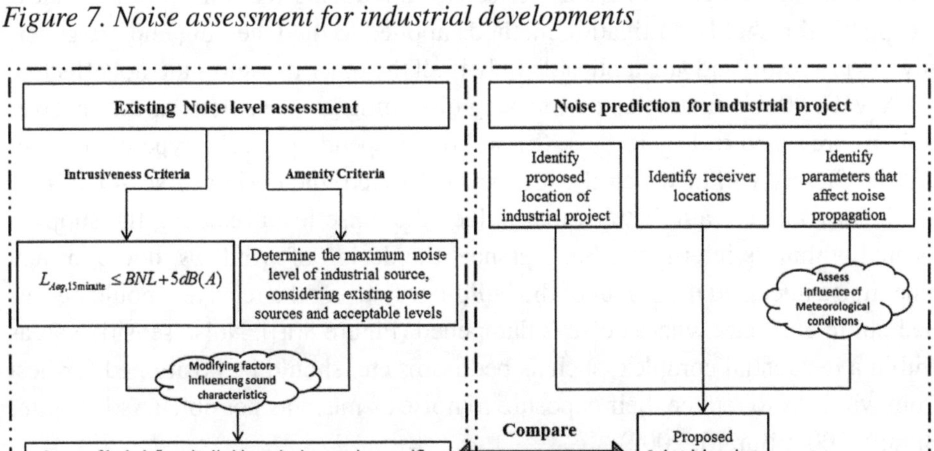

developments is to be kept. The result that ensues from these two assessments is known as the project-specific noise level. Before obtaining the project-specific levels, modifications may be required for each criterion to address the characteristics of noise that cause the greatest annoyance. Noise descriptor for the intrusiveness criteria is the L_{Aeq} measured over a period of 15 minutes, whereas the descriptor for amenity is the L_{Aeq} measured over a period corresponding to the hours within a day, evening or night time period (EPA, 2000).

While performing the above measurements, the expected noise level of the proposed industrial premises is calculated and is compared to the project specific noise level. In cases where the predicted noise levels exceed the project-specific level, i.e. the development does not satisfy threshold limits, efforts should be made to mitigate the excess noise produced. When predicting noise levels from a source, it is essential to identify every possible parameter that can directly impact the final sound level reaching the receiver. This includes meteorological effects that increase the sound levels due to exacerbation of propagation paths of sound waves (Marsh, 1982).

NOISE MITIGATION

Sound mitigation techniques may be used to reduce the level of noise produced within cities to acceptable limits. Each sound source is unique and hence requires a specific mitigation procedure (Manuel, 2005). Sound mitigation can be applied to the source itself, to the sound propagation path, or to the receiver surrounding the noise source. Noise should be preferably addressed at the source itself. The mitigation measures applicable to propagation path are used commonly when the source mitigation measures are not effective in reducing the noise level to below acceptable thresholds. Mitigation methods applied to the receiving end are generally the last resort and are applicable only if all the other measures are not effective (EPA VIC, 2008). Examples of various noise mitigation methods at the receiver end are presented in Figure 8(a), (b) and (c). Depending on the type of receiver and the facility within which the receiver is located, the design of such facilities can incorporate certain features that reduce the noise level reaching the subjects located within its internality. For instance, the design of windows, doors, or any other noise inlet, and the layout of the buildings housing the receiver should be carried out in a manner where noise is dampened (Figure 8(a)). Noise sensitive areas within a residential complex, such as bedrooms etc. should be positioned furthest from windows to reduce their exposure to noise as much as possible (Asdrubali & Buratti, 2005; Buratti, 2002).

Figure 8. Noise attenuation measures at receivers

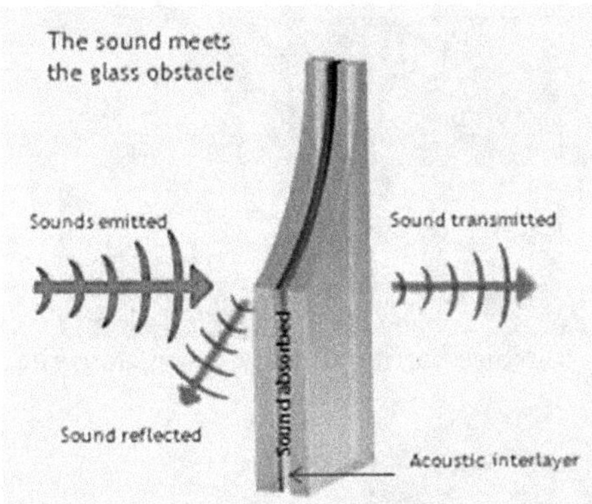

(a) Double glazed windows

(b) Barriers around building

(c) Noise insulation

Construction Sites Noise Mitigation

Construction noise controls that are applied directly at the source target the noise-emitting construction equipment. Some of the suggested controls include maintaining good working conditions, such as even haul roads etc., to ensure smooth operation of equipment, turning the engine off whenever the equipment is idle, or even performing regular engine check-ups and mechanical service routines (BSI British Standards, 2009). Furthermore, the use of enclosures which barricade the operation zone of the noise-emitting equipment is another common noise control approach to reduce the noise pollution from the construction site (Figure 9 (a)).

For noise control methods at the receiver end, the usual measures, including dampening of the noise levels through the use of concrete personal enclosures, noise absorption techniques and applications of noise reflectors, are adopted. Measures could be as simple as having buildings subjected to external disrupting sound in-

Figure 9. Noise attenuation measures on construction sites

(a) Insulated Noise enclosure

(b) Noise barrier around construction site

(c) Temporary noise barrier

ternally insulated with noise absorbing material such as fibreboard, plywood or fibre cement boards, all known to have good sound reduction indices (Standards Australia, 2010).

The application of noise attenuation measures at the transmission path is done mostly through ensuring a sufficient gap between source point and receiving point, or through the application of noise barriers that block the direct line of sight between the noise source and the receiver. Examples of such barriers are depicted in Figures 9 (b) and (c).

Road Noise Mitigation

The most effective traffic noise abatement measures are those implemented at the source, i.e. vehicles and around roads (Boer & Schroten, 2007). The methods to reduce the road noise at source include technical approaches and policy enforcement. The common technical approaches comprise the installation of noise mufflers on vehicles (Figure 10 (a)) and increasing the noise absorbance of the road pavement through the use of materials such as crumb rubber in the asphalt mix (Figure 10 (b)) (Losa, Leandri, & Rocchio, 2012). On the other hand, policies may also be enforced to reduce the road noise including setting limits for speed and flow of traffic at specific times during the day (Nielsen & Solberg, 1988).

Figure 10. Noise attenuation measures for traffic noise

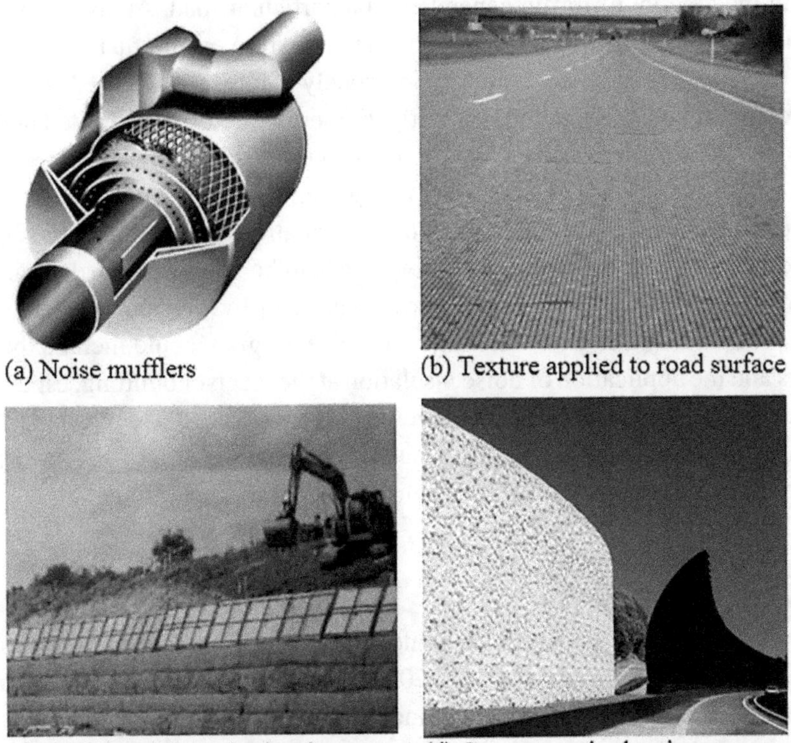

(a) Noise mufflers (b) Texture applied to road surface

(c) Earth bunds as noise barriers (d) Concrete noise barriers across roads

When a new road is to be constructed in a particular area, more flexibility is present in terms of changes to the planned location of the road to mitigate any exceeding sound levels (Sandberg, 1999). Changes to plans are usually less costly than dealing with the issues arising from high noise pollution after completion of the road construction. Alteration of the routes to prevent passage through noise sensitive regions should be considered. Alternatively, traffic management strategies can be applied to determine the routes and use of road by heavy vehicles such as trucks etc. (Department of Environment, Climate Change and Water NSW, 2011). Restrictions imposed on certain vehicles, during night time, targeting noise sensitive zones are also a plausible act that can be effectuated. Standards for noise emissions from vehicles which are to be implemented within a specified period of time could be proposed by the regulatory authorities (Ali & Tamura, 2003). Such strategies are more geared towards long term noise mitigation measures. Planning does not only involve the route selection and location of roads, but it also encompasses the location of facilities that are to be developed and whose operation is likely to cause an increase in the amount of traffic generated on a particular road. An assessment of the positioning of these facilities should be carried out to ensure that the increase in traffic noise level can be contained appropriately (Murphy & King, 2014).

For existing roads, the methods for mitigating noise are usually limited and are only applied in the form of noise barriers between the road and the noise receivers (Scholes & Sargent, 1971). In other words, targeting the noise propagation path is the most practicable method of handling noise attenuation on existing roads. Figure 10 (c) and (d) represent examples of barriers that can be applied to roadways.

At the receiving end, abatement programs can be set up by regulators to counteract the noise produced by traffic on existing roads. The program could include the use of barriers and the application of noise insulation at the receiver building, especially if the receiver happens to be residential dwellings (Naish, Tan, & Demirbilek, 2014; RTA, 2001). Other innovative measures, such as green roofs at facades, have been suggested in the literature (Van Renterghem & Botteldooren, 2009).

Railway Noise Mitigation

Several strategies can be deployed at the noise source to decrease the total sound levels dispersed from rail stations (EPA, 2013). The track path of the rail line may considerably affect rail noise generation and propagation. Therefore, the effects of alternative track paths on the noise pollution should be investigated during planning. Regulating the total number of vehicles passing the railway within a specified schedule and regulating the speed and lengths of the trains have also been suggested as alternative methods for controlling the noise pollution (Iwnicki, 2006). The speed of the train can be adjusted to a level where noise levels are within allowable limits. The length of the train passing by at certain instances can also be reduced by restricting the number and/or type of rolling stocks. In addition, the track on which trains traverse can also be modified through the use of special track forms and rail grinders (Figure 11 (a)). Operators can be asked to handle the train breaks in a certain way when passing through a noise sensitive area to avoid disruptions to receivers due to break and wheel squeals (Eadie & Santoro, 2006). Locomotives are a source of high noise levels due to their whistles and/or engines, and hence retrofits and noise mufflers can be applied to reduce noise (Xiaoan, 2006).

The topology around the railway line can also be used for noise attenuation purposes. As an example trenches can be dug to contain the noise at the source. Earth mounds and grass embankments can also be positioned on either side of the rail line to act as noise barriers (Figure 11 (b)). Ballast mats and vibration isolating baseplates, as shown in Figure 11 (c) and (d) respectively, can be installed to

Figure 11. Noise attenuation measures for railways

(a) Rail grinder for smoother contact between wheels and rails

(b) Railway earth mound/bund and trench for sound insulation

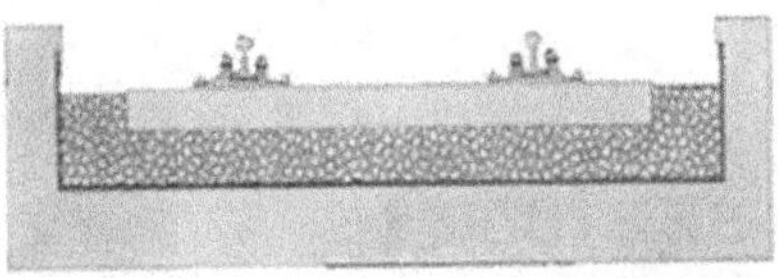

(c) Ballast mats for vibrations

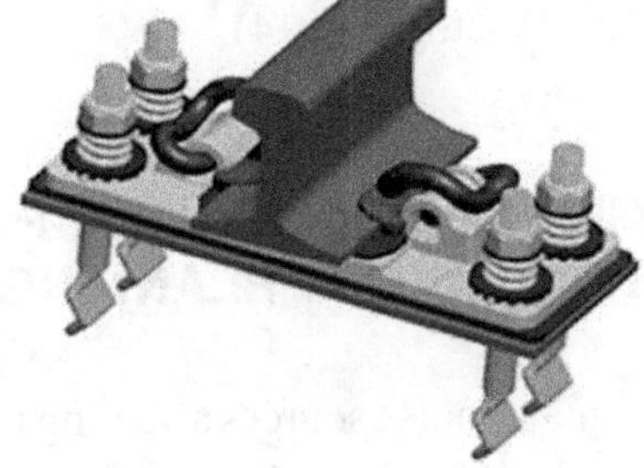

(d) Resilient baseplates

dampen ground-borne noise (RailCorp, 2012). When rail lines are very close to receivers, noise reduction measures targeting receivers may be required to bring down sound levels to acceptable thresholds.

Industrial Development Noise Mitigation

For noise mitigation strategies applied at industrial sources, the methodology can either be classified as a best practice management strategy or a best economical technology (EPA, 2000). For best practice strategies, simple operational measures such as scheduling noisy activities during daytime instead of night time, moving noisy equipment to areas that are well shielded due to barriers (Lan & Chiu, 2008), and servicing and maintaining equipment in good conditions such that noise due to mechanical faults are dealt with rapidly, constitute the measures that can be realised (Lyon, 2013). When technology is applied, the method usually involves the installation of noise mufflers, acoustic-range controlling equipment that limit the noise frequency, or pulse width modulation schemes that eliminate low order harmonics (Vijayraghavan & Krishnan, 1998). Efficient enclosures around the sound source and ensuring engines utilised are operated so that they are as quiet as can possibly be permitted without compromising productivity, are other reasonable actions that can be effected.

The propagation path between the noise source and the receiver can be treated in a number of ways to help reduce the total sound level travelling along the path. Effective measures are achieved when barriers are applied. Planning the land use is also an effective means for limiting total noise propagation (Barron, 2002). If buffer zones are allocated between noise sensitive land and industrial zones, then noise levels can be kept at a minimal as the distance between source and receiver increases. The scheme to be adopted for the layout of facilities around industrial areas and the allocation of space within these developments can also be designed in a way to satisfy an alignment that results in the least sound disruptions (EPA, 2000; EPA Victoria, 2014).

MINIMISING URBAN NOISE POLLUTION THROUGH FACILITY LAYOUT PLANNING: CASE STUDIES

The two major noise sources affecting noise-sensitive facilities in cities are the transport noise sources, including road and rail noise and the construction noise sources. Noise due to construction activities is especially a major concern in urban areas when multiple construction projects are overlapping in time. The noise pollution at sensitive receivers in an urban area can be significantly affected by the location of receivers with respect to facilities generating the noise (Hammad, Akbarnezhad, & Rey, 2016). Therefore, in sustainable urban planning where noise is a concern, service-based facilities such as hospitals, public parks etc. need to be strategically located within the city such that noise impacts from high-level sound producing sources are kept to a minimum. However, selecting the location of noise sensitive and noise generating facilities also requires considering the other economic, environmental, and social impacts of the selected locations. Specifically, a systematic approach is required to quantify the trade-offs between location decisions. Developing optimisation models that can account for the impact of noise pollution as well as other important factors such as operations costs in layout problems is thus an important step toward balanced decisions for facility location in urban areas (Hammad, Akbarnezhad, Rey, & Waller, 2015).

To illustrate the potential applications of noise minimisation models in facility layout planning, two case studies are presented where the aim is to produce a configuration of facilities such that overall noise is reduced while accounting for site layout design constraints. The first case presents a planning problem aimed at identifying the optimal locations of noise generating facilities on a construction site to minimise the noise at two existing noise sensitive buildings close to the construction site. The second case involves allocating different facility types to locations surrounding a railway line, such that sound level measured at these servicing facilities is minimised while at the same time the noise threshold limits corresponding to each facility is not violated.

Construction Site Noise Model

Figure 1 shows a construction site with a predetermined building location footprint. The aim is to identify the location of temporary construction facilities on the site so that the noise pollution at two neighbouring pre-existing noise-sensitive buildings located on the east and north sides of the construction site is minimised. Construction site facilities are grouped generally according to their functioning roles; temporary facilities represent those facilities that will be operating during a stage of the construction process and which will need to be positioned during the respective stage in which they are required (Andayesh & Sadeghpour, 2014; Li & Love, 1998). On the other hand, permanent facilities occupy one predetermined position throughout the entire project duration (Khalafallah & El-Rayes, 2008), *i.e.* they do not need to be allocated a position and their occupied space is deemed unavailable for other facility types. The temporary facilities considered in this case study are listed in Table 8 along with their dimensions. Each of these facilities can take on any shape; for simplicity purposes facility shapes are assumed to be squares. Permanent facilities include buildings under construction and any other objects that cannot be moved such as trees etc. Table 9 presents the dimensions of the locations available for locating the temporary facilities within the construction site. The noise emitted by different equipment utilised within each facility is shown in Table 10.

Table 8. Temporary facilities along with their dimensions and optimised locations on the construction site

Facility	Symbol	Width in *x*-Direction (m)	Length in *y*-Direction (m)	*x*-Coordinate Centroid	*y*-Coordinate Centroid
Offices	F1	10	10	110	90
False work	F2	10	10	10	50
Steel storage yard	F3	10	10	110	10
Formwork Assembly	F4	10	10	10	10
Soil stockpile	F5	10	10	10	30
Electrical generator	F6	10	10	30	10
Groundwater dewatering pump	F7	10	10	30	30

Table 9. Dimensions of predefined locations

Location	Width in x-Direction (m)	Length in y-Direction (m)
L1	100	80
L2	50	80
L3	50	120

Table 10. Sound source of equipment on construction site and their corresponding facilities

Sound Source/Equipment	Symbol	L_{Aeq} @ 10m (dB (A))	Associated Facility
Air Conditioner	S1	50	F1
Club Hammer	S2	79	F2, F4
Loading/Unloading of Steel	S3	72	F3
Material Hoist	S4	68	F3
Electric Saw	S5	81	F4
Electric Drill	S6	78	F4
Excavator	S7	86	F5
Loader	S8	68	F5
Electrical generator	S9	80	F6
Dewatering Pump	S10	78	F7

The methodology shown in Figure 4 is used to predict the total noise levels at two receivers assumed to be noise sensitive buildings surrounding the construction site, namely R1 and R2. R1 and R2 are located at a distance of 30 metres away from the Eastern and Northern side of the construction site boundary, respectively. To determine the total noise levels associated with moving plant on the construction site, mobile machinery that are involved with any facility operations are identified and linked to their corresponding facility. Noise levels are then measured at the point of interaction with the respective facilities between which the machinery moves. As an example, a loader that piles excavated soil is assumed to be emitting noise at its associated facility, being the soil stockpile, F5. Its noise levels are hence measured by taking the soil stockpile as its noise source reference point. A possible Mixed Integer Non Linear Programming (MINLP) optimisation model that can be utilised to minimise the noise pollution through construction site facility layout planning is represented through Equations 8-23. The objective function characterises the logarithmic summation of individual L_{Aeq} for each of the noise sources linked with the construction facilities. The objective function and constraints required are described in detail in the following sections.

Objective Function

$$Minimise \qquad \max_{r \in R} \left\{ 10\log_{10}\left(\frac{1}{10} \sum_{s} \sum_{i} t_s 10^{\frac{L_{sir}}{10}} \right) \right\} \tag{8}$$

The objective function presented in Equation 8 calculates the summation of the equivalent continuous sound pressure L_{sir} from different temporary facilities i, measured at receiver r due to noise source s. The mappings of the noise sources to the facilities are inputted to the model in the form of parameters, as shown in Table 10. The summation is carried out over the proportion of time each noise source operates within a daily 10-hour assessment period assumed for construction works. The maximum noise recorded at each receiver is calculated to ensure that the model provides a solution where the layout produced is one that results in minimal noise across all receivers.

Noise Estimation Constraints

To predict the noise level of each equipment involved in the construction work, Equations 9-11 are specified. In particular, Equation (9) calculates L_{sir} using L_{Aeqs} at a distance of 10 m for each noise source derived from relevant sound level data tables (see Table 2). Equation (10) is used to incorporate the attenuation due to the ground surface. It considers the distance between the noise receiver's abscissa and ordinate, RLX_r and RLY_r respectively, and the coordinates of the facilities' centroids, c_i^x and c_i^y, respectively. Noise intensification, G, is incorporated into Equation 9 and is assumed to be fixed at 3 dB (A) as specified in BS 5228 (BSI British Standards, 2009). Attenuation due to barrier effects is considered in Equation 11.

The attenuation depends on the distance between the receiver and the facility. For facilities that are far apart from the receiver, as determined by a reference chosen to be the farthest point on the construction site away from the respective receiver, r with coordinates f_r^x and f_r^y respectively, attenuation due to barriers is set at a maximum value of 10 dB (A). Attenuation value decreases with the distance between the receiver and the facility.

$$L_{sir} = LAeq_s - K_{ir} + G - \alpha_{ir} \quad \forall s \in N \quad \forall i \in F \quad \forall r \in R \tag{9}$$

$$K_{ir} = 20\log_{10}\sqrt{\left(RLX_r - c_i^x\right)^2 + \left(RLY_r - c_i^y\right)^2} - 8 \quad \forall i \in F \quad \forall r \in R \tag{10}$$

$$\alpha_{ir} = 10\frac{\sqrt{\left(c_i^x - RLX_r\right)^2 + \left(c_i^y - RLY_r\right)^2}}{\sqrt{\left(f_r^x - RLX_r\right)^2 + \left(f_r^y - RLY_r\right)^2}} \quad \forall i \in F \; \forall r \in R \tag{11}$$

Location Constraints

Equation (12) specifies that each facility i has to be located within an available location m. Equation (13) requires the use of each available location for locating the facilities. This is to prevent the site from getting compacted.

$$\sum_{m \in L} z_{im} = 1 \quad \forall i \in F_T \subseteq F \tag{12}$$

$$\sum_{i \in F_T} z_{im} \geq 1 \quad \forall m \in L \tag{13}$$

Boundary Constraints

Each facility positioned within a location cannot trespass onto an adjacent location or to regions that are outside the allocated site boundaries. To achieve this, Equations 14-17 are set up such that whenever the binary variable z_{im}, which assigns a facility i to location m, equals one, a single set of Equations out of 14 and 15 or 16 and 17 needs to be satisfied, so that the facilities are within the site boundaries.

$$cx_i + \left(0.5Wf_i\right) \leq \left(CLX_m + 0.5WL_m\right) z_{im} + W(1 - z_{im}) \quad \forall i \in F \;\; \forall m \in L \tag{14}$$

$$cx_i - \left(0.5Wf_i\right) \geq \left(CLX_m - 0.5WL_m\right) z_{im} \quad \forall i \in F \;\; \forall m \in L \tag{15}$$

$$cy_i + \left(0.5Lf_i\right) \leq \left(CLY_m + 0.5LL_m\right) z_{im} + B(1 - z_{im}) \quad \forall i \in F \quad \forall m \in L \tag{16}$$

$$cy_i - \left(0.5Lf_i\right) \geq \left(CLY_m - 0.5LL_m\right) z_{im} \quad \forall i \in F \quad \forall m \in L \tag{17}$$

WL_m and LL_m refer to the width and length of the locations in the x and y directions, respectively.

Non-Overlap Constraints

For facilities in the same location, the space that they occupy should be confined to that particular facility only and hence non-overlapping constraints between facilities assigned to the same location must be imposed. To achieve this, Equations (18)-(20) are assigned so that whenever two facilities are in the same location, either Equation (18) or (19) is satisfied but not both, as enforced by Equation (20).

$$\left|c_i^x - c_j^x\right| \geq 0.5(Wf_i + Wf_j) \cdot \mu_{ij}^x \quad \forall i, j \in F_T : i \neq j \tag{18}$$

$$\left|c_i^y - c_j^y\right| \geq 0.5(Lf_i + Lf_j) \cdot \mu_{ij}^y \quad \forall i, j \in F_T : i \neq j \tag{19}$$

$$1 + \mu_{ij}^x + \mu_{ij}^y \geq z_{jn} + z_{im} \quad \forall i, j \in F_T : i \neq j \ \forall m, n \in L : m = n \tag{20}$$

In these aforementioned equations, Wf_i and Lf_i refer to the width and length of facility i, respectively.

Domain of Variables

The domain of the binary variable z_{im} is stated through Equation (21). The variable can only be assigned one of 2 values; either 0 or 1. The domain of the continuous variables c_i^x and c_i^y is stated through Equation (22) and Equation (23), respectively.

$$z_{im} \in \left\{0,1\right\} \quad \forall i \in F, \forall m \in M \tag{21}$$

$$0 \leq c_i^x \leq W \quad \forall i \in F \tag{22}$$

$$0 \leq c_i^y \leq B \quad \forall i \in F \tag{23}$$

Optimisation Results

The optimisation model is solved using SCIP (Achterberg, 2008) and a global optimum solution is obtained, as shown in Table 8 and Figure 12. The layout produced indicates that facilities that are associated with the highest noise sources, such as formwork assembly yard, F4, and the soil stockpile, F5, are positioned at locations furthest apart from receivers R1 and R2. The site office, F1, being the least noise emitting facility is closely located to both receivers. In each location, the facilities are grouped towards the furthest corner away from the sides of the site adjacent to the receivers.

From the results obtained it seems legitimate to conclude that the utilisation of a MINLP model to plan the layout of the construction site, so that noise at specified receivers is minimised is an effective way of automating the planning of temporary facilities on construction sites in smart cities, all the while respecting sustainability goals.

Railway Noise Model

The second case study illustrates the application of layout planning in minimising the noise pollution due to operation of a heavy line railway. The heavy line railway was chosen to depict the concept of urban planning around a noise emitting source. The railway is assumed to have a fixed predetermined route. The optimal allocation of building facilities to predefined locations is sought. The noise level emitted from the railway is assumed to correspond to rolling stock noise. A representative train type, Tangara passenger train, is assumed to traverse the rail tracks (Sydney Trains, 2015). The noise level of this type of trains, as quantified by the sound exposure level L_{AE}, is around 89 dB (A) (SLR Consulting Australia, 2012). The outline of the route as well as the available locations for locating the different facilities, listed in Table 11, are shown in Figure 13. Each facility is assumed to be a unique receiver type; hence trigger levels for each facility are specified based on Tables 5 and 6.

Since the railway considered is an above-ground system, rather than an underground train system, the ground borne vibrations may be ignored. It is assumed that Steps 1-3 of the recommended procedure, Figure 6, have already been resolved. The railway model is formulated as a Binary Integer Model (BIP).

Figure 12. Optimum site layout configuration considering minimum noise levels at receivers R1 and R2

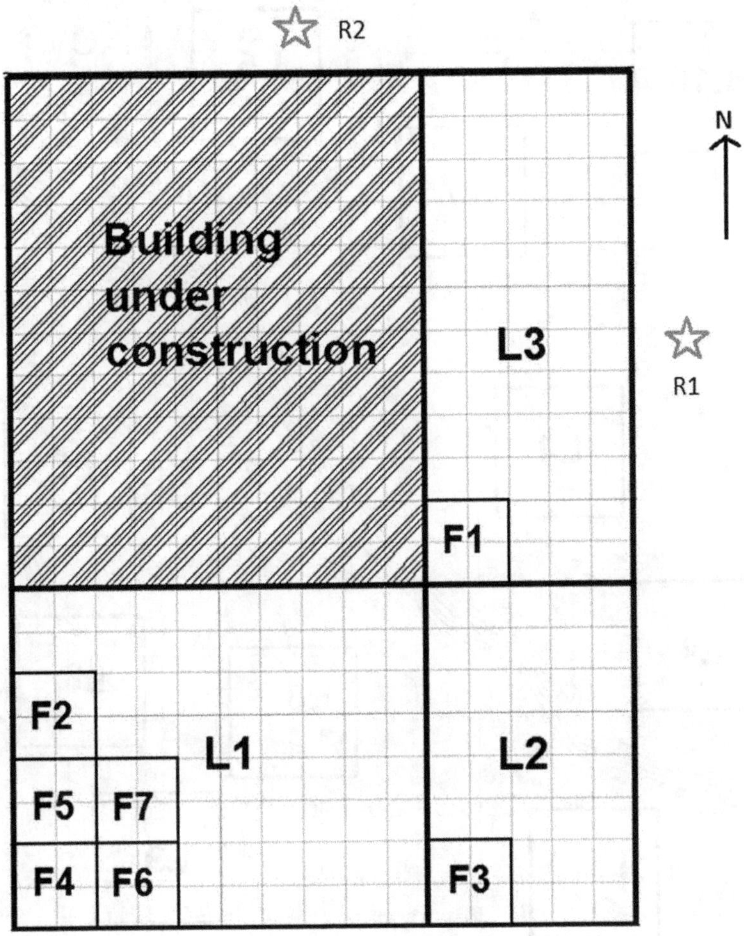

Table 11. Facilities to be located around the railway

Facility	Symbol	Size- *LxW* (m)
School	B1	100x100
Hospital	B2	100x100
Residential Complex	B3	100x100
Park	B4	100x100
Offices	B5	100x100
Restaurants	B6	100x100

Figure 13. Plan view of locations available around railway

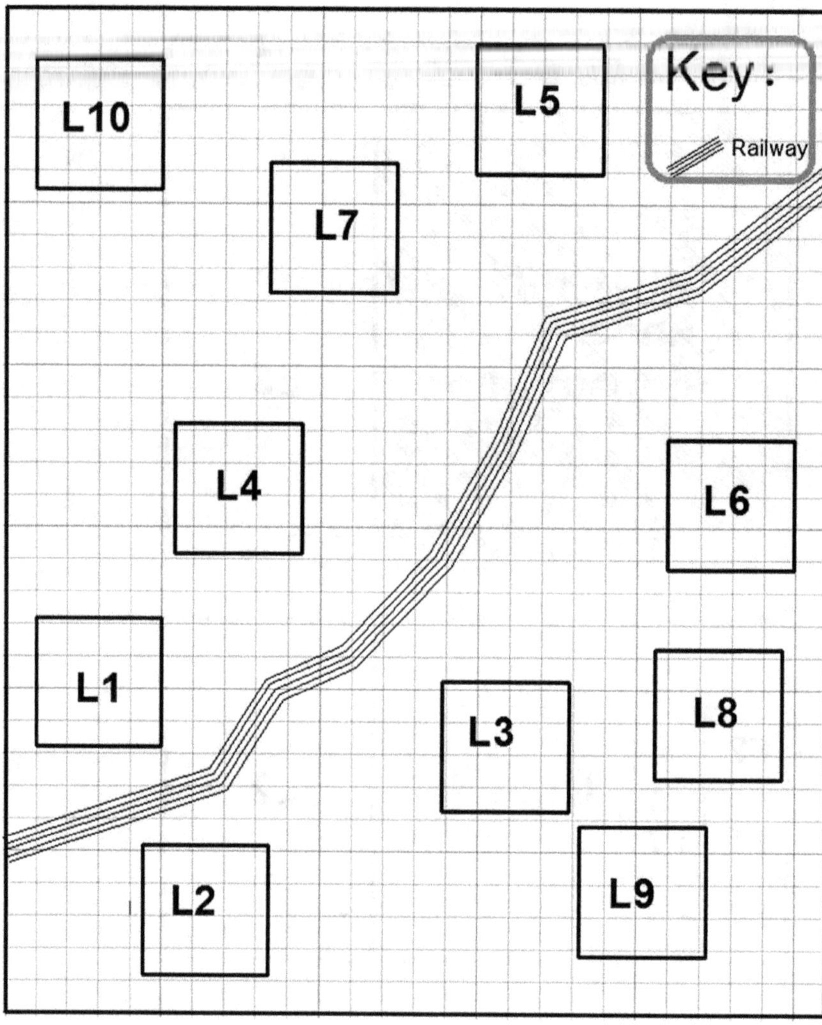

Pre-Processing

To predict the noise level at each location, Equation (24) is used, where the sound exposure level of the train L_{AE} is measured at a distance of 15 m from the noise source.

$$L_{Aeq,(period)} = 10\log_{10}\left(\frac{1}{T}n10^{L_{AE}}\right) - A_{Div} - A_{ground} \tag{24}$$

In Equation (24), *n*, which represents the total number of Tangara trains passing through the railway station in a day, is set to 480. This is derived based on the assumption that two trains pass across the city at a 15-minute interval and that the railway station has four service platforms. The assumed L_{AE} value of 89 dB (A) is converted to a L_{Aeq} figure, depending on the duration of the period adopted. Two periods including, day time period (15 hours), and the nosiest 1-hour period throughout the day are considered in this case study. As such, for each location shown in Figure 13 two L_{Aeq} parameters are defined (Table 12).

$$A_{Div} = 20\log\left(\frac{15.2}{d_m}\right) \tag{25}$$

$$A_{ground} = 4.8 - \left(\frac{2h}{d_m}\right)\left(17 + \frac{300}{d_m}\right) \tag{26}$$

To calculate the noise attenuation due to spherical dispersion of noise, Equation (25) is utilised (Wilson, 1989). Attenuation due to ground surface is calculated using Equation (26) (Harris, 1991). In Equations 25 and 26, the term d_m refers to the distance between the available locations, *m*, and closest point on the railway track to the location. The term *h* indicates the height of sound propagation measured from the ground, and is assigned a value of 2 metres in this case study. Table 12 displays the noise levels that correspond to each available location.

Table 12. Data for railway case study along with results

Location	d_m-Distance Away from Railway (metres)	$L_{Aeq(1h)}$ (dB(A))	$L_{Aeq(15h)}$ (dB(A))	Optimised Assignment of Facility
L1	15	80	68	-
L2	15	80	68	-
L3	25	75.5	63.5	-
L4	145	57.7	45.7	-
L5	155	56.7	44.7	B6
L6	165	55.82	43.8	B4
L7	400	47.6	36.6	B3
L8	800	41.43	29.4	B5
L9	1100	38.5	26.5	B1
L10	1700	34.6	22.6	B2

Objective Function

The objective function, Equation (27), minimises the summation of the L_{Aeq} levels that result due to the assignment of facilities, *s* and *r*, to locations, *m*, around the construction site. The indices *s* and *r* refer to those facilities where it is a requirement to assess noise levels based on the 1-hour L_{Aeq} and 15-hour L_{Aeq} descriptor respectively.

$$Minimise \sum_{m \in M} \sum_{s,r \in F} z_{sm} LAeq_m^{1hr} + z_{rm} LAeq_m^{15hr} \tag{27}$$

Allocation Constraints

Equation (28) specifies that each servicing facility *f* must be allocated to exactly one location.

$$\sum_{m \in M} z_{fm} = 1 \quad \forall f \in F \tag{28}$$

Trigger Level Constraints

Each servicing facility considered in this case study has a different threshold level for noise disruption. Equations (29)-(34) are used to ensure that the allocations that result from the model are in line with the regulations imposed for each building. The values on the right hand side of the equations correspond to the limits specified in (EPA, 2013)). For the school and the hospital their constraints are represented by Equations (29) and (30) respectively, whilst the rest of the facilities are represented through Equations (31)-(34).

$$z_{\mathrm{F1}m} LAeq_m^{1hr} \leq 40 \quad \forall m \in M \tag{29}$$

$$z_{\mathrm{F2}m} LAeq_m^{1hr} \leq 35 \quad \forall m \in M \tag{30}$$

$$z_{\mathrm{F3}m} LAeq_m^{15hr} \leq 60 \quad \forall m \in M \tag{31}$$

$$z_{\mathrm{F4}m} LAeq_m^{1hr} \leq 65 \quad \forall m \in M \tag{32}$$

$$z_{\mathrm{F5}m} LAeq_m^{15hr} \leq 65 \quad \forall m \in M \tag{33}$$

$$z_{\mathrm{F6}m} LAeq_m^{15hr} \leq 70 \quad \forall m \in M \tag{34}$$

Domain of Variables

The domain of the assigning variable z_{fm} is stated through Equation (35).

$$z_{fm} \in \{0,1\} \quad \forall f \in F, \forall m \in M \tag{35}$$

Optimisation Results

Results obtained from the BIP model are displayed in Table 12. CPLEX ("IBM CPLEX Optimizer - United States," 2014) was utilised as the solver for this model. As shown, none of the locations close to the railway line, namely L1-L4, are considered by the model. This is due to the fact that these locations result in high noise levels at the receivers, and in some instances, violate one of the constraints listed in Equations 29-34. The school and the hospital, being the most noise sensitive facilities, are allocated locations that are the furthest apart from the railway. The least noise sensitive facility, the restaurant, is assigned to location L5 which happens to be at a distance of only 155 m away from the rail tracks.

A Framework for Smart Urban Layout Planning by Considering Noise Pollution

An explicit framework is proposed, as depicted in Figure 14, to illustrate the possible avenues for linking the different optimisation models introduced in this chapter, together with other software that can be deployed to implement a smart urban planning solution, where the issue of noise pollution is considered. The framework shows a common noise prediction software, namely SoundPLAN, which can be used to predict the noise level estimates from developments such as railways and highways (SoundPLAN GmbH). This software has the ability to provide estimates of noise levels based on a set of links identifying the route of travel of the vehicles.

The noise model requires the input of Geographic information that lays out the topography surrounding the line of development, and this can be achieved through deploying Geographic Information Technology software such as ArcGIS (Esri, 2013). A platform could therefore be initiated to enable an association between linear optimisation solvers, such as CPLEX ("IBM CPLEX Optimizer - United States," 2014), where the noise optimisation model would be formulated, along with ArcGIS and SoundPLAN. The idea would then be to use the results generated from SoundPLAN, after modelling the terrain data and buildings surrounding the noise emitting development, using ArcGIS, as input into the optimisation model presented through formulations (1)-(20). An allocation of noise sensitive facilities can therefore be obtained where noise disruption is minimised. It is important to note that the framework is robust in the sense that for optimising facilities on a construction site, other software can be employed for construction noise prediction, in place of SounPLAN at step 3.

CONCLUSION

The layout of smart cities whereby the well-being of their residents is ensured through the mitigation of certain environmental and social nuisances is of utmost

Figure 14. The proposed framework for accounting for noise pollution through smart urban layout planning

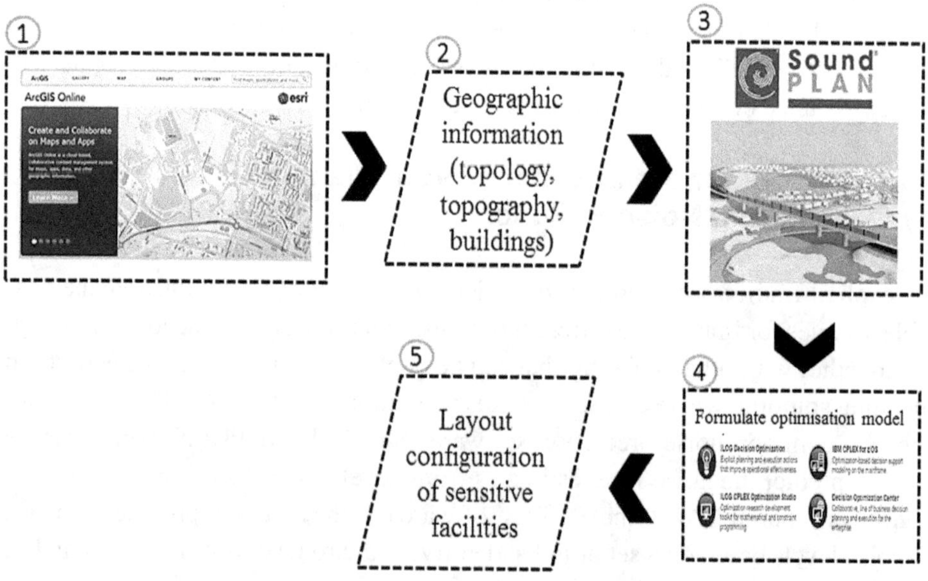

importance for development of sustainable urban areas. The ideas presented in this chapter highlight the importance of considering noise as a design variable when planning for smart cities. Specifically, the focus was placed on the noise pollution resulting from construction sites and from the public transport infrastructure. Noise sources typical of an urban setting were presented along with the necessary requisite noise assessments that are stated by different regulatory authorities in Australia. A novel mathematical model was then presented for urban layout planning to minimise the noise pollution at noise sensitive facilities. The results of two case studies presented in this chapter indicated that the layout of facilities around noise sources such as railways and construction sites as well as the layout of facilities within the noise generating sites can considerably affect the level of noise at the noise sensitive facilities. Therefore, the results of the case studies highlight the importance of considering noise as a decision parameter in planning of smart cities. In addition, a framework for implementation of the proposed optimisation method in conjunction with available noise prediction software was proposed. However, applying the proposed framework to smart urban planning requires development of suitable data transfer protocols to facilitate the transfer of data between different components of the proposed framework. It is also important to note that noise prediction models used as part of a smart solution proposal need to be updated as the layout of the city changes overtime. This temporal effect needs to be modelled and may require a certain set of assumptions to be made. Another important challenge in smart urban planning is accounting for the trade-off between noise pollution and other important environmental, social and economic impacts associated with the planning decision. This tends to turn the smart planning problem into a multi-objective layout problem where the location of facilities is optimised to minimise or maximise a number of usually conflicting objectives. The methodologies and optimisation models presented in this chapter can be helpful in developing the optimisation models needed to tackle such planning issues.

REFERENCES

Abo-Qudais, S., & Abu-Qdais, H. (2005). Perceptions and attitudes of individuals exposed to traffic noise in working places. *Building and Environment*, *40*(6), 778–787. doi:10.1016/j.buildenv.2004.08.013

Achterberg, T. (2008). SCIP: Solving constraint integer programs. *Mathematical Programming Computation*, *0*(1), 1–41. doi:10.1007/mpc.v0i1.4

Adams, M., Cox, T., Moore, G., Croxford, B., Refaee, M., & Sharples, S. (2006). Sustainable Soundscapes: Noise Policy and the Urban Experience. *Urban Studies (Edinburgh, Scotland)*, *43*(13), 2385–2398. doi:10.1080/00420980600972504

Ali, S. A., & Tamura, A. (2003). Road traffic noise levels, restrictions and annoyance in Greater Cairo, Egypt. *Applied Acoustics*, *64*(8), 815–823. doi:10.1016/S0003-682X(03)00031-8

Andayesh, M., & Sadeghpour, F. (2014). The time dimension in site layout planning. *Automation in Construction*, *44*, 129–139. doi:10.1016/j.autcon.2014.03.021

Anthopoulos, L. G., & Tsoukalas, I. A. (2006). The Implementation Model of a Digital City. The Case Study of the Digital City of Trikala, Greece. *Journal of E-Government*, *2*(2), 91–109. doi:10.1300/J399v02n02_06

Asdrubali, F., & Buratti, C. (2005). Sound intensity investigation of the acoustics performances of high insulation ventilating windows integrated with rolling shutter boxes. *Applied Acoustics*, *66*(9), 1088–1101. doi:10.1016/j.apacoust.2005.02.001

Australia, . (2013). *Locomotive Emissions Project-Scoping Study of Potential Measures to Reduce Emissions from New and In-Service Locomotives in NSW and Australia*. Australia: NSW EPA.

Babisch, W. (2008). Road traffic noise and cardiovascular risk. *Noise & Health*, *10*(38), 27. doi:10.4103/1463-1741.39005 PMID:18270405

Ballesteros, M. J., Fernández, M. D., Quintana, S., Ballesteros, J. A., & González, I. (2010). Noise emission evolution on construction sites. Measurement for controlling and assessing its impact on the people and on the environment. *Building and Environment*, *45*(3), 711–717. doi:10.1016/j.buildenv.2009.08.011

Barron, R. F. (2002). *Industrial Noise Control and Acoustics*. CRC Press. doi:10.1201/9780203910085

Belojević, G., Öhrström, E., & Rylander, R. (1992). Effects of noise on mental performance with regard to subjective noise sensitivity. *International Archives of Occupational and Environmental Health*, *64*(4), 293–301. doi:10.1007/BF00378288 PMID:1468799

Berg, R. E., & Stork, D. G. (2005). *The Physics of Sound*. Pearson Prentice Hall.

Black, D. A., Black, J. A., Issarayangyun, T., & Samuels, S. E. (2007). Aircraft noise exposure and resident's stress and hypertension: A public health perspective for airport environmental management. *Journal of Air Transport Management*, *13*(5), 264–276. doi:10.1016/j.jairtraman.2007.04.003

British Standards, B. S. I. (2009). *British Standards: Code of Practice for Noise and Vibration Control on Construction and Open Sites - Part 1: Noise, BS 5228*. British Standards Institution.

Brown, A. L., & Lam, K. C. (1987). Urban noise surveys. *Applied Acoustics*, *20*(1), 23–39. doi:10.1016/0003-682X(87)90081-8

Buratti, C. (2002). Indoor noise reduction index with open window. *Applied Acoustics*, *63*(4), 431–451. doi:10.1016/S0003-682X(01)00040-8

Busch-Vishniac, I. J., West, J. E., Barnhill, C., Hunter, T., Orellana, D., & Chivukula, R. (2005). Noise levels in Johns Hopkins Hospital. *The Journal of the Acoustical Society of America*, *118*(6), 3629–3645. doi:10.1121/1.2118327 PMID:16419808

Byrne, D. C., & Reeves, E. R. (2008). Analysis of Nonstandard Noise Dosimeter Microphone Positions. *Journal of Occupational and Environmental Hygiene*, *5*(3), 197–209. doi:10.1080/15459620701879438 PMID:18213533

Consulting Australia, S. L. R. (2012). *Northern Sydney Freight Corridor, North Strathfield Rail Underpass-Noise and Vibration Assessment*. Transport for New South Wales.

Craven, N., Bewes, O. G., Fenech, B. A., & Jones, R. (2013). Responding to the Environmental Noise Directive by demonstrating the benefits of rail grinding on the Great Britain's railway network. *Proceedings of the Institution of Mechanical Engineers, Part F: Journal of Rail and Rapid Transit*. doi:10.1177/0954409713494948

Crocker, M. J. (2007). *Handbook of Noise and Vibration Control*. John Wiley & Sons. doi:10.1002/9780470209707

den Boer, L., & Schroten, A. (2007). Traffic noise reduction in Europe. Report of CE Delft, Commissioned by the European Federation for Transport and Environment (T&E). Brussels.

Department of Environment. Climate Change and Water NSW. (2011). *NSW Road Noise Policy*. NSW, Australia. Retrieved from http://www.epa.nsw.gov.au/noise/traffic.htm

Department of Environment and Climate Change NSW. (2009). *Interim Construction Noise Guidline*. NSW, Australia: State of NSW and Department of Environment and Climate Change NSW.

Department of Environment and Resource Management. (2010). *State Planning Policy 5/10 Air, Noise and Hazardous Materials*. Queensland, Australia. Retrieved from http://www.ehp.qld.gov.au/noise/

Department of Infrastructure. Energy and Resources. (2011). *State Road Noise Strategy*. Tasmania, Australia. Retrieved from http://www.transport.tas.gov.au/__data/assets/pdf_file/0016/60514/State_Road_Noise_Strategy - Version_1.0,_June_2011.pdf

Ding, G. K. C. (2008). Sustainable construction—The role of environmental assessment tools. *Journal of Environmental Management*, *86*(3), 451–464. doi:10.1016/j.jenvman.2006.12.025 PMID:17289255

Eadie, D. T., & Santoro, M. (2006). Top-of-rail friction control for curve noise mitigation and corrugation rate reduction. *Journal of Sound and Vibration*, *293*(3–5), 747–757. doi:10.1016/j.jsv.2005.12.007

Egan, M. D. (2007). *Architectural Acoustics*. J. Ross Pub.

EPA. (2000). *NSW Industrial Noise Policy*. NSW, Australia: Environment Protection Authority. Retrieved from http://www.epa.nsw.gov.au/noise/industrial.htm

EPA. (2013). *Rail Infrastructure Noise Guideline*. NSW, Australia: State of NSW and Environment Protection Authority. Retrieved from http://www.epa.nsw.gov.au/noise/railnoisegl.htm

EPA VIC. (2008). NOISE CONTROL GUIDELINES. Victoria Government, Australia.

Falk, S. A., & Woods, N. F. (1973). Hospital Noise — Levels and Potential Health Hazards. *The New England Journal of Medicine*, *289*(15), 774–781. doi:10.1056/NEJM197310112891504 PMID:4728750

Fife, D., & Rappaport, E. (1976). Noise and hospital stay. *American Journal of Public Health*, *66*(7), 680–681. doi:10.2105/AJPH.66.7.680 PMID:937616

Fogari, R., Zoppi, A., Vanasia, A., Marasi, G., & Villa, G. (1994). Occupational noise exposure and blood pressure. *Journal of Hypertension*, *12*(4), 475–479. doi:10.1097/00004872-199404000-00019 PMID:8064173

Franssen, E. M., van Wiechen, C. M. A. G., Nagelkerke, N. J. D., & Lebret, E. (2004). Aircraft noise around a large international airport and its impact on general health and medication use. *Occupational and Environmental Medicine*, *61*(5), 405–413. doi:10.1136/oem.2002.005488 PMID:15090660

Fricke, F. (1985). Construction Site Noise. *Architectural Science Review*, *28*(1), 12–22. doi:10.1080/00038628.1985.9696566

Gan, W. Q., Davies, H. W., Koehoorn, M., & Brauer, M. (2012). Association of Long-term Exposure to Community Noise and Traffic-related Air Pollution With Coronary Heart Disease Mortality. *American Journal of Epidemiology, 175*(9), 898–906. doi:10.1093/aje/kwr424 PMID:22491084

Gehring, U., Tamburic, L., Sbihi, H., Davies, H. W., & Brauer, M. (2014). Impact of Noise and Air Pollution on Pregnancy Outcomes. *Epidemiology (Cambridge, Mass.), 25*(3), 351–358. doi:10.1097/EDE.0000000000000073 PMID:24595395

Gilchrist, A., Allouche, E. N., & Cowan, D. (2003). Prediction and mitigation of construction noise in an urban environment. *Canadian Journal of Civil Engineering, 30*(4), 659–672. doi:10.1139/l03-019

Goines, L., & Hagler, L. (2007). Noise pollution: A modem plague. *Southern Medical Journal, 100*(3), 287–294. doi:10.1097/SMJ.0b013e3180318be5 PMID:17396733

Hammad, A. W. A., Akbarnezhad, A., & Rey, D. (2016). A multi-objective mixed integer nonlinear programming model for construction site layout planning to minimise noise pollution and transport costs. *Automation in Construction*, 61, 73 – 85. http://dx.doi.org/10.1016/j.autcon.2015.10.010

Hammad, A. W. A., Akbarnezhad, A., Rey, D., & Waller, S. T. (2015). A computational method for estimating travel frequencies in site layout planning. *Journal of Construction Engineering and Management*, 142(5). http://dx.doi.org/10.1061/(ASCE)CO.1943-7862.0001086

Hammad, A. W. A., Rey, D., & Akbarnezhad, A. (2014). A mixed-integer nonlinear programming model for minimising construction site noise levels through site layout optimization. *Presented at the 31st International Symposium on Automation and Robotics in Construction and Mining, ISARC '14*(pp. 722–729).

Hanna, D., Blouin-Demers, G., Wilson, D. R., & Mennill, D. J. (2011). Anthropogenic noise affects song structure in red-winged blackbirds (Agelaius phoeniceus). *The Journal of Experimental Biology, 214*(21), 3549–3556. doi:10.1242/jeb.060194 PMID:21993783

Harris, C. M. (1991). *Handbook of acoustical measurements and noise control.* McGraw-Hill New York.

IBM CPLEX Optimizer - United States. (2014, December 30). Retrieved from http://www-01.ibm.com/software/commerce/optimization/cplex-optimizer/

Iwnicki, S. (2006). *Handbook of Railway Vehicle Dynamics*. CRC Press.

Jenkins, P., Smith, H., & Wang, Y. P. (2007). Planning and Housing in the Rapidly Urbanising World. London, UK: Routledge; Retrieved from http://eprints.gla.ac.uk/75688/

Katti, M., & Warren, P. S. (2004). Tits, noise and urban bioacoustics. *Trends in Ecology & Evolution, 19*(3), 109–110. doi:10.1016/j.tree.2003.12.006 PMID:16701237

Kenney, G. D., & Ayer, H. E. (1975). Noise Exposure and Hearing Levels of Workers in the Sheet Metal Construction Trade. *American Industrial Hygiene Association Journal, 36*(8), 626–632. doi:10.1080/0002889758507302 PMID:1227289

Khalafallah, A., & El-Rayes, K. (2008). Minimizing Construction-Related Security Risks during Airport Expansion Projects. *Journal of Construction Engineering and Management, 134*(1), 40–48. doi:10.1061/(ASCE)0733-9364(2008)134:1(40)

King, R. P., & Davis, J. R. (2003). Community noise: Health effects and management. *International Journal of Hygiene and Environmental Health, 206*(2), 123–131. doi:10.1078/1438-4639-00202 PMID:12708234

Korve, H. W., Farran, J. I., & Mansel, D. M. (1995). Integration of light rail transit into city streets. In *Transportation Research Board Conference Proceedings* (Vol. 1). Retrieved from http://trid.trb.org/view.aspx?id=457847

Lan, T.-S., & Chiu, M.-C. (2008). Optimal noise control on plant using simulated annealing. *Transactions of the Canadian Society for Mechanical Engineering, 32*(3-4), 423–437.

Lercher, P., Evans, G. W., Meis, M., & Kofler, W. W. (2002). Ambient neighbourhood noise and children's mental health. *Occupational and Environmental Medicine, 59*(6), 380–386. doi:10.1136/oem.59.6.380 PMID:12040113

Li, H., & Love, P. (1998). Site-Level Facilities Layout Using Genetic Algorithms. *Journal of Computing in Civil Engineering, 12*(4), 227–231. doi:10.1061/(ASCE)0887-3801(1998)12:4(227)

Losa, M., Leandri, P., & Rocchio, P. (2012). Crumb Rubber Modified Asphalt Concrete for Low Noise Surfaces. *Presented at the Asphalt Rubber Conference*, Munich. Retrieved from http://www.ra-foundation.org/crumb-rubber-modified-asphalt-concrete-for-low-noise-surfaces/

Lusk, S. L., Kerr, M. J., & Kauffman, S. A. (1998). Use of hearing protection and perceptions of noise exposure and hearing loss among construction workers. *American Industrial Hygiene Association Journal, 59*(7), 466–470. doi:10.1080/15428119891010217 PMID:9697294

Lyon, R. H. (2013). *Machinery Noise and Diagnostics*. Butterworth-Heinemann.

Manuel, J. (2005). Clamoring for Quiet: New Ways to Mitigate Noise. *Environmental Health Perspectives, 113*(1), A46–A49. doi:10.1289/ehp.113-a46 PMID:15631960

Marsh, K. J. (1982). The concawe model for calculating the propagation of noise from open-air industrial plants. *Applied Acoustics, 15*(6), 411–428. doi:10.1016/0003-682X(82)90003-2

Murphy, E., & King, E. A. (2014). Noise Mitigation Approaches. In E. M. A. King (Ed.), Environmental Noise Pollution (pp. 203–245). Boston: Elsevier. Retrieved from http://www.sciencedirect.com/science/article/pii/B9780124115958000070 doi:10.1016/B978-0-12-411595-8.00007-0

Næss, P. (2001). Urban Planning and Sustainable Development. *European Planning Studies, 9*(4), 503–524. doi:10.1080/09654310120049871

Naish, D. A., Tan, A. C. C., & Demirbilek, F. N. (2014). Simulating the effect of acoustic treatment types for residential balconies with road traffic noise. *Applied Acoustics, 79*, 131–140. doi:10.1016/j.apacoust.2013.12.021

Neirotti, P., De Marco, A., Cagliano, A. C., Mangano, G., & Scorrano, F. (2014). Current trends in Smart City initiatives: Some stylised facts. *Cities (London, England), 38*, 25–36. doi:10.1016/j.cities.2013.12.010

Neitzel, R., Seixas, N. S., Camp, J., & Yost, M. (1999). An Assessment of Occupational Noise Exposures in Four Construction Trades. *American Industrial Hygiene Association Journal, 60*(6), 807–817. doi:10.1080/00028899908984506 PMID:10635548

Ng, C. F.NG. (2000). Effects of building construction noise on residents: A quasi-experiment. *Journal of Environmental Psychology, 20*(4), 375–385. doi:10.1006/jevp.2000.0177

Nielsen, G., & Solberg, S. (1988). Costs and effects of traffic noise abatement measures: Basis for a national programme. *Applied Acoustics, 25*(3), 149–168. doi:10.1016/0003-682X(88)90091-6

Nissenbaum, M., Aramini, J., & Hanning, C. (2012). Effects of industrial wind turbine noise on sleep and health. *Noise & Health, 14*(60), 237. doi:10.4103/1463-1741.102961 PMID:23117539

Nurminen, T. (1995). Female noise exposure, shift work, and reproduction. *Journal of Occupational and Environmental Medicine / American College of Occupational and Environmental Medicine, 37*(8), 945–950.

Osada, Y. (1988). An overview of health effects on noise. *Journal of Sound and Vibration, 127*(3), 407–410. doi:10.1016/0022-460X(88)90363-X

Pedersen, E., & Waye, K. P. (2007). Wind turbine noise, annoyance and self-reported health and well-being in different living environments. *Occupational and Environmental Medicine, 64*(7), 480–486. doi:10.1136/oem.2006.031039 PMID:17332136

Penna, M., Pottstock, H., & Velasquez, N. (2005). Effect of natural and synthetic noise on evoked vocal responses in a frog of the temperate austral forest. *Animal Behaviour, 70*(3), 639–651. doi:10.1016/j.anbehav.2004.11.022

RailCorp. (2012). *SPC 235-RESILIENT BASEPLATES*. NSW Transport. Retrieved from http://www.asa.transport.nsw.gov.au/sites/default/files/asa/railcorp-legacy/disciplines/civil/spc-235.pdf

Razavi, H., Ramezanifar, E., & Bagherzadeh, J. (2014). An economic policy for noise control in industry using genetic algorithm. *Safety Science, 65*, 79–85. doi:10.1016/j.ssci.2013.12.010

Roads and Maritime Services NSW. (2009). *Road Design Guide*. Retrieved from http://www.rms.nsw.gov.au/business-industry/partners-suppliers/aus-roads-guides/road-design.html

Roads and Maritime Services. (2015). *Noise Criteria Guideline*. NSW, Australia. Retrieved from http://www.rms.nsw.gov.au/documents/about/environment/noise-criteria-guideline-book.pdf

Rossing, T. D., Wheeler, P. A., & Moore, F. R. (2002). *The Science of Sound*. Addison Wesley.

RTA. (2001). *RTA Environmental Noise Management Manual*. Roads and Traffic Authority of New South Wales. Retrieved from http://www.rms.nsw.gov.au/about/environment/reducing-noise/index.html

Sandberg, U. (1999, December 6-8). Abatement of Traffic, Vehicle and Tire/Road Noise - The Global Perspective. Proceedings of Internoise 99 - The 1999 International Congress On Noise Control Engineering, Fort Lauderdale, Florida, USA (Vol. 1). Retrieved from http://trid.trb.org/view.aspx?id=651529

Scholes, W. E., & Sargent, J. W. (1971). Designing against noise from road traffic. *Applied Acoustics, 4*(3), 203–234. doi:10.1016/0003-682X(71)90017-X

Sørensen, M., Lühdorf, P., Ketzel, M., Andersen, Z. J., Tjønneland, A., Overvad, K., & Raaschou-Nielsen, O. (2014). Combined effects of road traffic noise and ambient air pollution in relation to risk for stroke? *Environmental Research*, *133*, 49–55. doi:10.1016/j.envres.2014.05.011 PMID:24906068

Standards Australia. (2010). *Standards Australia 2010, Guide to noise and vibration control on construction, demolition and maintenance sites, AS2436.*

Standards Australia. (1990).

Sydney Trains. (2015). Tangara. Retrieved from http://www.sydneytrains.info/about/fleet/t_sets.htm

The Australian Noise Exposure Forecast (ANEF) System. (2014). Retrieved from http://www.infrastructure.gov.au/aviation/environmental/transparent_noise/expanding/app_a.aspx

Tiwari, R., Cervero, R., & Schipper, L. (2011). Driving CO2 reduction by Integrating Transport and Urban Design strategies. *Cities (London, England)*, *28*(5), 394–405. doi:10.1016/j.cities.2011.05.005

U.S. Department of Health and Human Services. (1988). Proposed National Strategies for the Prevention of Leading Work-related Diseases and Injuries: Noise-induced Hearing Loss (No. 89-135).

Vallet, M., Maurin, M., Page, M. A., Favre, B., & Pachiaudi, G. (1978). Annoyance from and habituation to road traffic noise from urban expressways. *Journal of Sound and Vibration*, *60*(3), 423–440. doi:10.1016/S0022-460X(78)80121-7

van Kempen, E., & Babisch, W. (2012). The quantitative relationship between road traffic noise and hypertension: A meta-analysis. *Journal of Hypertension*, *30*(6), 1075–1086. doi:10.1097/HJH.0b013e328352ac54 PMID:22473017

Van Renterghem, T., & Botteldooren, D. (2009). Reducing the acoustical façade load from road traffic with green roofs. *Building and Environment*, *44*(5), 1081–1087. doi:10.1016/j.buildenv.2008.07.013

Victoria EPA. (2014). Commercial and industrial noise. Retrieved from http://www.epa.vic.gov.au/your-environment/noise/commercial-and-industrial-noise#Introductiontonoisestandards

Vijayraghavan, P., & Krishnan, R. (1998). Noise in electric machines: a review. *Proceedings of theThirty-Third IAS Annual MeetingIndustry Applications Conference* (*Vol. 1*, pp. 251–258). Doi:<ALIGNMENT.qj></ALIGNMENT>10.1109/IAS.1998.732298

WHO. (2011). *Burden of disease from environmental noise, quantification of healthy years lost in Europe*. Copenhagen, Denmark: World Health Organisation.

Wilson, C. E. (1989). *Noise control: measurement, analysis, and control of sound and vibration*. Harper & Row Limited.

Xiaoan, G. (2006). Railway environmental noise control in China. *Journal of Sound and Vibration*, *293*(3–5), 1078–1085. doi:10.1016/j.jsv.2005.08.058

KEY TERMS AND DEFINITIONS

BIP (Binary Integer Programming): An integer programming model such that all variables included can take a value of either 0 or 1.

Construction Site Layout: Placing facilities on a construction site in a particular layout such that a certain objective function is respected.

dB (A): An A-weighted noise level decibel scale that accounts for the non-linear frequency response of the human ear.

MINLP (Mixed Integer Non-Linear Programming): A mathematical optimisation model that contains a non-linear objective function/ and or constraints where some of the variables involved are integer variables.

Noise Mitigation: The application of noise reduction techniques to sound sources, receivers and/or propagation paths such that overall noise levels reaching a receiver are reduced.

Receiver: People who are affected by noise levels propagating from a defined noise source.

Smart Urban Planning: The use of automated computational strategies to allocate different facilities and services within a delineated urban region based on certain objectives.

Chapter 7
Quality of Resilient Cities, the Issue of Urban Waste:
Waste Management as Part of Urban Metabolism

Elzbieta Dagny Rynska
Warsaw University of Technology, Poland

Anna Teresa Oniszk-Poplawska
Warsaw University of Technology Faculty of Architecture, Poland

Urszula Kozminska
Warsaw University of Technology Faculty of Architecture, Poland

ABSTRACT

This chapter focuses on the metabolic concept for the management and treatment of construction waste and organic fraction of municipal solid waste in urban areas. Analysis of related Dutch, German and Polish guidelines for environmental zoning of industrial plants, allows formulation of conditions for an optimal siting of waste infrastructure within urban unit. Protection zones are defined in accordance with specific requirements for waste facilities, which treat and recycle both municipal and construction waste. Distances from inhabited areas are related to environmental burdens generated by such facilities (incl. parameters such as odours, noise level, explosion impacts and emissions of other substances). Moreover, this chapter provides the analysis of a selected case studies of waste facilities processing. A comparison of European guidelines and implementation of practical solutions is described in the case study analysis, including the issues open for the discussion about sustainable siting for waste processing infrastructure within an urban unit.

DOI: 10.4018/978-1-5225-0302-6.ch007

INTRODUCTION

Waste is an underlying issue throughout Agenda 21, one of the first documents concerning sustainable development. It is pointed out as a source of many environmental problems caused more than often through anthropogenic activities. While some of the Chapters (20, 21) deal specifically with waste issues, others concentrate on the negative impacts and effects of waste. Even all those years back (1992), this very important document directly emphasized the need to manage waste, reduce generation of waste, or its effective integration into the 3R scheme. Basically, this procedure is based on the well-known Reuse – Reduce - Recycle pattern. In case of waste, this approach has been actually modified by Japanese with Reduce – Return - Reuse approach. Project started April 2014 and will run until March 2017. It focuses on reducing the use of imported goods and merchandise, encourages composting of biodegradable waste at home and reuses the waste. This project will work alongside the established Honiara Public-Private Cooperation Committee where people, agencies and administrators will collaborate to address and achieve the goal of the project in reducing household waste. The outcomes will decisively allow to formulate a case specific generally accepted redefinition of 3R in the waste sector.

Beside Agenda 21, Europe has introduced some particular Conventions. In case of the Basel Convention ("Ban Amendment", adopted in 1995 under Decision III/1), the emphasis was placed on the export prohibition of hazardous wastes, for any purpose, from countries listed in a proposed new annex ("Parties and other States which are members of OECD, EC, Liechtenstein"), to countries not listed in the annex. During subsequent meetings the Parties adopted lists of specific wastes characterized as hazardous or non-hazardous, thereby clarifying the scope of the Convention. In 1999, a Protocol on Liability and Compensation was adopted.

The second Convention is - The Rotterdam Convention constituted by a group of government-designated experts responsible for assessing whether banned, severely restricted chemicals or hazardous pesticide formulations should be made a subject to the Prior Informed Consent procedure. At its third meeting in February 2002, the interim committee recommended that three most widely-used pesticides and all forms of asbestos should be added to the Convention.

Similarly, the Stockholm Convention provides for the future establishment, by the Conference of the Parties, of a subsidiary body to be called the Persistent Organic Pollutants Review Committee.

Some of other relevant regional agreements also include air pollution or control of transboundary movement of hazardous waste *i.e.*:

- Convention on Long-range Transboundary Air Pollution (1979) including Aarhus Protocol on POPs (1998)www.unece.org/env/lrtap,

- Barcelona Convention for the Protection of the Mediterranean Sea Against Pollution (1976). Izmir Protocol (1996)www.rempec.org/barcelona.html,
- Waigani Convention to Ban the Importation into Forum Island Countries of Hazardous and Radioactive Wastes and to Control the Transboundary Movement and Management of Hazardous Wastes within the South Pacific Region (1995)www.forumsec.org.fj/docs/wc.htm & www.basel.int/Misclinks/waigani.html,
- Bamako Convention on the Ban of the Import into Africa and the Control of Transboundary Movement of Hazardous Wastes within Africa (1991)www.londonconvention.org/Bamako.htm
- Cartagena Convention for the Protection and Development of the Marine Environment in the Wider Caribbean Region (1983)www.cep.unep.org/law/cartnut.html,
- Central American Regional Agreement on Transboundary Movements of Hazardous Wastes (1992)www.basel.int/Misclinks/Centroamerica.html.

Even though, the number of people living in cities is steadily increasing and it has been estimated that in the near future 80% of the world's population will live in urbanised areas, the issue of urban waste remains unsolved. In Europe, despite the general tendency in population decrease, the process of urbanization will still continue as the urban sprawl takes over new open spaces. Even though some restrictive planning measures have been implemented in some countries, instead of revitalising brownfields, built up areas have been consuming more and more rural and forested areas. The urban metabolism concept (Wolman, 1965) is not a new idea, but the urban metabolism has increasingly intensified its presence (Kennedy, Pincetl, & Bunje, 2011) with the increasing consumption of materials and energy in urban areas (Barles, 2010). The ingredients of urban metabolism i.e. material and energy fluxes within a city depend on economic, environmental, social and infrastructural determinants (Kennedy et al., 2011), (Barles, 2010), (Anderberg, 1998), (Agudelo-Vera, Leduc, Mels, & Rijnaarts, 2012).

Simultaneously, the rising quality of life also means high rates of resource consumption and what follows a negative impact on the urban environment and surrounding areas. The most important issue is constant and unceasing generation of waste, far beyond the handling capacities of local governments. Even the richest and most influential cities are currently trying to solve the problems of high volumes of waste and costs involved development of disposal technologies and to elaborate adequate methodologies. Most local governments have identified waste management as a major problem that has reached proportions requiring drastic and quick measures. Still, this state has also provided an opportunity for some cities to find solutions - involving the community and the private sector; implementing innovative technologies and new disposal methods; prompting citizen behavior changes and raised awareness. These issues have been demonstrated by good practices found in various countries around the world. There is a need, therefore, for a complete rethinking of the idea - to analyze if waste is indeed a useless 'waste'. The currently supported solution is waste disposal provided by municipalities. Yet, current research proves that there is a need to move towards waste processing and waste recycling that involves public-private partnerships, aiming eventually at community driven waste minimization. Waste minimization programs should include deeper community participation, economic benefits coming from recovery of waste, There is a need to focus more on product life cycles (rather than end-of-pipe solutions), decentralized waste management, and life cycle costs perceived rather as long-term goals.

According to some sources the below key aspects should be considered [20]. The European waste hierarchy refers to the 5 steps included in the article 4 of the Waste Framework Directive [2008/98/EC] of the European Parliament and of the Council (European Parliament and Council, 2008):

- **Waste Prevention:** Preventing and reducing waste generation. Waste minimization is aimed at reducing the level of waste production through education and improved production process rather than technological means. It is a method of managing existing resources and technology in order to maximize the efficiency of available resource use. This process has the potential to reduce costs or increase profits by more effective use of resources and lower amount of waste. It has impact on individuals, industry as well as Local Governments.
- **Waste Reuse and Preparation for Reuse:** Giving the products a second life before they become waste. It also covers hazardous waste handling. Generally, there are two main groups of processes to be considered. Examples of reuse include: product, material reuse as well as durable packaging. Beyond environmental aspects, this choice also means employment opportunities in service and repair industries, better protection of products as well as social approach towards disposable products.
- **Waste Recycling:** Any recovery operation, by which waste materials are reprocessed into useful products, materials or substances whether used as original or for other purposes. It includes composting and it does not include incineration. The breaking down of materials from waste streams into raw materials, then reprocessed either into same material (closed loop) or a new product (open loop); this process generally includes waste separation and material reprocessing. There are various materials that are capable of being

recycled, and technology is advancing to allow the recycling of more materials. One of the most important benefits is that this process provides for reduction of virgin resources.
- **Other Waste Recovery Options:** Incineration, pyrolisis, gasification and other solutions which lead to recovery of energy.
- **Waste Disposal:** Processes to dispose of waste by landfilling.

One of the quality issues, which should be resolved in the context of resilient cities, is sustainable management of urban waste (both municipal and construction) as per the waste hierarchy mentioned above. The main task being, reduction of the waste volume, high reuse of materials and additional recovery of energy from municipal waste generated by city's metabolic processes. Cities are expected not only to consume goods and generate waste, but also to become a place, where materials and energy are recovered and reused.

Building materials make up the largest fraction in the global waste stream generated in urban areas and adequate urban planning and management may lead to lowering of their volume *(i.e.* through revitalisation instead of demolishment or recycling processes for various building materials). Waste generated by households is the second largest source of urban waste (Eurostat, 2015). Volumes of the organic fraction of municipal solid waste (OFMSW) have no chance of being reduced, but through organic recycling OFMSW may become a valuable fertiliser (digestate, compost) and additionally can be used as an alternative source of energy (biogas).

New waste infrastructure should be developed thoughtfully and sustainably. Intelligent urbanization requires new infrastructural solutions in the field of water and sewage, waste management systems and energy infrastructures. The waste hierarchy is the main principle for sustainable waste treatment in the European Union: *i.e.* reduction and recycling instead of landfilling. Therefore, a new, integrated approach to intelligent urbanisation is a future direction for an effective planning and sustainable development of cities. However, contemporary spatial planning procedures rarely include energy efficiency, effective reuse of building materials or waste management strategies. Yet, in various scientific researches it has been indicated that the issue of the reduction of waste volume, and its management should be analysed in the spatial planning context. This approach can be applied to waste management – both construction wastes, as well as OFMSW.

Inhabited areas generate waste, which - therefore, from the economic point of view (costs of logistics) should be treated in their proximity, *e.g.* on the city outskirts. However, the urbanisation process and urban-sprawl result in decreasing amounts of the available greenfield areas. Yet, at the same time there is growing tendency that more and more people move out of the city, where they expect to find a high quality space for living. On the other hand, the social acceptance level for environmentally friendly infrastructures is low. Not in my back yard (NIMBY) attitudes are commonly encountered in the peri-urban areas. For spatial planners' location of such environmental infrastructure facilities is, thus, challenging and requires procedures for trade-offs between various actors. Such trade-offs are usually expressed as minimum separation distances between technical infrastructures and inhabited areas.

A very broad approach, covering the social, economic, technology, political and administrative dimensions should be adopted. For example the social dimension of waste management involves waste minimization and involvement in source separated collection; the economic dimension - energy and material recovery and waste recycling; the technological dimension - waste processing; the spatial dimension- waste disposal and area demand for waste infrastructures; and finally the political and administrative cross-cutting issues throughout all dimensions of the waste hierarchy.

Furthermore, it is important that implemented management should provide for a gradual shift from "end-of pipe" solutions that focus on waste disposal, to a source based approach that is aimed at 'life-cycle' analysis". This places the responsibility also in manufacturers and retail businesses. Better education and higher awareness at the local and community level often forces businesses and industries to take a more environmentally friendly approach to their activities, including better management of wastes they produce. Therefore, an integrated approach to urban planning and multi-level actions concerning policies, regulations and economic incentives should be introduced to enable sustainable metabolism of urban waste. Adequate waste management and processing infrastructure may be successfully introduced in urban areas without detriment of natural environment and human health - if all local conditions social, environmental and economic aspects of waste processing are taken into consideration. This article also suggest policy, legal and administrative measures to design the idea of sustainable waste metabolism in a given area.

WASTE MANGEMENT AS A COMPONENT OF URBAN METABOLISM

Focus on Construction and Demolition Waste

Construction waste makes up a significant fraction of global waste streams, especially in urban areas: 10-15% of all waste generated globally (UNEP, 2011) and 34% of the EU waste in 2010 (Eurostat, 2015). Moreover, it has been estimated that with current trends the global raw material consumption is due to triple – from 49 billion Mg/y in 2000 to 141 billion Mg/y in 2050 (UNEP, 2011). The concept of sustainable urban metabolism and closed circulation of materials is in line with European directives and policy documents (European Parliament and Council, 2008), (Commission of the European Communities, 2014) and may have a significant impact on waste reduction and its further reuse.

Construction and demolition wastes (code 17, acc. to: the European Waste Catalogue) (Commission of the European Communities, 2000) originate from construction, renovation and demolition of buildings and road infrastructure. This kind of waste materials can be recovered for new materials or energy, which conforms to principles of sustainable urban metabolism. The concept of intelligent urbanization relies on modern systems and installations, which aim to reduce amounts of waste and environmental impacts. In countries, such as the Netherlands or Germany the rate of recovered construction waste is high, for example the level of recycling of construction and demolition waste in the Netherlands in 2010 amounted to 98.1%, in Germany – 86.3%. In post communist countries, such as Poland, only 28.3% of such waste is recovered (Tojo & Fischer, 2011). However, recirculation process of construction waste within an urban unit needs to incorporate necessary infrastructures and management (incl. spatial planning) procedures.

Continuous material and waste cycles are promoted by German and Dutch policies and development strategies (German Federal Ministry of Education and Research, 2011), (Ministerie van Infrastructuur en Milieu, 2014) which see sustainability as a cross sector task and where the development is based on the conservation of natural resources, preventing environmental burdens, increased material efficiency, life cycle approach and innovation. Successful waste management system and high recovery rates for construction and demolition waste depends on national and regional policies and regulations. In some countries, such as Germany waste legal acts (Bundesregierung, 2012) do not only conform to European directives, but also incorporate sustainable urban metabolism principles into the waste management. Their objective is to create closed substance cycles (Germany) and material chains (the Netherlands) by promoting prevention, reuse, recycling and other forms of recovery of waste over its incineration and landfilling. Moreover, a life cycle approach to waste management requires that prevention is analysed earlier – before a product or material enters the waste phase. Furthermore, closed circulation of building materials can be enhanced by establishing minimum standards (a required type of recovery) and targets (recycling levels) as the ones introduced by German construction and demolition waste ordinance (Bundesregierung, 2002) or Dutch waste policy (Ministerie van Infrastructuur en Milieu, 2014). These policies are accompanied by multiple programs, institutions and tools, which aim to enable effective use of building materials and their closed circulation in urban areas. There are national programs concerning sustainable innovation and waste (*e.g.* German Environmental Innovation Programme or Dutch Waste as a Resource Programme) and institutions, which investigate the issues of material efficiency, Cradle-to-

Cradle and sustainable technologies (*e.g.* DEMEA, Circular Economy Foundation). Moreover, such instruments as producer responsibility, voluntary agreements for SMEs, pro-environmental taxes, green investments, sustainable procurement and eco-design are introduced to close building material flows. Lacking data concerning location, availability and properties of waste is a crucial problem for the process of construction and demolition waste recovery. Thus, in Germany an online networking platform (MaRess - The Resource Efficiency Network) was introduced and in the Netherlands - an online database as a harvest map (Oogstkaart Nederland). Furthermore, there are some measuring tools, which facilitate material flows monitoring (Sustainability Monitor) and forecasting (Flow Calculator), which are used also for sustainable spatial planning (Urban Metabolism Analysis) and pro-environmental construction (LCA, Environmental Impact Calculator).

Sustainable construction waste management has been a topic of debates but in practice closed circulation of building materials has so far not been achieved in real life. The organisation of a waste collection system is fundamental for successful reuse of materials. The commitment to the life-cycle approach, selective waste collection, recycling and preparation of waste for further reuse (including: demolition waste) requires a lot of organisational effort and change of attitudes – a shift from seeing it as a problem to perceiving it as a resource. Waste can be successfully reused in the construction sector, which was demonstrated with success in some architectural works worldwide, *e.g.*: as recycled concrete (Superuse Studios, Kringloop Zuid, Maastricht, 2014), bricks (D.Chipperfield, Am Kupfergraben, Berlin, 2007), rubble (W.Shu, Historic Museum, Ningbo, 2008), prefabricated panels (Wiewiorra Hopp Architekten, Plattenpalast, Berlin, 2009), aluminium (Karo Architekten, Open-Air Library, Magdeburg, 2009), steel (Superuse Studios, Dordyart, Dordrecht, 2012), wood (Superuse Studios, Willa Welpeloo, Enschede, 2009), ceramic tiles (Superuse Studios, Earthship, Zwolle, 2009), plastics (Superuse Studios, WORM, Rotterdam, 2012), carpets, rubber floors, glass powder, gypsum, composites made out of recycled components or wind turbines (Superuse Studios, Re-wind, Rotterdam) and road signs (Gigon Guyer, Transport Museum, Lucerne, 2009). Construction waste is utilized to build easily assembled, reusable systems of temporary shelters in developing countries (Habitat for Humanity, Engineers Without Borders, S.Ban, Paper Log Houses in Japan, Indie, Turkey, Philippines, 1995-2014).

Such reclaimed materials as tyres, concrete, window frames, roofs, wood or installations are commonly used in alternative architecture (*e.g.* M. Reynolds, Earthships, New Mexico) - for both structure and finishing (Nunan, 2010). Moreover, reused bricks, reclaimed timber, steel structures, stone, metalwork, windows and old machines are present in adaptations of industrial buildings for housing (Atelier Zeinstra van der Pol, Aquartis, Amsterdam, 2001), public (A.Franco, J.Garcia Roldan, M.Langrita, V.Navarro, A.Garcia Abril, Matadero, Madrid, 2007-2011) or commercial uses (Neri & Hu, Waterfront Hotel, Shanghai, 2010). Construction waste can be used to create such elements as: a building's structure (steel machine as a load-bearing system of Villa Welpeloo in Enschede), a façade (multi-material rubble on the façades of the Historic Museum in Ningbo), windows (frames as openings in the Esspressobar in Delft), stairs and lifts (truck steel sheets as staircase in Dordyart in Dordrecht), finishing (perforated desktops as wall and floor panels in WORM in Rotterdam), furniture (wind turbine as outdoor bench in Rotterdam) or whole rooms (plastic containers as toilet units in WORM in Rotterdam) (see Figure 1).

Focus on Organic Fraction of Municipal Solid Waste

In the case of OFMSW closed circulation of nitrogen and phosphorus, energy recovery and the bioeconomy understood as re-source-efficient recycling have been prioritised. In Germany the Federal Government has long supported the above mentioned ideas by adopting policies, taking the concept of natural cycles of materials as its point of orientation:

- The National Policy Strategy on Bioeconomy (2013) (Bundesministerium für Ernährung und Landwirtschaft, 2014),
- The National Raw Materials Strategy (2010) (German Federal Ministry of Education and Research, 2011),
- The German Resource Efficiency Programme (2012) (Bundesministerium für Umwelt Naturschutz und Reaktorsicherheit, 2012).

Changing requirements in the area of waste management – ever demanding limits for the recovery and recycling on one hand, and waste storage restrictions on the another, are a driving force for the development of new and retrofits of existing waste treatment facilities. Waste hierarchy and limitation on the fossil fuel energy consumption seem to be the most crucial environmental criteria. Anaerobic digestion (AD) plants can be considered as the most desired option from the point of view of sustainable waste metabolism: they recycle nutrients and produce energy at the

Figure 1. a-e Open air library, Magdeburg; Am Kupfergraben, Berlin; Kringloop Zuid, Maastricht; Plattenpalast, Berlin; Villa Welpeloo, Enschede
Source: photographs by U. Koźmińska

same time. However, the final choice of a technology for processing of OFMSW will depend not only on the environmental factors. For new investments, social and economic limitations must be seriously taken into consideration. Barriers for the possible application of AD plants using OFMSW lie mainly in the organizational aspects (source separation, OFMSW quality, spatial conditions, and involvement of local authorities) and in the economic constraints (investment and O&M costs, the level of the gate fee, the possibility of selling energy). The economic efficiency of the AD process is mainly affected by the gate-fee. Other components of the economic evaluation are remuneration for the green energy generation (income) as well as O&M costs for waste pre-treatment and digestate after treatment. However, income generated by the renewable energy sell was discovered to be far less significant than the gate fee. In the economic accounting the gate fee is the core position.

The criterion analysing technological alternatives includes boundary conditions related to the economics of certain technological options, expressed by the size of population within the region and the gate fee threshold values. For a population of more than 300 thousand inhabitants, technological option to consider is an incineration plant. For AD plants it is essential to evaluate the minimum required plant capacity, which based on the study should not be lower than 20 thousand Mg/year.

The question which needs to be answered - is the urban waste metabolism economically feasible under all conditions? It can be estimated using waste accumulation indicators, population density, and assumed effectiveness of selective collection as per urban unit (with type and size residential area). In the case of dispersed buildings in rural areas (rural and rural-urban municipalities) an minimum threshold value for SSO collection has been defined at 150 inhab./km^2 (EPEA i VHE, 2008) and the area from which waste is transported to the waste processing facility in the analysed cases usually did not exceed 40 km.

SITING OF URBAN WASTE INFRASTRUCTURES

The perception of waste facilities and expectations towards separation distances between technical infrastructure and inhabited areas vary depending on actors involved. Inhabitants tend to maximise their expectations in order to push away an undesired waste processing facility from their dwelling place. Local authorities and developers see economic interest in minimising distances and reserving as much land as possible for future developments (*e.g.* housing estates).

The problem of waste treatment is a significant social challenge in urban areas. The proximity to waste generation is a desired option from the economic point of view (Eiselt & Marianov, 2014), while inhabitants tend to prefer much further distances due to NIMBY (not in my back yard) phenomena. Such circumstances translate into difficulties to find suitable locations for waste treatment facilities in the vicinity of medium and large cities, where there is a particularly high demand for such investments (Achillas et al., 2011), (Khadivi & Fatemi Ghomi, 2012).

The simplest, but least accurate method to evaluate the environmental impact and indicate the width of protection zones around an industrial facility is to draw an R radius circle around the site. In some countries adequate guidelines for the estimation of distances to human settlements and other sensitive areas are available. The *Dutch guidelines for the environmental zoning of industrial plants* (Vereniging van Nederlandse Gemeenten, 2009) analyse different types of production and processing facilities which are split-up into four categories: exposition to emissions of odour, site specific substances, noise and danger of explosion. Consequently, adequate separation distances between industrial plants and sensitive areas (including housing, educational, recreational and health entities) are stipulated. Similarly, the *German regional guidelines for industrial sites by the Federal Land North Rhine-Westphalia* (MUNLV, 2007) provide minimum distances between selected industrial facilities and living areas. On the national level in Germany the *Technical Instructions on Air Quality Control – TA Luft* (Bundesministerium für Umwelt (BMU), 2002) stipulate requirements in the area of air emission impacts.

However, if the protection zones were determined roughly following the above mentioned simplified procedures, distances from an industrial site would not account for real onsite impacts *e.g.* dispersion of pollutants in time, including impact of varying meteorological and topographical conditions in its immediate vicinity. Therefore, in specific situations protection zones should be determined with more sophisticated tools (such as dispersion or acoustic modelling). Zones determined in the process of calculation are smaller than R-radius determined zones and take elliptic forms. The method to estimate the width of a protection zone, in order to define the optimal location of a waste management facility within a city, should be chosen after consideration of cost effectiveness of the planning process (Petts & Gev, 1994). Due to the fact that dispersion methods are expensive (a few thousand of EUROs) and time consuming, it is always a matter of a trade-off between quality of end results and costs to produce them.

Focus on Construction and Demolition Waste

Recycling of construction waste requires adequate technical infrastructure. The problem of siting of waste infrastructures has been examined since 1970's (Germany) and 1980's (the Netherlands). Building materials can be reused without processing (*e.g.* radiators, installations), with partial transformations (*e.g.* reused steel beams as structural elements, wooden planks or window frames as facade systems) or can be completely processed (*e.g.* recycled glass, concrete). Reused materials can be firmly incorporated into a building (*e.g.* bricks combined with mortar in Am Kupfergraben Gallery in Berlin) or may be flexibly assembled for future reuse (*e.g.* bolted steel structure in Villa Welpeloo in Enschede). Thus, selective waste collection, on-site separation, cleaning processes, suitable organisation of demolition companies and processing infrastructure is required. Building elements can be either reused as whole parts, which require adequate storage space; or slightly repurposed, which requires place for dedicated workshops. Construction waste may be processed in specific installations or returned to a manufacturer, where secondary materials are converted into brand new products. Construction waste recycling is feasible when recycling facility is located close to a waste material generation site - acc. to Harvest Map concept the optimal distance between the construction site, material extraction place and processing plant is limited to 50km concept (Hinte, E. van, Peeren, C., & Jongert, 2007). Thus, to achieve sustainable circulation of construction waste it is crucial to incorporate necessary facilities within the urban tissue. Due to environmental impacts such as: odours, noise level, other pollution and explosion danger, visual impact, traffic, accidents, other health risks (Petts & Gev, 1994) this kind of facilities need to be carefully planned and located in a predefined distance from inhabited areas (both existing and stipulated in master zoning plans).

The Dutch guidelines (Vereniging van Nederlandse Gemeenten, 2009) stipulate environmental impacts for selected industrial plants in four chosen categories: odours, noise level, other pollution and danger of explosion. They define distances to sensitive areas (incl. inhabited) in four categories and suggest the highest distance value for planning purposes. For recycling facilities such distances are stipulated at: for metals (greatest distance: 500m), aggregates (700m), rubble (300m) or for other waste systems (300m). Significant burdens related to building materials production are odour emissions (for metals, mineral wool, bitumen) and noise levels (for cement, concrete, aggregates and metals). The emissions by other substances and the explosion danger can become an issue during the processing of following materials: plasters, cement and metals. Wood production places should be located in a distance ranging from 100m (for sawmills, veneers and plywood production plants) to 200m (for wood protection companies, which use creosote and salt). Rubble processing plants should keep the distance of 100m and the ones transforming plastics – of 200m (plastics without the use of phenols) and 300m (with use the of phenol raisins). Glass factories, depending on their capacity, should be located between 100m (for a plant with <5thou. Mg/y capacity) to 300m (>5thou. Mg/y) and glass processing companies 50m to inhabited areas. Production plants should keep the distance of 1500m for steel and iron, 1,000m for cement (>100 thou. Mg/y), 700m for concrete (>100Mg/d), 500m for mineral and fiberglass (>5thou. Mg/y), calcium, bitumen, cement (<100 thou. Mg/y), 200m for brick and ceramic tiles, 300m for plaster, thermal insulations, asphalts and 100m for mineral products. Repair of metal products requires significantly smaller distances than its production – workshops in closed buildings should keep the distance of 100m and the open-air ones – of 300m. The companies stamping, pressing and forming metals need to be away by 200m, welding and metal surface enterprises – by 100m. Concrete processing plants require 50m of a protection zone, but when processing involves crushing, screening, drying in large quantities the distance rises up to 700m.

Similarly, regional guidelines of the German North Rhine-Westphalia (MUNLV, 2007) define distances between industrial plants and inhabited areas. The assessment of a required distance is based on overall impact levels. Therefore, in the North Rhine-Westphalia open and closed waste collection points should be away from dwellings by respectively 500m and 200m. Production and processing plants for aluminium, concrete (capacity of 1 Mg/h), metals (incl. steel processing), fibreboards and wood-based materials should keep the distance of 1,000m from inhabited areas. Facilities producing and processing cement should be located 700m from housing, the ones with bitumen, glass, fiberglass, rubble – 500m, with aggregates, gravels, sands, bricks and ceramic products (combustion chamber volume above $4m^3$, stocking density of the combusted material above 300 kg/m^3), felts and wood (sawmills, veneer) – 300m and the ones processing plastics (without phenols) – 100m.

The distances indicated by the Dutch and German guidelines are proportional to the environmental burdens imposed by a specific type of material. The greatest distance should be kept between inhabited (or other sensitive) areas and facilities processing steel, metals, concrete, cement, aggregates and bricks (500-1500m). The smallest distance is required for production or processing plants for plastics, ceramic products and wood (100-300m), which is presented in Table 1.

In Poland, according to the former guidelines (PCUGP, 1967) (withdrawn), for environmental zoning in urban planning, industrial plants were grouped in five categories, which required a minimum protection zone of 50m, 100m, 300m, 500m and 1000m. In the case of building materials, the largest distance of 1000m was recommended for industrial plants producing: steel, aluminium, cement, calcium, plywood, fibreboards; 500m for smaller cement producers (capacity up to 150 thou. Mg/y), 300m for glass wool, aggregates, felts, cables in rubble insulation, wood preservation; 100m for asbestos, concrete, natural and artificial stone, mortars, porcelain, wooden elements and 50m for gypsum and clay products. Moreover, the distances for solid waste disposal (1,000m) and utilization points (500m) were indicated. Later in 1980's (MAGTiOŚ, 1982) (also withdrawn legislation) some zoning requirements were added, *e.g.*: the distance of 500m for industrial plants producing plastics, synthetic fibres, glass, pitch, bitumen; 300m for paints, varnishes, PVC; 100m for synthetic raisins, furniture, building materials' storages and 50m for gravel, carpets and welding workshops.

Table 1. Separation distances between communal infrastructure processing construction waste and housing areas

Distance	Holland Facility Type	Germany Facility Type (production and processing)
1500m	steel and iron production	-
1000m	cement production (>100 thou. Mg/y)	aluminum concrete (cap. of 1 Mg/h) metals (incl. steel processing) fiberboard wood-based materials
700m	recycling facilities for aggregates concrete production (>100 Mg/d) concrete processing (with crushing, screening, drying)	cement
500m	recycling facilities for metals cement production (<100 thou. Mg/y) fiberglass production calcium production bitumen production	bitumen glass fiberglass rubble waste collection points (open)
300m	glass production (cap. >5 thou. Mg/y) plaster production thermal insulation production asphalt production plastics processing (with phenol raisins) repair of metal products (open-air) other waste systems	aggregates, gravels, sands bricks, ceramic products (com.vol.> 4m^3, stock. dens.> 300 kg/m^3) felts wood (sawmills, veneer)
200m	plastics processing (without phenols) wood production (with creosote and salt) bricks production ceramic tiles production stamping, pressing and forming metals	wood concrete natural and artificial stone mortars porcelain
100m	glass production (cap. <5 thou. Mg/y) rubble processing mineral products production wood production (sawmills, veneers, plywood) repair of metal products (indoor) welding and metal surface enterprises	plastics (without phenols)
50m	glass processing concrete processing	-

Source: after (Vereniging van Nederlandse Gemeenten, 2009), (MUNLV, 2007)

Focus on Organic Fraction of Municipal Solid Waste

The procedures for siting facilities processing organic waste are usually expressed by air quality standards. Odours in the vicinity of any facility processing organic waste are in fact the most common cause of protests from local communities (Petts & Gev,

1994). Protection zones have been determined from the 1960's in countries such as Austria, Belgium, Canada, Germany, the Netherlands, Poland, Sweden and Switzerland for both agricultural facilities (such as stables), and municipal infrastructure (such as waste processing plants or WWTPs) (Oniszk Popławska & Kulig, 2014).

In Poland protection zones were already specified by the law in the 1960's but later withdrawn. The width of a protection zone around a waste water treatment plant (WWTP) varied from 100m to 1,000m depending on the plant's size, 500m for organic waste processing facilities and for landfilling of waste with organic content from 500m (below 10 ha) to 1,000m. The separation distances in other countries are presented in Table 2.

Table 2. Separation distances between communal infrastructure processing organic material and housing areas

Specification	Germany		Holland	
	[m]	Plant Capacity		[m]
Waste water treatment plants (WWTP)				
Medium	300	< 100 thou. IE*	< 100 thou. IE*	200
Big		-	100-300 thou. IE*	300
Very big	500	> 500 thou. IE*	< 300 thou. IE*	500
Processing, storage, landfilling of waste				
Storing of waste	500	>10 Mg/d or with the total capacity>100 Mg	-	-
Open landfills	500	Landfilling time > 1 year	-	-
Sorting of waste	300	>10 Mg/d	-	300
Processing of waste	700 500	>50 Mg/d 10-50 Mg/d	-	-
Burning of waste and fuel	500	>1 Mg/d	Waste incineration plants, thermal power> 75 MW Bioenergy systems electrical power <50 MW_e:	300 100
Drying of waste	300			
Biological Treatment Plants				
Composting plants	500 300	Open >3 thou. Mg/a <3 thou. Mg/a Enclosed>20 thou. Mg/a	- Non-aerated <5 thou. Mg/a - Non-aerated 5 - 20 thou. Mg/a - Aerated <20 thou. Mg/a - Aerated > 20 thou. Mg/a - Green waste, closed	300 700 100 200 200
Anaerobic digestion plants	500 300	>10 Mg/d Open Closed	-	-

Source: after (Vereniging van Nederlandse Gemeenten, 2009), (MUNLV, 2007), (Bundesministerium für Umwelt (BMU), 2002)

*IE - inhabitant equivalent

CASE STUDY: INTERGRATED PROCESSING OF CONSTRUCTION AND OFMSW

Two case studies were analysed to illustrate the concept of integrated waste metabolism, both of them in Germany one in Bayern (Kirchstockach) the other one in Schleswig-Holstein (Trittau) Data were collected through interviews with plant operators and on site research visits (Oniszk-Popławska, 2014), (Oniszk-Popławska, 2015), additional literature review, analysis of online information (company websites, geoportals). Evaluation of gathered data gave a valuable insight into the problem of the optimal distances between urban waste infrastructure and inhabited (and other sensitive) areas as well indication for optimal waste policy to be carried out by the local/regional authorities.

An integrated approach to urban waste management has been applied in the German federal state of Bavaria, near Munich in Kirchstockach. Two waste processing facilities are located on premises belonging to the company M. Ganser GmbH & Co. KG (Figure 2). The company's goal to protect nature and raw materials resources aims to enable long-term and sustainable urban waste processing – for both OFMSW and construction and demolition waste. The plant was chosen as a case study, because both investigated waste streams are processed in the facility, which is located close to inhabited areas. Thus, it was an adequate example to examine in the context of environmental zoning and sitting of urban waste infrastructure. The analysis of the case study in Kirchstockach underlined the importance of social aspects in urban planning (Figures 2 and 3).

Figure 2. Distances of waste processing sites to inhabited areas in the Kirchstockach M. Ganser GmbH & Co. KG and Ganser Baustoffe facilities
Source: own elaboration after Bayern Geoportal (http://geoportal.bayern.de/bayernatlas)

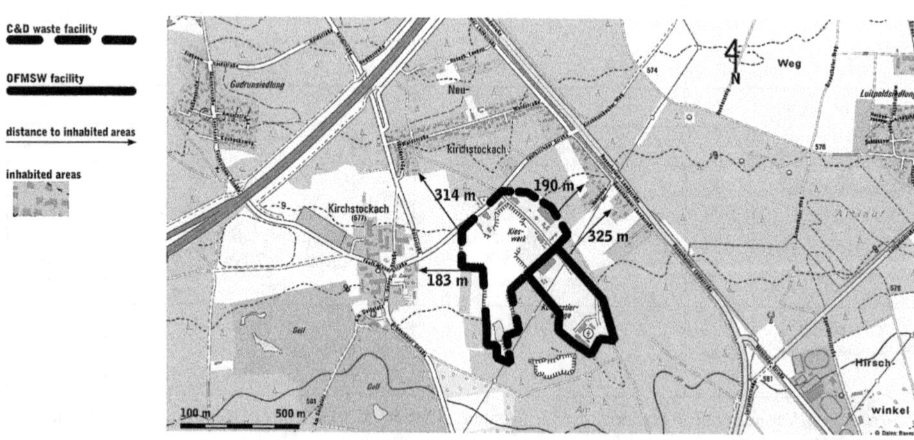

Figure 3. Distances of waste processing sites to inhabited areas in Trittau, BUHCK Gruppe
Source: own elaboration after Google Earth

In Trittau the closest distance to the inhabited areas is 490 meters, however the company also pays due care to public acceptance issues for two reasons: 1) in order to increase the level of selective collection of food and green waste among the inhabitants from the surrounding villages and also 2) to alleviate possible NIMBY impacts.

Focus on Construction and Demolition Waste

In case of the Ganser plant in Kirchstockach the concept of urban metabolism finds its application in the production and processing of sand, gravel and concrete. Construction waste is washed, screened and crushed. It is treated in a washing unit, vertical crusher and pipe rod mills for its future reuse aggregates (in accordance with a DIN-standard), sands and gravels. Those materials are later processed to create high quality concrete, cement or other building products. Ganser Baustoffe policy meets also other principles of urban metabolism – it produces repair systems and leads constant research to improve properties of building materials. The company promotes the idea of closed circulation of building materials by processing approx. 500 thou. Mg of construction waste per year. Two plants in Kirchstockach (dry mining) and Feldkirchen (wet mining) process all kinds of gravel, sand, aggregates as well as products for wall construction, floors and finishing. The offer of Ganser Baustoffe includes: gravel, sand, aggregates (for mortars, concrete and asphalt), concrete (high quality concrete and finishing), masonry products (mortars, waterproofing), facade elements (bases, plasters, thermal insulation, paints), finishing for walls and ceilings, floors (screeds, supplements), tiles (bases, finishing's, acoustic insulations) and other products (adhesives, mortars, grouts). The Ganser Kirchstockach plant is located at the distance of *c.* 180m to inhabited areas (Figure 4), which is closer than the distance indicated in the German and Dutch guidelines for the environmental zoning (according to German guidelines a minimum separation distance for aggregates, sands, gravels and concrete is 300m and for concrete - 500m and for cement 700m). Gathered data does not provide quantities of specific building materials processed in Ganser Kirchstockach plant. Therefore, it is difficult to evaluate optimal protection zone between Ganser Kirchstockach plant and inhabited areas. According to Ganser Gruppe manager, there were some complaints concerning odour and noise during the transport. However, due to small amount of inhabitants affected (3 residents) and company's effort to collaborate with the local community (*e.g.* through round-table meetings organised twice a year) the existence of processing plant in the close vicinity of inhabited areas continues undisturbed.

Focus on Organic Fraction of Municipal Solid Waste

Within the Munich metropolitan region two installations operate fuelled by the OFMSW. One is located within the city of Munich and operated by the municipal utility company AVM, where source selected organics (SSO) of the city of Munich are processed. The other one (in operation since 1997) is located in Kirchstockach,

Figure 4. C&D waste collection and processing, the last row depicts the facility in Trittau
Source: photographs by A. Oniszk-Popławska

it takes in SSO from areas surrounding the city ring administered by the Landkreis München (the province of Munich). Waste is transported to the plant from a distance of even up to 40 km towards the North and directed to the processing plant in the South (even though the other OFMSW plant is much closer). However, a well-developed network of roads and highways around Munich allows for efficient logistics. SSO to Kirchstockach is delivered from the area of *c.* 667 km^2 (service area of the Munich province), from over 400 thou. residents, with the annual capacity of the plant *c.* 30 thou. Mg per year (Figure 5).

The concept of material metabolism is realised by organic recycling. OFMSW kitchen waste is processed in an anaerobic digestion (AD) plant, green waste is added to the digested waste stream in a later composting stage. 12 thou. Mg of AD by-products *i.e.* digestate (after dehydration), is mixed with the same amount of green waste, and later used as a fertiliser or earth compound for example for horticultural purposes. Impurities make up approx. 20% of the total amount of organic waste. The residual waste after separation from the organic fraction is used as a road construction material. In addition, metals are recovered by a magnetic separator and sold for further recycling.

The energy metabolism concept is realised by power production units fuelled by biogas produced in a process of anaerobic digestion, with the help of 3 cogeneration units CHP 3x310 kW_e. Biogas production results in 4.5 MWh/y electricity generations, of which less than 50% is fed to the power grid (the rest is used for the plant's own processes). The heat is used either for plant's own needs (like sustaining the digestion temperature) or for drying of wood (ordered by an external sawmill).

Figure 5. OFMSW processing, the final stage of organic recycling: the first row the facility in Kirchstockach, the second row in Trittau
Source: photographs by A.Oniszk-Popławska

At the AD plant at Kirchstockach odour problem has been recorded as a nuisance for the local residents - the nearest are located at the distance of 325m. An air-tight hall, where the AD processing takes place is equipped with a mechanical ventilation system and a biofilter, which cleans exhaust air. However, the main source of odour emissions is the entrance gate opening approx. 20 times a day, when SSO waste is delivered by lorries. However, in order to sustain a high public acceptance, operation and maintenance aspects have been discussed twice a year among the plant operator, inhabitants, municipality, and a construction supervisor.

The plant operator stressed (Oniszk-Popławska, 2014) the need to introduce proactive educational measures in close cooperation with local authorities. Such activities include open days, organized in Kirstockach already a couple of times, with a wide public representation (age range 10-60 years). The facility has also been exhibited during the "Energy days" festival, during which local people could learn about specifics of the AD technology. A consistent educational policy has contributed to a high participation of citizens in separate selection of kitchen and garden waste (in Germany called Biotonne). It should be emphasized that at the beginning of the implementation of the system social resistance and doubts about its effectiveness were common. Today, residents are well informed on the environmental advantages and willingly participate in the separate collection system. Such a success however resulted from the commitment on the part of local authorities. The results of interviews with stakeholders showed that a separate collection of organics is a long-term project, effects of which can be seen sometimes after a decade (Oniszk-Popławska, 2014).

The second combined facility is located in Trittau, a waste processing facility, a joint venture of the waste disposal companies BUHCK and Remondis (Oniszk-Popławska, 2015). Since 1998 AWT has been operating a composting facility with green waste and source separated kitchen waste as well as old grocery's food from the surrounding districts of Lauenburg and Stormarn. Since 2013 an anaerobic digestion plant has been coupled with the composting plant and allows the additional production of biogas. Each year *c.* 30 thou. Mg of biowaste provide *c.* 2.5 million m^3 of biogas and 15 thou. Mg of quality compost. The investment is private and thanks to the utilisation of the existing brownfields infrastructure (former composting sites) the costs borne by the local inhabitants are much smaller than in facilities with comparative sizes (60 EUR per Mg in Trittau compared to 90 EUR in Kirchstockach).

Both companies have actively participated in the information campaigns, info days, invited school trips and had telephone advisory service on "how to separate effectively". This has done for a different reason in Trittau (than in Kirchstockach) because Biotonne is collected from 60 villages and the inhabitants are not used to source separation of food waste, therefore so far despite of the Biotonne obligation (Bundesregierung, 2012) the level of food waste in the organic material has been relatively small. It is hoped that it can be increased by educational measures (Oniszk-Popławska, 2015). The public acceptance measures are also performed to sustain the uptake of the produce digestate in order to recalculate the nitrogen and phosphorous contained in food waste back to the fields.

CONCLUSION AND RECOMMENDATIONS

The new, holistic approach to the city metabolism and treatment of construction and OFMSW waste will influence the way cities are planned, waste management systems organised and inhabited areas delimitated. The process of urban planning will need to incorporate adequate siting procedures for waste processing facilities. Waste management system is the main determinant for organisation of selective collection points and transportation facilities. In urban spatial planning, it will be necessary to take under consideration life-cycles of urban infrastructure and buildings as well as related environmental impacts and environmental burdens such as odours, noise levels, emissions of other substances and danger of explosion. It is assumed that in the future also buildings will be designed to account for urban metabolism issues (*e.g.* to enable their easy maintenance, servicing and dismantling close to separate collection and recycling sites).

The urban planning, which aims to create a closed circulation of materials, requires adequate organisational measures (siting and management of deconstruction companies, production, cleaning and processing facilities, storage and repurposing workshops). Such infrastructures should be incorporated into urban tissue to facilitate sustainable approach to materials used in built environment and waste generated in inhabited areas. However, due to significant burdens, which are generated during processing of urban waste infrastructures need to be placed at safe distances from inhabited (and other sensitive) areas. For building construction waste, according to the Dutch and German guidelines, the largest distances are required for the production and processing plants of such building materials as: metals, aggregates, gravels, sands, concrete, cement, bitumen and fiberglass. Smaller protections zones are indicated for infrastructures transforming wood, ceramics and plastics.

For the organic fraction of municipal solid waste the distances depend on the facility's size and kind of treatment the biggest distances are required for open, non aerated composting pants (700 m) the smallest (200m) for air-tight closed facilities. Usually the distance should not be smaller than 200-300m for facilities with capacity exceeding 3 thou. Mg of annual output.

The analysis of a selected case study - the Ganser Kirchstockach plant showed that effective and sustainable processing of construction and OFMSW waste in urban areas is possible. Successful recycling of such building materials as gravel, aggregates, cement and concrete as well as SSO processing is possible in a close proximity to inhabited areas. But, this specific case study demonstrated also that an industrial plant processing building materials was sited closer than recommended by the Dutch and German guidelines. Thanks to additional efforts and clear information policies for inhabitants of neighbouring urban areas the project can still be evaluated as successful.

In case of the plant processing organic waste in Kirchstockach separation the distance for a closed facility is not smaller than recommended (300m), however complaints are still present. In the Trittau facility the distances are bigger but the operator still pays attention to public communication and avoidance of NIMBY effect. Therefore, it is recommended that the evaluation of impacts and distances during a siting procedure is conducted individually for each plant. During such tailor made evaluation it is evident that local social, cultural, topographical and meteorological criteria should be taken into account. The distances indicated in the guidelines should be treated with due care and used only in the preliminary, scoping procedure.

To create sustainable material flows within a city, it is necessary to undertake multi-level actions at the level of policy making, regulations, economic incentives and education. The principles of closed circulation of organic and inorganic waste need to be incorporated in national and regional legal acts as well as in development strategies in all sectors of economy. The optimal waste recovery treatment and recycling levels should be defined for each waste stream (*e.g.* in A form of minimum standards and targets).

The first step to plan a sustainable metabolism is to set priorities and establish a legal framework, followed by the application of given policies for example by the erection first pilot investments: the Kirchstockach facility processing source separated organics was erected in 1997 and it was the first anaerobic digestion plant using OFMSW in Germany, currently there are *c.* 200 of them all over the country.

The third step is to enforce legislation, like in the case of separate selection of kitchen and garden waste (Biotonne) has been made obligatory starting January 2015 (Bundesregierung, 2012). Also, economic incentives for new investments as well as measures for sustaining their economic feasibility are important. The citizens have to be aware of the fact that the new quality of waste recycling and the realisation of urban metabolism needs to be paid for by the participants of the metabolic chain. In case of the organic fraction the enforcement of such investments is covered by a higher gate fee, *i.e.* by the inhabitants themselves paying for waste collection and disposal. If such a radical change is made without the knowledge and participation of citizens in the process, it is probably due to fail. Therefore, increasing the public acceptance of higher costs, and on the other hand the proximity to such investments in the so-far rural surrounding is a foundation for their active participation in the process. In order to become a part of bioeconomy transformation they not only need to accept the investment but they have to participate actively in source separated collection. Thus, information campaigns, open days and other means leading to higher involvement of citizens in the process are prerequisite to the success of the bioeconomy transition. In case of poorer countries where part of the society actually scavenges the waste as a form of living – neither information campaigns nor open days will lead to a change of attitude. It is an interesting thought, that a society has to b sufficiently rich to understand analysed approach.

In the case of construction and demolition waste sustainable procurement, pro-environmental taxes, green investments, producer responsibility, voluntary agreements for SMEs and eco-design are needed for successful recovery of urban waste. Moreover, tools for measuring, monitoring and forecasting material flows are required to enable closed circulation of urban waste. A common problem concerning waste, the lack of data about its location and availability, may be solved through introduction of adequate, open access, online databases.

Waste treated as a part of city management is not a standalone feature, it is more of an urban comportment connected directly or indirectly, to a number of issues such as urban lifestyles, resource consumption patterns, jobs and income levels. There is a whole culture of waste management that needs to be put in place - from the micro-level of household, neighbourhood and community levels up to the macro levels of city, state and nation. The main level – is not even the city level, it starts much lower, at the level of households and neighbourhoods.

Different approaches to the management of waste within a city are possible. The most common approach for the operation of OFMSW anaerobic digestion plants is the public private partnership (PPP), for in most of the cases it is the local government which is a responsible body for the waste management. In such a PPP agreement different administrative measures can be agreed upon: specification of who is responsible for waste delivery its amounts, plot ownership issues, equities in the investment etc. The involvement of the local authorities has a great importance for the successful implementation of AD facility (social acceptance and system efficiency). In the case study analysis of 10 AD plants in Europe most of the anaerobic digestion plants were in some way governed by the local authorities. Usually the plant operator was an entity with both private and public (city or region owned) equity. In some cases, authorities owned the land, where the plant was located (*e.g.* lease agreement). In most of the European countries municipalities own the MSW streams and decide on treatment methods. Moreover, the public equity boosts the investment process. Another issue is the public perception; investments with the participation of a local authority tend to have a higher acceptance level by the local community.

Moreover, an integrated approach to urban planning is required to enable sustainable flows of organic and construction and demolition waste. Processing infrastructure should be planned in urban areas after detailed analysis of social, environmental and economic conditions. Guidelines for sitting of urban waste infrastructure enable optimal location of processing facilities in urban areas. However, the evaluation of local context and impacts needs to be conducted individually for each plant. Facilities, according to their size, the type of the treatment and materials processed, should be located in a safe distance from inhabited (or other sensitive) areas, preferably in the existing brownfields. Sometimes, additional efforts and clear information policies for inhabitants of urban areas, where processing plants are introduced, are required – for undisturbed existence of processing facility.

ACKNOWLEDGMENT

The research leading to these results has received funding from the Polish-Norwegian Research Programme operated by the National Centre for Research and Development under the Norwegian Financial Mechanism 2009-2014 in the frame of Project Contract No [Pol-Nor/210329/46/2013]. URL: http://mbist.arch.pw.edu.pl/index.php/en/

REFERENCES

Achillas, C., Vlachokostas, C., Moussiopoulos, N., Banias, G., Kafetzopoulos, G., & Karagiannidis, A. (2011). Social acceptance for the development of a waste-to-energy plant in an urban area. *Resources, Conservation and Recycling*, *55*(9-10), 857–863. doi:10.1016/j.resconrec.2011.04.012

Agudelo-Vera, C. M., Leduc, W. R. W., Mels, A. R., & Rijnaarts, H. H. M. (2012). Harvesting urban resources towards more resilient cities. *Resources, Conservation and Recycling*, *64*, 3–12. doi:10.1016/j.resconrec.2012.01.014

Anderberg, S. (1998). Industrial metabolism and the linkages between economics, ethics and the environment. *Ecological Economics*, *24*(2-3), 311–320. doi:10.1016/S0921-8009(97)00151-1

Barles, S. (2010). Society, energy and materials: The contribution of urban metabolism studies to sustainable urban development issues. *Journal of Environmental Planning and Management*, *53*(4), 439–455. doi:10.1080/09640561003703772

Bundesministerium für Ernährung und Landwirtschaft. (2014). Nationale Politikstrategie Bioökonomie. National Policy Strategy on Bioeconomy - Renewable resources and biotechnological processes as a basis for food, industry and energy.

Bundesministerium für Umwelt (BMU). (2002). Technical Instructions on Air Quality Control – TA Luft.

Bundesministerium für Umwelt Naturschutz und Reaktorsicherheit. (2012). German Resource Efficiency Programme, Deutsches Ressourceneffizienz- programm (ProgRess), 124.

Bundesregierung. Law on Life-Cycle Management, Kreislaufwirtschaftsgesetz, Gesetz zur Förderung der Kreislaufwirtschaft und Sicherung der umweltverträglichen Bewirtschaftung von Abfällen (KrWG) (2012).

Bundesregierung. The Commercial Waste Ordinance, Verordnung über die Entsorgung von gewerblichen Siedlungsabfällen und von bestimmten Bau- und Abbruchabfällen (Gewerbeabfallverordnung - GewAbfV) (2002).

Commission of the European Communities. (2000). Commission Decision on the European List of Waste (COM 2000/532/EC). *Official Journal of the European Communities, 2000D0532*(01.01.2002). doi:(2000/532/EC)

Commission of the European Communities. (2014). 15 Final. Communication from the Commission to the European Parliament, the Council, the European economic and social committee and the committee of the regions a policy framework for climate and energy in the period from 2020 to 2030.

Eiselt, H., & Marianov, V. (2015). Location modeling for municipal solid waste facilities. *Computers & Operations Research, 62*, 305–315. doi:10.1016/j.cor.2014.05.003

EPEA i VHE. (2008). *Ökologisches Leistungsprofil von Verfahren zur Behandlung von biogenen Reststoffen Kompass für die Entscheidungsfindung vor dem Hintergrund der geplanten Überarbeitung*.

European Parliament and Council. Directive 2008/98/EC of the European Parliament and of the Council of 19 November 2008 on waste and repealing certain directives (Waste framework directive)., L 312/3 3–30 (2008). doi:2008/98/EC.; 32008L0098

Eurostat. (2015). Generation of waste [env_wasgen]. Retrieved from http://appsso.eurostat.ec.europa.eu/nui/show.do?dataset=env_wasgen&lang=en

German Federal Ministry of Education and Research. (2011). The German Government's raw materials strategy.

Kennedy, C., Pincetl, S., & Bunje, P. (2011). The study of urban metabolism and its applications to urban planning and design. *Environmental Pollution,* 159(8-9), 1965–73. doi:10.1016/j.envpol.2010.10.022

Khadivi, M. R., & Fatemi Ghomi, S. M. T. (2012). Solid waste facilities location using of analytical network process and data envelopment analysis approaches. *Waste Management (New York, N.Y.)*, *32*(6), 1258–1265. doi:10.1016/j.wasman.2012.02.002 PMID:22382043

MAGTiOŚ. Zarządzenie Ministra Administracji, Gospodarki Terenowej i Ochrony Środowiska z dnia 9 listopada 1982 r. w sprawie szczegółowych zasad wyznaczania granic i obszarów stref ochronnych oraz orientacyjnych wskaźników ich szerokości. (1982).

Ministerie van Infrastructuur en Milieu. National Waste Management Plan (NWMP), Landelijk afvalbeheerplan 2009-2021 (2014). Den Haag.

MUNLV. (2007). *Bauleitplanung*. Düsseldorf: Abstandserlass.

Nunan, J. (2010). *The complete guide to alternative home building materials & methods: including sod, compressed earth, plaster, straw, beer cans, bottles, cordwood, and*. Atlantic Publishing Group Inc.

Oniszk-Popławska, A. (2014). *Site visit to Kirstockach within an M-BIST project.*

Oniszk-Popławska, A. (2015). *Site visit to Trittau within an M-BIST project.*

Oniszk Popławska, A., & Kulig, A. (2014). Application of Odour Predictions to Spatial Planning, the Case of Agricultural Biogas. *Chemical Engineering Transactions*, *40*, 277–282. doi:10.3303/CET1440047

PCUGP. Zarządzenie Prezesa Centralnego Urzędu Gospodarki Wodnej z dnia 30 maja 1967 r. w sprawie szerokości stref ochronnych ustanawianych dla ochrony powietrza atmosferycznego przed zanieczyszczeniem. (1967).

Petts, J., & Gev, E. (1994). *Environmental impact assessment for waste treatment and disposal facilities*. West Sussex: John Wiley&Sons.

Tojo, N., & Fischer, C. (2011). Europe as a Recycling Society. European Recycling Policies in relation to the actual. *ETC/SCP Working Paper*, (ETC/SCP working paper 2/2011).

UNEP. (2011). *Waste and Climate Change*. Global Trends and Strategy Framework.

van Hinte, E., Peeren, C., & Jongert, J. (2007). *Superuse: Constructing New Architecture by Shortcutting Material Flows*. Rotterdam, the Netherlands: Nai Publishers.

Vereniging van Nederlandse Gemeenten. Bedrijven en milieuzonering' uit (2009). Holland.

Wolman, A. (1965). The metabolism of cities. *Scientific American*, *3*(213). PMID:14338520

Chapter 8
Utilization of SVM, LSSVM and GP for Predicting the Medical Waste Generation

J. Jagan
VIT University, India

Yıldırım Dalkiliç
Erzincan University, Turkey

Pijush Samui
National Institute of Technology Patna, India

ABSTRACT

The prediction of wastes generated in the hospital will help their management for several activities like storage, transport and disposing. This chapter adopts Support Vector Machine (SVM), Least Square Support Vector Machine (LSSVM) and Genetic Programming (GP) in order to estimate the rate of medical waste generation. In the event of predicting the rate, type of hospital, capacity and bed occupancy has been used as inputs of SVM, LSSVM and GP. SVM is based on statistical learning theory, which provides an elegant tool for nonlinear system modeling. LSSVM is the re-formulation to the general SVM. GP, a best part of evolutionary algorithm and also the specification of Genetic Algorithm (GA). These SVM, LSSVM and GP have been used as the regression techniques. The results show the performance of the developed SVM, LSSVM and GP models were elegant and outstanding.

DOI: 10.4018/978-1-5225-0302-6.ch008

INTRODUCTION

The development of industries, urban growth together with population and population densities will automatically increase the production of wastes. World Health Organization (WHO) has advocated that the hospital wastes should be treated as the particular wastes (Rushbrook et al., 2000). These hospital wastes are more hazardous and are rapidly increasing in the society. Inadequate hospital waste management will cause environmental pollution, generation of germs and insects, foul smell, etc, will lead to the contagious diseases like typhoid, cholera and AIDS through the wound from syringes and needles contaminated with the human blood (Henry & Heinke, 1996). This research has the better scope in the smart cities with more industries and population. Potentially dangerous hospital wastes can be predicted for the prevention of some hazardous diseases.

Through inventory, questionnaire survey and formal and informal interviews, data has been collected for empirical field observation in Dhaka city, Bangladesh (Hassan et al., 2008). Very deep interviews were carried out to obtain the knowledge of earlier and current management of medical wastes. Their examination provides the outcome of average waste generation 1.9 kg/bed/day. Hassan et al., (2008) delivers the reasons for improper management of medical wastes like, lack of awareness, appropriate policy and laws, and willingness. Omar et al., (2012) has considered the medical waste generation in Malaysia and determined the factors affecting management of medical wastes. They detected and detailed specifically flaw in segregation process of medical waste. Omar et al., (2012) has adopted Microsoft Excel and its pertinent statistical gadget like ANOVA for evaluating the management of medical waste. Maha et al., (2015) had thrown their limelight work distinctly on dental clinics at Lebanon. They considered 242 dental clinics arbitrarily and made survey with 32 framed questions. According to their scrutiny, the process of isolating the waste was vanished and it is mingled with the municipality wastes. Their audit provides the necessity of adopting efficient and responsible medical waste management, in order to prevent disaster in public health and pollution at Lebanon. Hence the adequate management of hospital waste is required to reduce the health and environmental risks. Unfortunately, collection, transport and disposal of waste have no proper strategy or planning behind its implementation. It urgently needs to that needs to be regulated by definitive rules (Askariana et al., 2004).

In order to determine the rate of waste generation, the data like (i) number of bed/patient available at the hospital; (ii) type of specialization (type and range of care); and (iii) number, kind, and size of departments have been utilized (Award et al., 2004). The goal of this chapter is to predict the rate of waste generation based on the mentioned criteria's. Artificial Neural Network (ANN) and Multiple Linear

Regression (MLR) have been adopted for this determination (Jahandideh et al., 2009). Since ANN & MLR have some disadvantages such as the low accuracy of MLR and the black box approach, over fitting issues, lack of generalization capability and slow convergence rate of ANN (Park & Rilett, 1999; Kecman, 2001), various effective methods have been utilized.

The methodology of pursuing the medical waste generation includes the upcoming procedures. A total of 50 hospitals private, governmental, educational and non-educational, but university-related hospitals in Fars province, Sharz city, Iran have been considered for the collection of dataset (Jahandideh et al., 2009). The dataset contains information about type of hospital, Capacity(C), Bed Occupancy (B_0) and Waste generation rates (kg/day) (w_t). This chapter accommodates the machine learning techniques like Support Vector Machine (SVM), Least Square Support Vector Machine (LSSVM) and Genetic Programming (GP) for determination of rate of medical waste generation.

SUPPORT VECTOR MACHINE (SVM)

The SVM have been formulated by Vapnik (1998, 1999) based on statistical learning theory. Initially, SVM was utilized for two class classifications, later it was developed to multi-classification, regression, estimation and other similar works. Support Vector machines can be defined as systems which use hypothesis space of a linear functions in a high dimensional feature space, trained with a learning algorithm from optimization theory that implements a learning bias derived from statistical learning theory. The formulation uses the Structural Risk Minimization (SRM) principle, which has been shown to be superior, (Burges, 1998) to traditional Empirical Risk Minimization (ERM) principle. SRM reduces an upper bound on the expected risk, whereas ERM minimizes the error on the training data. This differentiation equips SVM with a greater ability to generalize, which is the goal in statistical learning. Since SVM follows the Structural Risk Minimization (SRM) principle, it reduces the occurrence of risk during the training phase as well as enhances its generalization capability. Due to its high performance and detecting the global optimums in real applications, it is most widely used technique in many fields ranging.

Details of SVM

Let us consider the typical regression formulation for predictive learning (Cherkassky & Mulier, 1998; Hastie, Tibshirani, & Friedman, 2001; Vapnik, 1999). The intention is to assess unknown real-valued function in the relationship

$$y = r\left(x\right) + \delta \tag{1}$$

where x is a multivariate input, y is the output and δ is the noise data which is uniformly distributed zero mean random error. The estimation is based on a finite number of samples i.e. training data $\left(x_i, y_i\right), \left(i = 1, \ldots, n\right)$. The uniformly distributed training data samples are engendered according to the probability density function.

$$p\left(x, y\right) = p\left(x\right) p\left(y \mid x\right) \tag{2}$$

The unknown function in equation (1) is the mean of the output regression function

$$r\left(x\right) = \int yp\left(y \mid x\right) dy \tag{3}$$

An estimation procedure identifies the best model $f\left(x, \varpi_0\right)$ from the set of approximating functions $f\left(x, \varpi\right)$. It is then parameterized by a set of parameters $\varpi \in \Omega$. The eminence of the approximation can be determined by the inconsistency measure $L\left(y, f\left(x, \varpi\right)\right)$. The ultimate aim of learning this is to pick the best model reducing estimation risk.

$$R\left(\varpi\right) = \int L\left(y, f\left(x, \varpi\right)\right) p\left(x, y\right) dxdy \; f\left(x, \varpi\right) \tag{4}$$

The regression function is equation (3) is the one minimizing estimation risk (4) with the squared loss function

$$L\left(y, f\left(x, \varpi\right)\right) = \left(y - f\left(x, \varpi\right)\right)^2 \tag{5}$$

It should be noted that, carried by the learning technique may or may not have the regression function (3). Hence we have to determine the regressor function $f\left(x, \varpi_0\right)$ that reduces the prediction risk function using only the training data.

$$R\left(\varpi\right) = \int \left(y - f\left(x, \varpi\right)\right)^2 p\left(x, y\right) dxdy \tag{6}$$

The risk functional ascertains the accuracy of the unknown target function $r\left(x\right)$. SVM maps the input x to the n dimensional space using some non-linear mapping and then the linear mapping will be developed in the feature space (Cherkassky & Mulier, 1998; Smola & Scholkopf, 1998; Vapnik, 1998, 1999).

$$f\left(x,\varpi\right)=\sum_{i=1}^{n}\varpi_i g_i\left(x\right)+b \tag{7}$$

where $g_i\left(x\right)$, i=1,.....,n indicates the non-linear transformations and b is the bias.

Regression forecasts can be procured by the reduction of the empirical risk on the training data. Typical loss functions used for minimization of empirical risk include squared error and absolute value error. SVM regression uses a type of loss function called ε-insensitive loss proposed by Vapnik (1998, 1999).

$$L_\varepsilon\left(y,f\left(x,\varpi\right)\right)=\begin{cases}0 & \text{if } \left|y-f\left(x,\varpi\right)\leq\varepsilon\right| \\ \left|\left|y-f\left(x,\varpi\right)\leq\ \varepsilon\right|\right. & \text{otherwise}\end{cases} \tag{8}$$

The empirical risk is defined in the following equation (9)

$$R_{emp}\left(\varpi\right)=\frac{1}{n}\sum_{i=1}^{n}L_\varepsilon\left(y_i,f\left(x_i,\varpi\right)\right) \tag{9}$$

SVM performs linear regression in a high dimensional feature space by utilizing the ε insensitive loss. Also simultaneously it engages to deduce the complexity of the model by minimizing $\left\|\varpi\right\|^2$. This can be illustrated by setting up a non-negative slack variable ξ_i,ξ_i^* $i=1,...,n$ in order to determine the divergence training samples away from ε insensitive region. Thus SVM regression is formulated as minimization of the following function:

Minimize: $\frac{1}{2}\left\|\varpi\right\|^2+C\sum_{i=1}^{n}\left(\xi_i+\xi_i^*\right)$

$$\text{Subjected to: }\begin{cases} y_i - f(x_i, \varpi) - b \le \varepsilon + \xi_i^* \\ f(x_i, \varpi) + b \le \varepsilon + \xi_i^* \\ \xi_i, \xi_i^* \ge 0 \end{cases} \tag{10}$$

where C is the regularization parameter which is a positive constant. This optimization formulation can be transformed into the dual problem (Vapnik, 1998, 1999).

$$f(x) = \sum_{i=1}^{n} (\alpha_i - \alpha_i^*) K(x_i, x) + b \tag{11}$$

The dual variables are subjected to constraint $0 \le \alpha_i, \alpha_i^* \le C$ and $K(x_i, x)$ is the kernel function. The points that come into view with non-zero coefficients are called support vectors. SVM generalization performance depends on the better setting of hyper parameters and the kernel parameters. The problem of determining depends upon the above mentioned hyper and kernel parameters.

The parameter C measures the tradeoff whereas the degrees of deviation are determined by ε. For example, when C is too large i.e. infinity, then the empirical risk alone can be minimized without regards to model complexity. Parameter ε controls the width of the ε -insensitive zone, used to fit the training data (Cherkassky & Mulier, 1998; Vapnik, 1998, 1999). The value of ε can affect the number of support vectors used to construct the regression function. Larger ε - value result in fewer support vectors selected, and results in more 'flat' (fewer complexes) regression estimates. Hence, both C and ε -values affect model complexity (but in a different way).

Radial basis function has been adopted for this study.

$$K(x_i, x) = \exp\left\{-\frac{(x_i - x)^T (x_i - x)}{2Ã^2}\right\} \tag{12}$$

x_i and x are the training and test patterns, T is the transpose and σ is a the global basis function width.

The total datasets have been divided into the following two groups:

- **Training Dataset:** This is used to construct the SVM model. In this article, 35 datasets out of 50 have been adopted as training dataset.

- **Testing Dataset:** This is used to evaluate the developed SVM. This article uses the remaining 15 datasets as testing dataset.

The datasets are scaled between 0 and 1. The programming of SVM has been done by using MATLAB.

LEAST SQUARE SUPPORT VECTOR MACHINE (LSSVM)

LSSVM was proposed by Suykens et al (2002). LSSVM is the statistical learning theory which adopts the least squares linear system as the loss function to the original SVM. LSSVM retains the SRM principle which leads to the improvement of speed than the classical SVM. It was achieved due to the varying inequality constraints into equations and takes the squared loss function. Hence, it solves system of equations instead of quadratic problems. LSSVM adopts the same inputs as that of SVM.

Details of LSSVM

Consider a given training set of N data points $\left\{x_k, y_k\right\}_{k=1}^{N}$ with input data $x_k \in R^N$ and output $y_k \in r$ where R^N is the N-dimensional vector space and r is the one-dimensional vector space.

In feature space LSSVM models take the form

$$y\left(x\right) = w^T \phi\left(x\right) + b \tag{13}$$

where the nonlinear mapping $\phi\left(.\right)$ maps the input data into a higher dimensional feature space; $w \in R^n$; $b \in r$; w - adjustable weight vector; b - the scalar threshold.

In LSSVM for function estimation the following optimization problem is formulated:

$$\text{Minimize: } \frac{1}{2} w^T w + \gamma \frac{1}{2} \sum_{k=1}^{N} e_k^2$$

$$\text{Subjected to: } y\left(x\right) = w^T \phi\left(x_k\right) + b + e_k \text{, k=1,...,N.} \tag{14}$$

where γ is the regularization parameter, determining the trade-off between the fitting error minimization and smoothness and e_k is error variable.

The Lagrangian ($L(w,b,e;\alpha)$) for the above optimization problem (14) is

$$L(w,b,e;\alpha) = \frac{1}{2}w^T w + \gamma \frac{1}{2}\sum_{i=1}^{N} e_i^2 - \sum_{i=1}^{N} \alpha_i \left\{ w^T \phi(x_i) + b + e_i - y_i \right\} \tag{15}$$

with α_k Lagrange multipliers. The conditions for optimality are given by

$$\frac{\partial L}{\partial w} = 0 \Rightarrow w = \sum_{k=1}^{N} \alpha_k \phi(x_k)$$

$$\frac{\partial L}{\partial b} = 0 \Rightarrow \sum_{k=1}^{N} \alpha_k = 0$$

$$\frac{\partial L}{\partial e_k} = 0 \Rightarrow \alpha_k = \gamma e_k, \text{ k=1,...,N.}$$

$$\frac{\partial L}{\partial \alpha_k} = 0 \Rightarrow w^T \phi(x_k) + b + e_k - y_k = 0, \text{ k=1,...,N.} \tag{16}$$

After elimination of e_k and w, the solution is given by the following set of linear equations

$$\begin{bmatrix} 0 & 1^T \\ 1 & \Omega + \gamma^{-1} I \end{bmatrix} \begin{bmatrix} b \\ \alpha \end{bmatrix} = \begin{bmatrix} 0 \\ y \end{bmatrix} \tag{17}$$

where $y = [y_1, \ldots, y_N]$, $1 = [1, \ldots, 1]$, $\alpha = [\alpha_1, \ldots, \alpha_N]$ and Mercer's theorem (Vapnik, 1998; Smola and Scholkopf, 1998), is applied within the Ω matrix,

$$\Omega = \phi(x_k)^T \phi(x_l) = k(x_k, x_l), \text{ k,l=1,...,N}$$

where $k\left(x_k, x_l\right)$ is the kernel function. Choosing $\gamma>0$, ensures the matrix $\Phi = \begin{bmatrix} 0 & 1^T \\ 1 & \Omega + \gamma^{-1} I \end{bmatrix}$ is invertible. Then the analytical of α and b is given by

$$\begin{bmatrix} b \\ \alpha \end{bmatrix} = \Phi^{-1} \begin{bmatrix} 0 \\ y \end{bmatrix} \tag{18}$$

Radial basis function has been used as a kernel in this analysis. Radial basis function is given by

$$K\left(x_k, x_l\right) = \exp\left\{ -\frac{\left(x_k - x_l\right)\left(x_k - x_l\right)^T}{2\sigma^2} \right\} \text{k,l=1,...,N} \tag{19}$$

where σ is the width of radial basis function.

The resulting LSSVM model for rate of medical waste generation prediction becomes then

$$y\left(x\right) = \sum_{k=1}^{N} \alpha_k K\left(x, x_k\right) + b \tag{20}$$

LSSVM utilizes the same dataset which was used for developing SVM.

- **Training Dataset:** This is used to build the LSSVM model. In this article, 35 datasets out of 50 have been adopted as training dataset.
- **Testing Dataset:** This is used to verify the built LSSVM. This article uses the remaining 15 datasets as testing dataset.

The datasets are scaled between 0 and 1. The programming of LSSVM has been done by using MATLAB.

GENETIC PROGRAMMING (GP)

Genetic Programming (GP) was proposed by Koza (1992). GP is one of the best parts of the evolutionary algorithm and also the specification of genetic algorithms

(GA) (Goldberg, 1989). It helps in producing the clear prediction equations which satisfies the inabilities of some soft computing techniques (Adil et al., 2008). GP adopts Darwins concept of "survival of the fittest" and also some appropriate combination of computer programs. The computer program can be exposed in terms of parse tree. The explanation of parse tree was depicted in Figure 1.

Figure 1 explains the parse tree diagram of computer program for a simple equation $\left[\left(2a-7b\right)+\left(3x+6y\right)\right]$.

Details of GP

This section conveys the details of the GP. In 1st step, random populations of equations were created. Then the fitness value of each equation is determined in the sequential step. 'Parents' are selected in this step. 3rd step makes 'offspring's are created through the procedure of reproduction, mutation and crossover. In 4th step, the best equation is obtained. The following Figure 2 explains the technical procedure of GP.

Crossover is one of the genetic operators used to combine the population. This operator gently selects preferable parent's chromosome and couple them to produce an offspring. The crossover point was elected randomly. The following Figure 3 explains the process of crossover.

Mutation is another genetic operator which is identical to the crossover only in selecting the chromosomes of the parent's (i.e.) in the mutation selection point only.

Figure 1. Example of parse tree

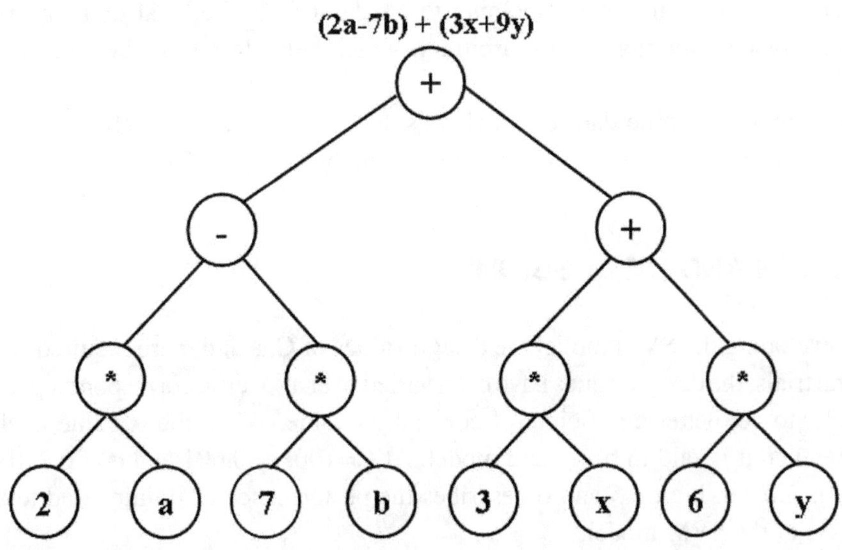

Figure 2. Flowchart of a simplified GP

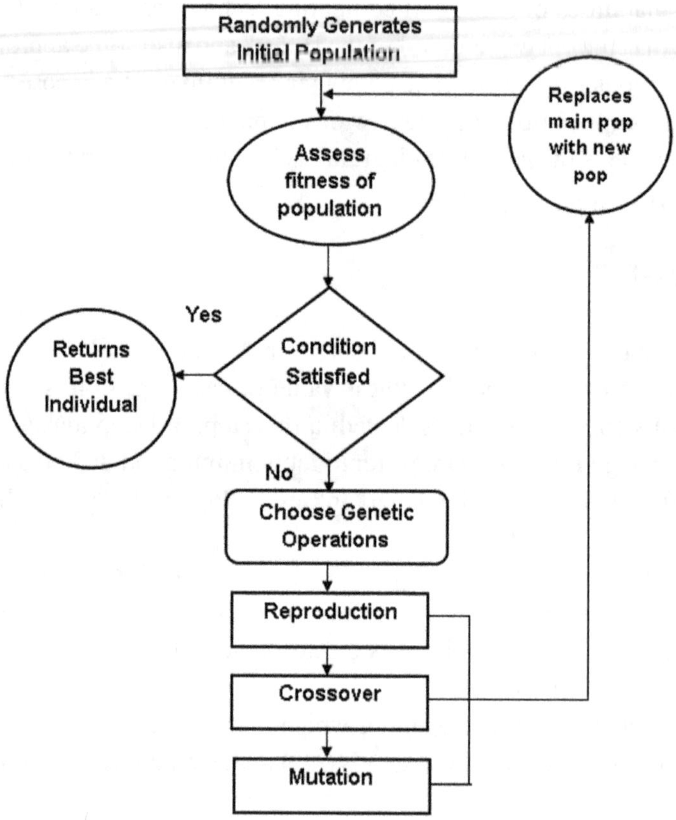

It changes the building blocks helping to breakout from the local minimal traps. The size of the mutation is also arbitrary, which helps to get the best individual (Figure 4).

GP adopts the same dataset which was used for developing SVM & LSSVM models. The programming of GP has been done by using MATLAB.

RESULTS AND DISCUSSION

For developing the SVM model the design values of C, ε and σ are required. After several trials, the design values have been varied over a specific corresponding range in order to get better coefficient of correlation value. When the R value is close to one, then it is said to be a good model. At C=100, ε=0.001 and σ = 0.7, the R value is greater. Figure 5 and 6 describes the performance of training and testing datasets of the SVM model.

Figure 3. Process of crossover

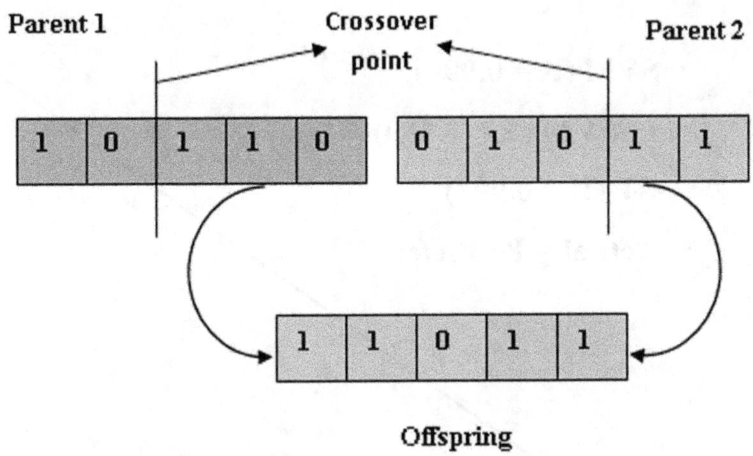

Figure 4. Methodology of mutation

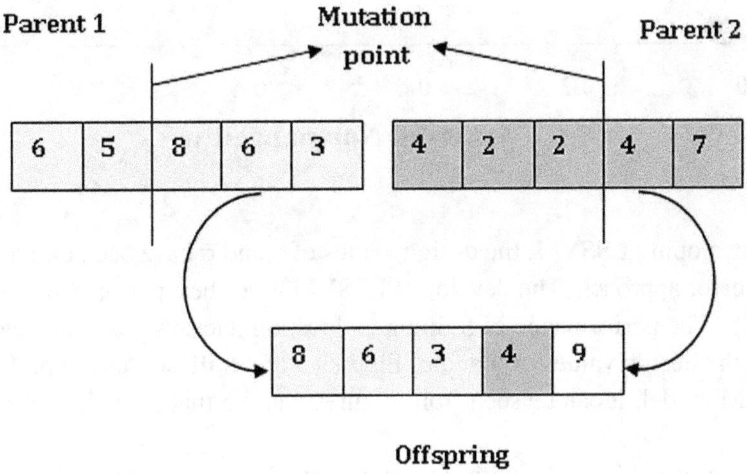

Therefore, the equation to determine w_t is as follows:

$$y = \sum_{i=1}^{31} w_i \exp\left\{-\frac{\left(x_i - x\right)^T \left(x_i - x\right)}{0.98}\right\} \tag{21}$$

The values of $\left(\alpha_i - \alpha_i^*\right)$ is shown in the following Figure 7.

Figure 5. Performance of training dataset of SVM model

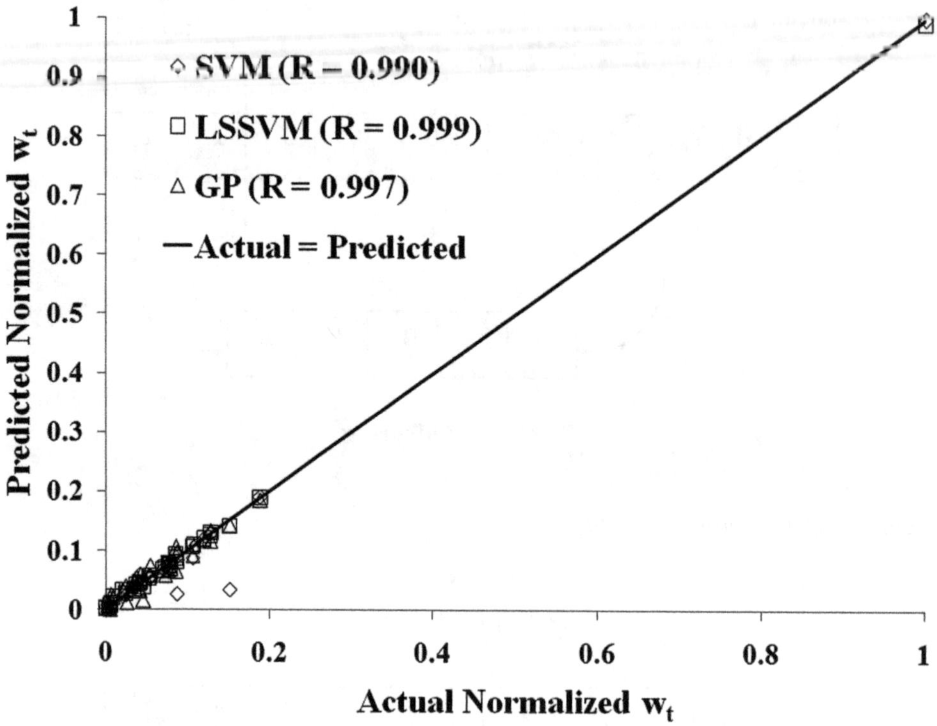

For developing LSSVM, the design values of γ and σ have been determined by trial and error approach. The developed LSSVM gives best performance at γ=100 and σ=0.2. The performance of training and testing datasets has been determined by using the design values of γ and σ. Figures 5 and 6 illustrates the performance of LSSVM model. It can be seen from Figures 5 and 6 that the value of R is close to one.

The developed LSSVM give the following equation (by putting N=31, b=0.4258, σ=0.2, $y(x)=w_t$ and $K\left(x,x_k\right)=\exp\left[\frac{-\left(x-x_k\right)\left(x-x_k\right)}{2\sigma^2}\right]$ in equation 20) for prediction of w_t.

$$d=\sum_{i=1}^{31}\alpha_k \exp\left[\frac{-\left(x-x_k\right)\left(x-x_k\right)^T}{0.08}\right] \tag{22}$$

Figure 6. Performance of testing dataset of SVM model

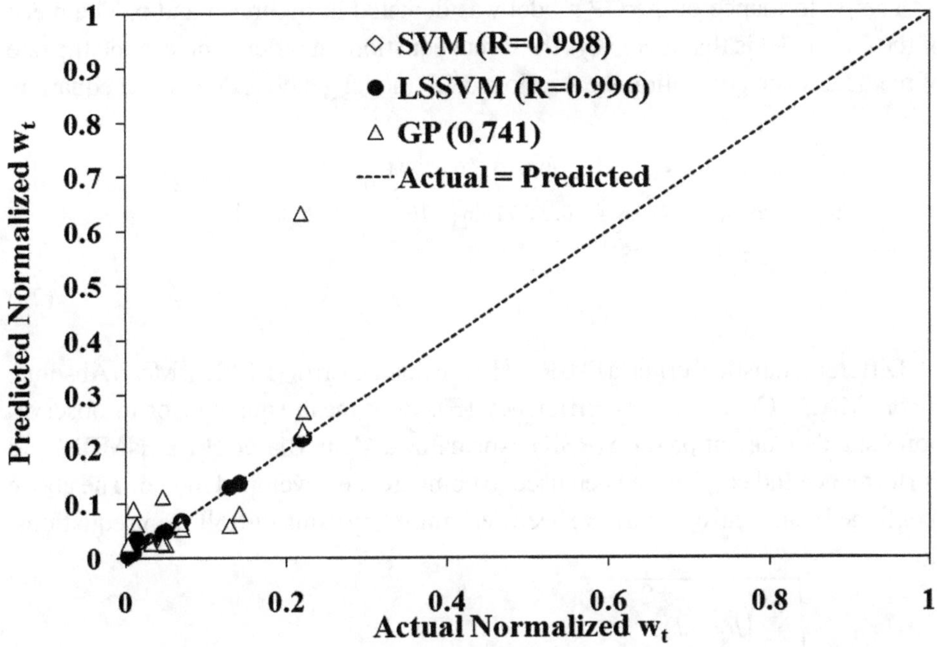

Figure 7. Values of $\left(\alpha_i - \alpha_i^*\right)$

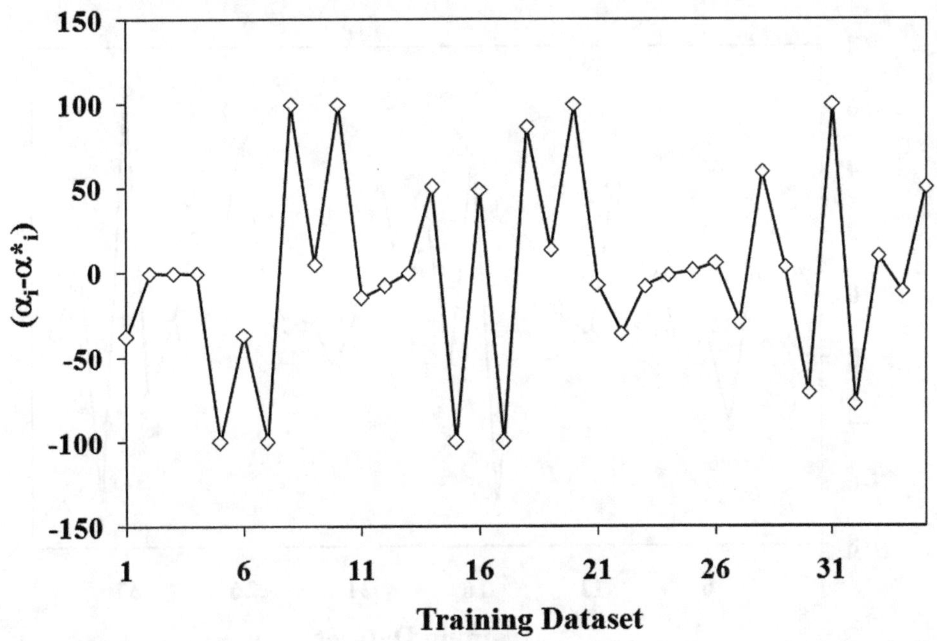

Figure 8 shows the value of α_k.

The performance of the GP model was depicted in Figures 5 and 6. The merit of the GP model is that it conveys the direct equation. In order to determine the rate of medical waste generation the developed GP model generates the direct equation.

$$0.2366\left(2x_1\right)^2 - 0.02179\cos\left(-x_1 - 6.299x_2\right) - 0.01204x_6 + 0.05986\sin\left(2x_2\left(x_5 - 8.176\right)\right) - 0.02338\sin\left(-8.729x_2 - 8.729x_6\right) - 0.03271\sin\left(-16.35x_2 - 16.35x_5\right) + 0.02408x_2\left(x_2 + x_5\right) + 0.04439 \tag{23}$$

Different statistical criteria like Root Mean Square Error (RMSE), Mean Absolute Error (MAE), Coefficient of Efficiency (E), Root mean square error to observation's standard deviation ratio (RSR), Normalized Mean Biased Error (NMBE) and Performance Index (ρ) have been used to compare the developed model. The above mentioned statistical criteria have been determined by using the following equations.

$$RMSE = \sqrt{\frac{\sum_{i=1}^{n}\left(h_i - t_i\right)^2}{n}} \tag{24}$$

Figure 8. Values of α_k

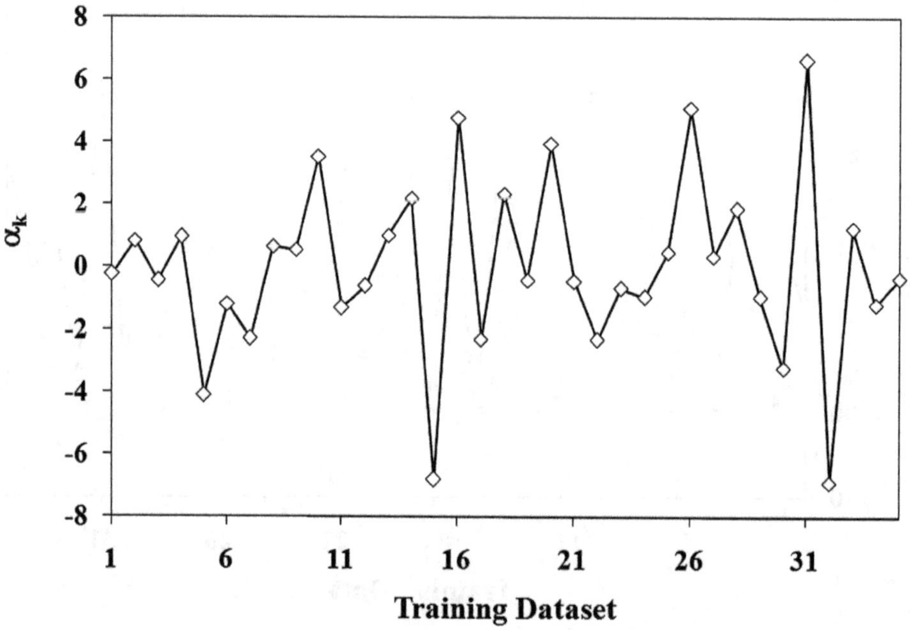

$$MAE = \frac{\sum_{i=1}^{n}\left|h_i - t_i\right|}{n} \tag{25}$$

$$E = 1 - \frac{\sum_{i=1}^{N}\left(h_i - t_i\right)^2}{\sum_{i=1}^{N}\left(hi - \bar{h}_i\right)^2} \tag{26}$$

$$RSR = \frac{RMSE}{\sqrt{\frac{1}{N}\sum_{i=1}^{N}\left(hi - \bar{h}_i\right)^2}} \tag{27}$$

$$NMBE = \frac{\sum_{i=1}^{n}\left(t_i - h_i\right)}{\bar{h}i\left(n - p\right)} \tag{28}$$

$$\rho = \frac{\frac{RMSE}{\bar{h}_i}}{R+1} \tag{29}$$

where h_i and t_i are the actual and predicted values for the ith output, $\bar{h}i$ and $\bar{t}i$ are the average of the measured and predicted outputs, n is the total number of data and p is the number of predictor variables.

By using the above equations the values have been determined and it is tabulated in the following Table 1.

Table 1. Statistical criteria's for the developed models

	SVM		LSSVM		GP	
	Training	Testing	Training	Testing	Training	Testing
RMSE	0.0229	0.0211	0.0043	0.0304	0.0125	0.3296
MAE	0.0079	0.0088	0.0030	0.0146	0.0106	0.2483
E	0.9804	0.9962	0.9993	0.9922	0.9941	0.0811
RSR	0.1401	0.0613	0.0263	0.0884	0.0767	0.9586
NMBE(%)	-4.5931	1.2095	0.0362	-0.5208	-0.4643	-59.5678
ρ	0.1333	0.0267	0.0249	0.0385	0.0727	0.4790

SOLUTIONS AND RECOMMENDATIONS

The regression modeling procedures utilized in this chapter convey the equations for future estimation on rate of medical waste generation. It depicts the path for easy management on the generated medical wastes. With any number of inputs, the predicted equation will provide the outcomes directly without any conditions. By predicting the rate of waste generation in smart cities, the authorities can perform flawless planning for storing, transportation and disposal of medical wastes. This will lead to the prevention of harm effects to human and environmental pollution.

FUTURE RESEARCH DIRECTIONS

The variation in the dataset ranges may lead to the inappropriate results. This can be resolved by the researchers in upcoming years. Various datasets can be collected and utilize the different procedures for foretelling the rate of medical waste generations. It is also suggested for the scientists to adopt various the adopted techniques for foretelling the generation of some other wastes like solid wastes, electronic wastes, agricultural wastes, radioactive wastes, construction and demolition wastes, etc. to escape from the ill effects for the smart cities and its citizens.

CONCLUSION

This chapter examines the capability of SVM, LSSVM and GP for prediction of rate of medical waste generation based on type of hospital, capacity and bed occupancy. The dataset plays a prominent role in the regression techniques. The dataset which goes beyond the adopted range may lead to inappropriate result which could be the practical difficulty. The researchers can try to overcome this difficulty in future. The developed SVM, LSSVM and GP give elegant and encouraging performances. GP uses four tuning parameters, whereas SVM and LSSVM use 3 and 2 parameters. SVM and LSSVM outperform the GP model. The developed equations can be used practically. This chapter issues the robust models based on SVM, LSSVM and GP for prediction of rate of medical waste generation.

REFERENCES

Baykasoglu, A., Gullu, H., Canakci, H., & Ozbakir, L. (2008). Prediction of compressive and tensile strength of limestone via genetic programming. *Expert Systems with Applications*, *35*(1-2), 111–123. doi:10.1016/j.eswa.2007.06.006

Burges, C. (1998). A tutorial on support vector machines for pattern recognition. In *Data Mining and Knowledge Discovery*. Boston: Kluwer Academic Publishers.

Cherkassky, V., & Mulier, F. (1998). *Learning from data: Concepts, theory, and methods*. New York: Wiley.

Goldberg, D. E. (1989). *Genetic Algorithms in Search, Optimization, and Machine Learning* (p. 432). Massachusetts: Addison-Wesley Professional.

Hassan, M. M., Ahmed, S., Rahman, K. A., & Biswas, T. (2008). Pattern of medical waste management: Existing scenario in Dhaka City, Bangladesh. *BMC Public Health*, *8*(1), 36–45. doi:10.1186/1471-2458-8-36 PMID:18221548

Hastie, T., Tibshirani, R., & Friedman, J. (2001). *The elements of statistical learning: Data mining, inference and prediction*. Berlin: Springer. doi:10.1007/978-0-387-21606-5

Henry, G., & Heinke, G. W. (1996). *Environmental Science and Engineering* (2nd ed.). Englewood, NJ: Prentice-Hall.

Jahandideh, S., Jahandideh, S., Asadabadi, E. B., Askarian, M., Movahedi, M. M., Hosseini, S., & Jahandideh, M. (2009). The use of artificial neural networks and multiple linear regression to predict rate of medical waste generation. *Waste Management (New York, N.Y.)*, *29*(11), 2874–2879. doi:10.1016/j.wasman.2009.06.027 PMID:19643591

Koza, J. R. (1992). *Genetic Programming: On the Programming of Computers by Means of Natural Selection*. Cambridge, MA: MIT press.

Maha, H. D., Rita, K., Samar, K., & Darine, M. (2015). Current status of dental waste management in Lebanon. *Environmental Nanotechnology. Monitoring & Management, 4*, 1–5.

Omar, D., Nazli, S. N., & Karuppannan, S. A. L. (2012). Clinical Waste Management in District Hospitals of Tumpat, Batu Pahat and Taiping. *Procedia: Social and Behavioral Sciences*, *68*, 134–145. doi:10.1016/j.sbspro.2012.12.213

Rushbrook, Ph., Chandra, C., & Gayton, S., (2000). *Starting healthcare waste management in medical institution, practical approach.* World Health Organization (WHO) Healthcare Practical Information Series No. 1.

Smola, A., & Scholkopf, B. (1998). A tutorial on support vector regression. NeuroCOLT Technical Report NC-TR-98-030, Royal Holloway College, University of London, UK.

Smola, A., & Scholkopf, B. (1998). On a kernel based method for pattern recognition, regression, approximation and operator inversion. *Algorithmica*, *22*(1-2), 211–231. doi:10.1007/PL00013831

Vapnik, V. N. (1998). *Statistical learning theory*. New York: Wiley.

Vapnik, V. N. (1999). *The nature of statistical learning theory* (2nd ed.). Berlin: Springer.

ADDITIONAL READING

Abdulla, F., Abu Qdais, H., & Rabi, A. (2008). Site investigation on medical waste management practices in northern Jordan. *Waste Management (New York, N.Y.)*, *28*(2), 450–458. doi:10.1016/j.wasman.2007.02.035 PMID:17507209

Akpieyi, A., Tudor, T. L., & Dutra, C. (2015). The utilisation of risk-based frameworks for managing healthcare waste: A case study of the National Health Service in London. *Safety Science*, *72*, 127–132. doi:10.1016/j.ssci.2014.08.014

Alavi, A. H., Gandomi, A. H., Mollahasani, A., Bazaz, J. B., & Talatahari, S. (2013). 12 - Linear and Tree-Based Genetic Programming for Solving Geotechnical Engineering Problems. *Metaheuristics in Water, Geotechnical and Transport Engineering*, 289-310.

Ali Ghorbani, M., Khatibi, R., Aytek, A., Makarynskyy, O., & Shiri, J. (2010). Sea water level forecasting using genetic programming and comparing the performance with Artificial Neural Networks. *Computers & Geosciences*, *36*(5), 620–627. doi:10.1016/j.cageo.2009.09.014

Anagnostopoulos, T., Kolomvatsos, K., Anagnostopoulos, C., Zaslavsky, A., & Hadjiefthymiades, S. (2015). Assessing dynamic models for high priority waste collection in smart cities. *Journal of Systems and Software*, *110*, 178–192. doi:10.1016/j.jss.2015.08.049

Babanajad, S. K., Gandomi, A. H., Mohammadzadeh S, D., & Alavi, A. H. (2013). Numerical modeling of concrete strength under multiaxial confinement pressures using linear genetic programming. *Automation in Construction*, *36*, 136–144. doi:10.1016/j.autcon.2013.08.016

Birpınar, M. E., Bilgili, M. S., & Erdoğan, T.(2009). Medical waste management in Turkey: A case study of Istanbul. *Waste Management (New York, N.Y.)*, *29*(1), 445–448. doi:10.1016/j.wasman.2008.03.015 PMID:18550354

Blenkharn, J. I. (2007). Standards of clinical waste management in hospitals—A second look. *Public Health*, *121*(7), 540–545. doi:10.1016/j.puhe.2006.12.005 PMID:17296210

Bokhoree, C., Beeharry, Y., Makoondlall Chadee, T., Doobah, T., & Soomary, T. (2014). Assessment of Environmental and Health Risks Associated with the Management of Medical Waste in Mauritius. *APCBEE Procedia*, *9*, 36–41. doi:10.1016/j.apcbee.2014.01.007

Carmona, C. J., Ruiz Rodado, V., del Jesus, M. J., Weber, A., Grootveld, M., González, P., & Elizondo, D. (2015). A fuzzy genetic programming-based algorithm for subgroup discovery and the application to one problem of pathogenesis of acute sore throat conditions in humans. *Information Sciences*, *298*(20), 180–197. doi:10.1016/j.ins.2014.11.030

Castro, A., Pérez, J. L., Rabuñal, J. R., & Iglesias, G. (2015). Genetic programming and floating boom performance. *Ocean Engineering*, *104*(1), 310–318. doi:10.1016/j.oceaneng.2015.05.023

Cazzani, A., & Ruge, P. (2014). Symmetric matrix-valued transmitting boundary formulation in the time-domain for soil–structure interaction problems. *Soil Dynamics and Earthquake Engineering*, *57*, 104–120. doi:10.1016/j.soildyn.2013.11.002

Chen, Y., Liu, L., Feng, Q., & Chen, G. (2012). Key Study on the Operation Management of Medical Waste Incineration Disposal Facilities. *Procedia Environmental Sciences*, *16*, 208–213. doi:10.1016/j.proenv.2012.10.029

Chen, Y., Zhao, R., Xue, J., & Li, J. (2013). Generation and distriution of PAHs in the process of medical waste incineration. *Waste Management (New York, N.Y.)*, *33*(5), 1165–1173. doi:10.1016/j.wasman.2013.01.011 PMID:23462270

Chou, J.-S., Yang, K.-H., Pampang, J. P., & Pham, A.-D. (2015). Evolutionary metaheuristic intelligence to simulate tensile loads in reinforcement for geosynthetic-reinforced soil structures. *Computers and Geotechnics*, *66*, 1–15. doi:10.1016/j.compgeo.2015.01.001

Coker, A., Sangodoyin, A., Sridhar, M., Booth, C., Olomolaiye, P., & Hammond, F. (2009). Medical waste management in Ibadan, Nigeria: Obstacles and prospects. *Waste Management (New York, N.Y.)*, *29*(2), 804–811. doi:10.1016/j.wasman.2008.06.040 PMID:18835151

Shiri, J., Kişi, Ö., Landeras, G., López, J. J., Nazemi, A. H., & Stuyt, L. C. (2012). Daily reference evapotranspiration modeling by using genetic programming approach in the Basque Country (Northern Spain). *Journal of Hydrology (Amsterdam)*, *414–415*(11), 302–316.

Danandeh Mehr, A., Kahya, E., & Yerdelen, C. (2014). Linear genetic programming application for successive-station monthly streamflow prediction. *Computers & Geosciences*, *70*, 63–72. doi:10.1016/j.cageo.2014.04.015

Ding, M., Cao, Y., & Wu, Q. (2013). Novel approach of crater detection by crater candidate region selection and matrix-pattern oriented least squares support vector machine. *Chinese Journal of Aeronautics*, *26*(2), 385–393. doi:10.1016/j.cja.2013.02.016

Elhenawy, M., Chen, H., & Rakha, H. A. (2014). Dynamic travel time prediction using data clustering and genetic programming. *Transportation Research Part C, Emerging Technologies*, *42*, 82–98. doi:10.1016/j.trc.2014.02.016

Fallah Mehdipour, E., Bozorg Haddad, O., & Mariño, M. A. (2013). Prediction and simulation of monthly groundwater levels by genetic programming. *Journal of Hydro-environment Research*, *7*(4), 253–260. doi:10.1016/j.jher.2013.03.005

Gandomi, A. H., Mohammadzadeh S, D., Pérez-Ordóñez, J. L., & Alavi, A. H. (2014). Linear genetic programming for shear strength prediction of reinforced concrete beams without stirrups. *Applied Soft Computing*, *19*, 112–120. doi:10.1016/j.asoc.2014.02.007

Gandomi, A. H., & Roke, D. A. (2015). Assessment of artificial neural network and genetic programming as predictive tools. *Advances in Engineering Software*, *88*, 63–72. doi:10.1016/j.advengsoft.2015.05.007

Garg, A., & Lam, J. S. L. (2015). Improving environmental sustainability by formulation of generalized power consumption models using an ensemble based multi-gene genetic programming approach. *Journal of Cleaner Production*, *102*(1), 246–263. doi:10.1016/j.jclepro.2015.04.068

Ghotbi Ravandi, E., Rahmannejad, R., Feili Monfared, A. E., & Ghotbi Ravandi, E. (2013). Application of numerical modeling and genetic programming to estimate rock mass modulus of deformation. *International Journal of Mining Science and Technology*, *23*(5), 733–737. doi:10.1016/j.ijmst.2013.08.018

Güllü, H. (2012). Prediction of peak ground acceleration by genetic expression programming and regression: A comparison using likelihood-based measure. *Engineering Geology*, *141–142*(19), 92–113.

Güllü, H. (2014). Function finding via genetic expression programming for strength and elastic properties of clay treated with bottom ash. *Engineering Applications of Artificial Intelligence*, *35*, 143–157. doi:10.1016/j.engappai.2014.06.020

Gusca, J., Kalnins, S. N., Blumberga, D., Bozhko, L., Khabdullina, Z., & Khabdullin, A. (2015). Assessment Method of Health Care Waste Generation in Latvia and Kazakhstan. *Energy Procedia*, *72*, 175–179. doi:10.1016/j.egypro.2015.06.025

Duan, H., Huang, Q., Wang, Q., Zhou, B., & Li, J. (2008). Hazardous waste generation and management in China: A review. *Journal of Hazardous Materials*, *158*(2-3), 221–227. PMID:18353544

Brusco, J.M., & Ogg, M. (2010). Health Care Waste Management and Environmentally Preferable Purchasing. *AORN Journal*, *92*(6), S62–S69. doi:10.1016/j.aorn.2010.10.001

Insa, E., Zamorano, M., & López, R. (2010). Critical review of medical waste legislation in Spain. *Resources, Conservation and Recycling*, *54*(12), 1048–1059. doi:10.1016/j.resconrec.2010.06.005

Islam, T., Rico-Ramirez, M. A., & Han, D. (2012). Tree-based genetic programming approach infer microphysical parameters of the DSDs from the polarization diversity measurements. *Computers & Geosciences*, *48*, 20–30. doi:10.1016/j.cageo.2012.05.028

Jafarian, Y., Kermani, E., & Baziar, M. H. (2010). Empirical predictive model for the vmax/amax ratio of strong ground motions using genetic programming. *Computers & Geosciences*, *36*(12), 1523–1531. doi:10.1016/j.cageo.2010.07.002

Jang, Y.-C., Lee, C., Yoon, O.-S., & Kim, H.(2006). Medical waste management in Korea. *Journal of Environmental Management*, *80*(2), 107–115. doi:10.1016/j.jenvman.2005.08.018 PMID:16338054

Ji, J., Zhang, C., Kodikara, J., & Yang, S.-Q. (2015). Prediction of stress concentration factor of corrosion pits on buried pipes by least squares support vector machine. *Engineering Failure Analysis*, *55*, 131–138. doi:10.1016/j.engfailanal.2015.05.010

Ji, J., Zhang, C., Kodikara, J., & Yang, S.-Q. (2015). Prediction of stress concentration factor of corrosion pits on buried pipes by least squares support vector machine. *Engineering Failure Analysis*, *55*, 131–138. doi:10.1016/j.engfailanal.2015.05.010

Kara, I. F. (2011). Prediction of shear strength of FRP-reinforced concrete beams without stirrups based on genetic programming. *Advances in Engineering Software*, *42*(6), 295–304. doi:10.1016/j.advengsoft.2011.02.002

Kayadelen, C. (2011). Soil liquefaction modeling by genetic expression programming and neuro fuzzy. *Expert Systems with Applications*, *38*(4), 4080–4087. doi:10.1016/j.eswa.2010.09.071

Khatibinia, M., Javad Fadaee, M., Salajegheh, J., & Salajegheh, E. (2013). Seismic reliability assessment of RC structures including soil–structure interaction using wavelet weighted least squares support vector machine. *Reliability Engineering & System Safety*, *110*, 22–33. doi:10.1016/j.ress.2012.09.006

Kisi, O. (2012). Modeling discharge-suspended sediment relationship using least square support vector machine. *Journal of Hydrology (Amsterdam)*, *456–457*(16), 110–120.

Kisi, O. (2015). Pan evaporation modeling using least square support vector machine, multivariate adaptive regression splines and M5 model tree. *Journal of Hydrology (Amsterdam)*, *528*, 312–320. doi:10.1016/j.jhydrol.2015.06.052

Kiyoumars, R., Dominique, M., & Jalal, S. (2014). Evaluation of genetic programming-based models for simulating friction factor in alluvial channels. *Journal of Hydrology (Amsterdam)*, *517*(19), 1154–1161.

Komilis, D., Fouki, A., & Papadopoulos, D. (2012). Hazardous medical waste generation rates of different categories of health-care facilities. *Waste Management (New York, N.Y.)*, *32*(7), 1434–1441. doi:10.1016/j.wasman.2012.02.015 PMID:22444895

Kuepouo, G. (2013). Estimating environmental release of mercury from medical-thermometers and potential "hot spot" development: Case study of need for improved waste management capacity in Cameroon. *Resources, Conservation and Recycling*, *71*, 48–52. doi:10.1016/j.resconrec.2012.10.004

Kumari, R., Srivastava, K., Wakhlu, A., & Singh, A. (2013). Establishing biomedical waste management system in Medical University of India – A successful practical approach. *Clinical Epidemiology and Global Health*, *1*(3), 131–136. doi:10.1016/j.cegh.2012.11.004

Lee, C. C., & Huffman, G. L. (1996). Medical waste management/incineration. *Journal of Hazardous Materials*, *48*(1–3), 1–30.

Li, B., Li, D., Zhang, Z., Yang, S., & Wang, F. (2015). Slope stability analysis based on quantum-behaved particle swarm optimization and least squares support vector machine. *Applied Mathematical Modelling*, *39*(17), 5253–5264. doi:10.1016/j.apm.2015.03.032

Liu, H.-C., You, J.-X., Lu, C., & Chen, Y.-Z. (2015). Evaluating health-care waste treatment technologies using a hybrid multi-criteria decision making model. *Renewable & Sustainable Energy Reviews*, *41*, 932–942. doi:10.1016/j.rser.2014.08.061

Lowe, J. J., Gibbs, S. G., Schwedhelm, S. S., Nguyen, J., & Smith, P. W. (2014). Nebraska Biocontainment Unit perspective on disposal of Ebola medical waste. *American Journal of Infection Control*, *42*(12), 1256–1257. doi:10.1016/j.ajic.2014.10.006 PMID:25465251

Manga, V. E., Forton, O. T., Mofor, L. A., & Woodard, R. (2011). Health care waste management in Cameroon: A case study from the Southwestern Region. *Resources, Conservation and Recycling*, *57*, 108–116. doi:10.1016/j.resconrec.2011.10.002

Marinković, N., Vitale, K., Holcer, N. J., Džakula, A., & Pavić, T. (2008). Management of hazardous medical waste in Croatia. *Waste Management (New York, N.Y.)*, *28*(6), 1049–1056. doi:10.1016/j.wasman.2007.01.021 PMID:17451931

Markus, D., Sophie, L. G., David, S., Murray Lark, R., Heather, A. S., & Dayton, D. (2014). Mapping seabed sediments: Comparison of manual, geostatistical, object-based image analysis and machine learning approaches. *Continental Shelf Research*, *84*(1), 107–119.

Masum, A. P., William, T. H., & Mosharraf, H. S. (2011). Assessment of occupational and environmental safety associated with medical waste disposal in developing countries: A qualitative approach. *Safety Science*, *49*(8–9), 1200–1207.

Mato, R. R. A. M., & Kassenga, G. R. (1997). A study on problems of management of medical solid wastes in Dar es Salaam and their remedial measures. *Resources, Conservation and Recycling*, *21*(1), 1–16. doi:10.1016/S0921-3449(97)00010-1

Windfeld, E.S., & Brooks, M.S. (2015). Medical waste management – A review. *Journal of Environmental Management*, *163*(1), 98–108. PMID:26301686

Meshgi, A., Schmitter, P., Chui, T. F. M., & Babovic, V. (2015). Development of a modular streamflow model to quantify runoff contributions from different land uses in tropical urban environments using Genetic Programming. *Journal of Hydrology (Amsterdam)*, *525*, 711–723. doi:10.1016/j.jhydrol.2015.04.032

Moreira, A. M. M., & Günther, W. M. R. (2013). Assessment of medical waste management at a primary health-care center in São Paulo, Brazil. *Waste Management (New York, N.Y.)*, *33*(1), 162–167. doi:10.1016/j.wasman.2012.09.018 PMID:23122204

Muduli, P. K., & Das, S. K. (2015). Model uncertainty of SPT-based method for evaluation of seismic soil liquefaction potential using multi-gene genetic programming. *Soil and Foundation*, *55*(2), 258–275. doi:10.1016/j.sandf.2015.02.003

Najafzadeh, N., & Barani, A. (2011). Comparison of group method of data handling based genetic programming and back propagation systems to predict scour depth around bridge piers. *Scientia Iranica*, *18*(6), 1207–1213. doi:10.1016/j.scient.2011.11.017

Nasseri, M., Moeini, A., & Tabesh, M. (2011). Forecasting monthly urban water demand using Extended Kalman Filter and Genetic Programming. *Expert Systems with Applications*, *38*(6), 7387–7395. doi:10.1016/j.eswa.2010.12.087

Ozbek, A., Unsal, M., & Dikec, A. (2013). Estimating uniaxial compressive strength of rocks using genetic expression programming. *Journal of Rock Mechanics and Geotechnical Engineering*, *5*(4), 325–329. doi:10.1016/j.jrmge.2013.05.006

Ozgur, K., Ali Hosseinzadeh, D., Mesut, C., & Jalal, S. (2012). Suspended sediment modeling using genetic programming and soft computing techniques. *Journal of Hydrology (Amsterdam)*, *450–451*(11), 48–58.

Pérez, J. L., Cladera, A., Rabuñal, J. R., & Abella, F. M. (2010). Optimal adjustment of EC-2 shear formulation for concrete elements without web reinforcement using Genetic Programming. *Engineering Structures*, *32*(11), 3452–3466. doi:10.1016/j.engstruct.2010.07.006

Pérez, J. L., Cladera, A., Rabuñal, J. R., & Martínez-Abella, F. (2012). Optimization of existing equations using a new Genetic Programming algorithm: Application to the shear strength of reinforced concrete beams. *Advances in Engineering Software*, *50*, 82–96. doi:10.1016/j.advengsoft.2012.02.008

Qiang, C., Jale, T., & Jie, C. (2014). Confidence and prediction intervals for semi-parametric mixed-effect least squares support vector machine. *Pattern Recognition Letters*, *40*(15), 88–95.

Stonemetz, J., Pham, J.C., Necochea, A.J., McGready, J., Hody, R.E., & Martinez, E.A. (2011). Reduction of Regulated Medical Waste Using Lean Sigma Results in Financial Gains for Hospital. *Anesthesiology Clinics*, *29*(1), 145–152. doi:10.1016/j.anclin.2010.11.007 PMID:21295759

Salazar, F., Toledo, M. A., Oñate, E., & Morán, R.(2015). An empirical comparison of machine learning techniques for dam behaviour modelling. *Structural Safety*, *56*, 9–17. doi:10.1016/j.strusafe.2015.05.001

Sarıdemir, M. (2010). Genetic programming approach for prediction of compressive strength of concretes containing rice husk ash. *Construction & Building Materials*, *24*(10), 1911–1919. doi:10.1016/j.conbuildmat.2010.04.011

Sarıdemir, M. (2011). Empirical modeling of splitting tensile strength from cylinder compressive strength of concrete by genetic programming. *Expert Systems with Applications*, *38*(11), 14257–14268.

Sarıdemir, M. (2014). Effect of specimen size and shape on compressive strength of concrete containing fly ash: Application of geneticprogramming for design. *Materials & Design*, *56*, 297–304. doi:10.1016/j.matdes.2013.10.073

Satishkumar, S. K., & Rajib, M. (2012). Prediction of monthly rainfall on homogeneous monsoon regions of India based on large scale circulation patterns using genetic programming. *Journal of Hydrology (Amsterdam)*, *454–455*(6), 26–41.

Segura Muñoz, S. I., Da Silva Oliveira, A., Nikaido, M., Trevilato, T. M. B., Bocio, A., Takayanagui, A. M. M., & Domingo, J. L. (2006). Metal levels in sugar cane (Saccharum spp.) samples from an area under the influence of a municipal landfill and a medical waste treatment system in Brazil. *Environment International*, *32*(1), 52–57. doi:10.1016/j.envint.2005.04.008 PMID:15990169

Singh, Z., Bhalwar, R., Jayaram, J., & Tilak, V. W.(2001). An introduction to essentials of bio-medical waste management. *Medical Journal Armed Forces India*, *57*(2), 144–147. doi:10.1016/S0377-1237(01)80136-2

Sivapragasam, C., Muttil, N., Catherin, J., & Visweshwaran, S. (2015). Infilling of Rainfall Information Using Genetic Programming. *Aquatic Procedia*, *4*, 1016–1022. doi:10.1016/j.aqpro.2015.02.128

Sreekanth, J., & Datta, B. (2010). Multi-objective management of saltwater intrusion in coastal aquifers using genetic programming and modular neural network based surrogate models. *Journal of Hydrology (Amsterdam)*, *393*(3–4), 245–256. doi:10.1016/j.jhydrol.2010.08.023

Sundell-Bergman, S., de laCruz, I., Avila, R., & Hasselblad, S. (2008). A new approach to assessment and management of the impact from medical liquid radioactive waste. *Journal of Environmental Radioactivity*, *99*(10), 1572–1577. doi:10.1016/j.jenvrad.2007.12.005 PMID:18242799

Ta Kang, L., Meng Wei, W., & Ping, C. (2013). Influence of waste management policy on the characteristics of beach litter in Kaohsiung, Taiwan. *Marine Pollution Bulletin*, *72*(1), 99–106. PMID:23673204

Tanyildizi, H., & Çevik, A. (2010). Modeling mechanical performance of lightweight concrete containing silica fume exposed to high temperature using genetic programming. *Construction & Building Materials*, *24*(12), 2612–2618. doi:10.1016/j.conbuildmat.2010.05.001

Tinoco, J.,, Gomes Correia, A., & Cortez,, P. (2014). A novel approach to predicting Young's modulus of jet grouting laboratory formulations over time using data mining techniques. *Engineering Geology*, *169*(4), 50–60.

Tudor, T. L., Barr, S. W., & Gilg, A. W. (2007). Linking intended behaviour and actions: A case study of healthcare waste management in the Cornwall NHS. *Resources, Conservation and Recycling*, *51*(1), 1–23. doi:10.1016/j.resconrec.2006.06.009

Tudor, T. L., Woolridge, A. C., Phillips, C. A., Holliday, M., Laird, K., Bannister, S., & Rushbrook, P. et al. (2010). Evaluating the link between the management of clinical waste in the National Health Service (NHS) and the risk of the spread of infections: A case study of three hospitals in England. *International Journal of Hygiene and Environmental Health*, *213*(6), 432–436. doi:10.1016/j.ijheh.2010.08.002 PMID:20833586

Yantao, X. (2015). Comparison of hospital medical waste generation rate based on diagnosis-related groups. *Journal of Cleaner Production*, *100*(1), 202–207.

Yeganeh, B., Shafie Pour Motlagh, M., Rashidi, Y., & Kamalan, H. (2012). Prediction of CO concentrations based on a hybrid Partial Least and Square and support vector machine model. *Atmospheric Environment*, *55*, 357–365. doi:10.1016/j.atmosenv.2012.02.092

Yong, Z., Gang, X., Guanxing, W., Tao, Z., & Dawei, J. (2009). Medical waste management in China: A case study of Nanjing. *Waste Management (New York, N.Y.)*, *29*(4), 1376–1382. doi:10.1016/j.wasman.2008.10.023 PMID:19157838

You, Lv., Tingting, Y., & Jizhen, L. (2015). An adaptive least squares support vector machine model with a novel update for NOx emission prediction. *Chemometrics and Intelligent Laboratory Systems*, *145*(15), 103–113.

Zaji, A. H., & Bonakdari, H. (2015). Application of artificial neural network and genetic programming models for estimating the longitudinal velocity field in open channel junctions. *Flow Measurement and Instrumentation*, *41*, 81–89. doi:10.1016/j.flowmeasinst.2014.10.011

KEY TERMS AND DEFINITIONS

Artificial Neural Network: Artificial Neural Network is a structure framed with densely interconnected with artificial neurons that performs the parallel computations for data processing and knowledge representations.

Estimation: The action of forecasting or predicting something is called as estimation.

Genetic Programming: Genetic programming (GP) is a systematic, domain-independent, evolutionary computation technique for computational problem solving.

Least Square Support Vector Machine: Least Square Support Vector Machine (LSSVM) least squares version of SVM. In this, the solutions can be obtained by solving the set of linear equations.

Medical Waste: Medical waste is all waste materials generated at health care facilities, such as hospitals, clinics, physician's offices, dental practices, blood banks, and veterinary hospitals/clinics, as well as medical research facilities and laboratories.

Support Vector Machine: Support Vector Machine (SVM) is supervised learning algorithm used for analyzing data and recognizes designs, also used for classification and regression analysis.

Chapter 9
Smart City Solutions for India:
Learnings from Best Practices around the World

Shekhar Vishnu Nagargoje
National Institute of Construction Management and Research, India

Mehul Manjibhai Sutaria
National Institute of Construction Management and Research, India

Gaurvav Krishna Somani
National Institute of Construction Management and Research, India

Navneet Jha
National Institute of Construction Management and Research, India

ABSTRACT

Indian Government has proposed to develop 100 Smart Cities and 500 AMRUT (Atal Mission for Rejuvenation and Urban Transformation) Cities in its overall commitment of sustainable development. The proposed study aims at documenting the potential and need of developing Smart cities in India, to understand smart city principles and various dimensions of smart city adopted in various parts of the world which will further help in deriving recommendations and lessons for India's future smart cities. The objective of this study is to generate learning and exploration that will improve confidence in civic authorities, experts, city planners, developers; and help stimulate further private sector investment in developing Smart cities of India.

DOI: 10.4018/978-1-5225-0302-6.ch009

INTRODUCTION

The global environment in which people live is undergoing significant changes. In 1900, only 13% of the world's population lived in cities. By 2050, that number will reach to 70% (UN-DESA report, 2011). India with its 377 million urban population, living in 7936 cities and towns, is World's second largest country in terms of urbanization. (Keshava, 2014) It provides an opportunity for social transformation, economic development, employment generation, infrastructure and urban development (Jain, 2015). However, with haphazard urban growth and deficit in energy, transport, housing and other basic infrastructure services, Indian cities are continuously under pressure of rapid urban expansion, gap in infrastructure services, growing carbon emissions and dangers of climate change. Cities are not only the engines of economic growth, but also the development of civil culture and vice-versa. Cities in India are growing indiscriminately which is disintegrating their eco-system and social fabric, trampling the rich heritage. India is heading towards the post-industrial economy, where knowledge economy is taking over. The existing employment sectors, market and government are shrinking and the GNP (Gross National Product) is rising faster than the birth rate. It means redistribution and reorganization or work, economy and time. Knowledge economy and micro-communication are giving a new meaning to work, leisure and creative civic culture (Jain, 2015). Indian cities need to address the specific issues of the local population and ecological context. A city with circular metabolism can give as much to the environment as it takes out (World Future Council, 2012). It reduces its ecological impact and safeguards its future sources of supply. To facilitate the efficient use of natural energy the form of cities and buildings has to be different (Ahuja and Tatsutani, 2009). They should be planned and designed to incorporate both old and new ecological strategies, learning from traditional forms, as well as modern technology.

The city of future has to be smart and sustainable, which integrates the peri-urban and rural communities around it for an overall regional development. Technology can play a larger role in making cities and communities smarter and resilient (De Angelis, 2015). The new age technology can create smarter and more sustainable cities. It is important to understand and analyze the potential of making smart and resilient city that can best maximize the social, economic and environmental sustainability of a city.

Physical and electronic communication brings opportunities of change and exchange. The ability to innovate to constantly changing ideas will ensure appropriate response in time and space context. Sustainable development is giving birth to a new aesthetic, where the nature (sun, wind, water and earth) integrates with technology (nano-technology, renewable, biotechnology, eco-efficiency, etc.) to create an efficient, comfortable, peaceful and beautiful built environment (Jain, 2015).

It is a colossal task to integrate technology and address all the issues of sustainability, but some successful examples of transformation of existing cities or design of new cities as smart, sustainable, energy-efficient and green are evident and can be used as learning lessons for future smart cities. These include Plan IT valley (Portugal), Fuiisawa (Japan), Amsterdam, Cairo, Dubai, Masdar (UAE), Dongton (China), Singapore, Edinburg, Kochi Business Park, Lyon, Malaga (Spain), Malta, Santander (Spain) Songdo (South Korea), Kazan smart city (Tatarstan), Holyoke (Massachusetts), Dublin (lreland), Shenyang (China) and Southampton.

Next few decades will be critical in shaping the urban infrastructure in India as investments flow to cities for expanding their infrastructure networks. India faces a unique challenge of closing the deficit in urban infrastructure and services, while at the same time making a transition to a more sustainable path. However, this deficit may well be a boon in disguise, as it could serve as an opportunity to leapfrog to more sustainable, less energy and resource-intensive forms of infrastructure (Wankhede, 2013).

This chapter attempts to address the need for smart cities in India and assess the smart city solutions adopted in various parts of the world which will further help in deriving recommendations and lessons for India's future smart cities.

BACKGROUND

The decadal growth of population in urban India is greater than rural population leading to the increase in urban population from around 27.8% (286 million) in 2001 to 31.2% (377 million) in 2011(Census 2011) and is estimated to be 40% by 2030 and more than 50% by 2050. The growth of urban population in the country is not only explosive but also highly skewed. The number of census towns increased by 53.7% from 5161 in 2001 to 7935 in 2011 with number of Class I towns increased from 441 to 468 and the million plus cities increased from 35 to 53 over the past decade from 2001 to 2011 (Census 2011). The urbanization in India is unplanned and haphazard in general, this in itself is a root cause of almost all the problems (Jawaid and Khan, 2015). The condition of infrastructure and services have deteriorated over the time, leading to the increasing demand for better quality of infrastructure and services such as water supply, sewerage, drainage, public health and sanitation, roads, streets, city transport and primary education. Urban local bodies are the primary agencies for providing, maintaining and administrating the infrastructural needs of the people. But dependence on state and national governments for large capital funds for infrastructure development leads to delay in actual implementation leading to further gap in providing basic infrastructure services. The NIUA report of 2011 clearly point out that the poor delivery of urban service in the country is the result of the neglect of urban planning and infrastructure by state governments, the fragmented and overlapping institutional responsibilities of the state government, urban local bodies, Development Authorities and Parastatal agencies in different states. This lack of infrastructure and poor service further adds pressure on ecology. Cities alone accounts for 50% of world's population (31.8% in India), 75% of energy consumption and 80% of carbon emission (Parsuraman, 2013). The problems of pollution, congestion, deteriorating quality of life and infrastructure and rising cost coupled with challenges of environmental degradation, climate change, resource depletion, alteration in the ecological cycle, intensifies the need for up-gradation in the approach cities are planned for future. Cities of future need to become more energy efficient, sustainable and liveable.

Smart cities addresses the need of energy efficiency and sustainability, however it is important to note that there is no single and specific definition of a "smart city". Will the real smart city stand up? The question asked by Hollands in his article. As per Hollands, cities often claim to be smart, but do so without defining what this means, or offering any evidence to support such proclamations, like the digital city, intelligent city, hi-tech city and neighborhoods of smart communities abound, they all fail to convey what it means to be smart and why it is important for cities to be defined in such terms (Hollands, 2008). The validity of any city's claim to be smart has to be based on something than their use of information and communication technologies (ICT). Smart city is described as a city that pushes the quality of resource management and service provision to the limit possible at the time (Kang, Zang et al, 2014). Following this concept, smart city projects are part of city's renewal and modernisation. Another version is that it is "a space for coexistence among people who, based on the available technologies, can thrive and develop, while taking into account economic, social and environmental sustainability", (Branchi et al, 2014). (Cavada, Hunt and Rogers, 2014), maintain that definitions hitherto available revolve around business strategy or resilience but smart refers to services, security and growth of stakeholders (people, academics, government and commercial) in an urban environment. (Lee et al, 2012), introduced the layered concept of the basic smart city architecture that integrates aspects of 1) integration of city infrastructures 2) governance 3) intelligence and sustainability 4) urban openness 5) urban innovativeness, and 6) service innovation as main aspects. They listed the six domains mentioned above and eighteen perspectives that fall within the domains. Alternatively some scholars have identified six key dimensions a little differently as that of (Lee et al, 2010) as follows: natural resource preservation, community development, housing, economic development, transportation choice, and planning for smart growth (Hawkins, Christopher V., 2011; Ye, Mandpe and Meyer, 2005).

In the recent COP21 Paris Summit, on Climate Change, India has set an ambitious goal of receiving 40% of its power from renewable resources by 2030 and in recent weeks launched a solar power alliance aimed at growing solar power production in the developing world. The country also recently set a target to develop 100 GW of solar power capacity by 2022, a huge ramp up from current capacity (www.mnre.gov.in) On 1 October 2015, India submitted its Intended Nationally Determined Contribution (INDC), including the targets to lower the emissions intensity of GDP by 33% to 35% by 2030 below 2005 levels, to increase the share of non-fossil based power generation capacity to 40% of installed electric power capacity by 2030 (equivalent to 26–30% of generation in 2030), and to create an additional (cumulative) carbon sink of 2.5–3 GtCO2e through additional forest and tree cover by 2030. For 2020, India has earlier put forward a pledge to reduce the emissions intensity of GDP by 20% to 25% by 2020 below 2005 levels.

Indian Government has also proposed to develop 100 Smart Cities and 500 AMRUT Cities in its overall commitment of sustainable development. The Smart Cities Mission is a bold initiative by the Government of India to drive economic growth and improve the quality of life of people by enabling local development and harnessing technology as a means to create smart outcomes for citizens (www.Smartcities.gov.in, 2015). In the guidelines for smart cities prepared by government, it is observed that; in the imagination of any city dweller in India, the picture of a Smart City contains a wish list of infrastructure and services that describes his or her level of aspiration. To provide for the aspirations and needs of the citizens, urban planners ideally aim at developing the entire urban eco-system, which is represented by the four pillars of comprehensive development — institutional, physical, social and economic infrastructure. This can be a long term goal and cities can work towards developing such comprehensive infrastructure incrementally, adding on layers of 'smartness'. In the approach to the Smart Cities Mission, the objective is to promote cities that provide core infrastructure and give a decent quality of life to its citizens, a clean and sustainable environment and application of 'Smart' Solutions. The focus is on sustainable and inclusive development and the idea is to look at compact areas, create a replicable model which will act like a light house to other aspiring cities. The core infrastructure elements in a Smart City would include: adequate water supply, assured electricity supply, sanitation, including solid waste management, sanitation, including solid waste management, efficient urban mobility and public transport, affordable housing, especially for the poor, robust IT connectivity and digitalization, good governance, especially e-Governance and citizen participation, sustainable environment, safety and security of citizens, particularly women, children and the elderly, and health and education. This could be further categorised and discussed as smart growth, smart economy, smart governance, smart mobility, smart living and smart people.

Smart Growth

In the USA, according to (Adams and Gerard, 2000), cities in the US like Austin in Texas, has developed a local plan to manage growth known as the 'Smart Growth Initiative' (Gerard, 2000). The initiative covers a variety of programmes viz. undertake renewal in the urban core with intensified land use, preservation of environmentally sensitive areas, and address larger regional strategies for future growth. Smart growth at local levels include: a) In-fill development projects on vacant or underutilized parcels of land and b) Planning "Transit-oriented development" (TOD), in which cities use subway and commuter rail stations as hubs for mixed-use developments within a half-mile radius, including housing, retail, and day care (Adams and Gerard, 2015). Other growth initiatives that can be dubbed in the 'smart' category include permitting systems that help retain or develop environmentally sensitive areas, land pooling and banking, and the use of priority funding areas (PFAs) that direct development toward existing community centers (Smart Growth Network, 2002). In case of smart growth, policy choices are shaped by the difference in responsiveness of locally elected officials to interest group demands (Feiock, 2004). Vienna considered as a topmost smart city in the world has the highest scores due to its public transport network, extensive social housing activities and social services with high penetration and affordability for everyone, and also others factors (Smart City Wien Initiative Report, 2013). In another example of Seoul under Mayor Myung Bak Lee from 2001 -2006, has transformed the city into an easy commuting, working and living hub through new infrastructure construction while providing greater opportunities to business to function efficiently (Cervero, 2000).

Smart Economy

Cities are agglomerations of population and economic activity. European researchers, as per the Espon 1.1.1 project criteria, state that a 'smart' economy has characteristics of innovativeness, entrepreneurship, economic image and trademarks, productivity, flexibility of labour markets, international embeddedness and ability to transform (Giffinger, Fertner et al, 2012). According to (Adam, 2014), a city may deplete critical resource within its own boundaries or its hinterlands, lose its comparative advantage in cross-border trade, or suffer severe social ills. It may also be subjected to external shocks from natural and man-made disasters. The smart economy is concerned with how attractive and competitive the region is with regard to factors such as the stimulation of innovation, entrepreneurship, productivity and international appeal. In addition, new collaboration models often result in new revenue models. Amsterdam has adopted some innovative models, such as Budget Monitoring is a method that facilitates citizens to screen, assess, and actively participate in the decisions on public policy-making and government expenditure. It offers citizens the power, knowledge, and self belief to take action for the right to live in better environments. The Centre for Budget monitoring binds diverse community-led organizations, allowing valuable encounters among them as well as dialogue between citizens, organizations, and the government in mapping out problems and their solutions (Amsterdam 2015). In Amsterdam, due to several reasons, people were leaving for their offices at about the same time and were spending about 30% of their time travelling. Several past experiences of implementing Smart Work Centres show that once people use their cars on their way to the office, they don't want to change their behaviour. The first pilot for this project was introduced as *Smart Work@ IJburg*. Almost every day several kilometers of traffic jams block the entrances and exits to the island making IJburg a very suitable location to get people out of their cars. Amsterdam Smart City is therefore aiming at a Smart Work Center near the homes of the employees that are normally stuck in these traffic jams. ASC aims to make it as easy as possible. No hassle with payments for the employees. This is the reason; a payment system is currently being developed. Another feature is that the Smart Work centres will have a telepresence system, enabling the highest quality of video-conferencing currently available.

Smart Governance

To make cities enabled, empowered and smart, changes in traditional governance are very important. The traditional "closed and top down" approach will have to be replaced by a more "open model" for the management of its stakeholders. City officials need to encourage innovative and inclusive, open and transparent governance systems. Use of open and inclusive networks, and data infrastructure, citizen engagement, and integrated management structures enable the same. In the new millennium environmental concern of citizens at the local level have led them to increase their support for environmental and conservation goals. (Shah et al., 2015)

Mobile Seoul (m.Seoul) makes use of Mobile Web technology and mobile applications to provide Seoul's citizens with 62 unique services over 11 types of mobile device. A wide range of public information is available over the Mobile Web, but the inconvenience of navigating is averted by mobile apps concentrating on the provision of the most commonly-demanded information. m.Seoul apps support location-based services pinpointing nearby public offices, restrooms, hospitals, supermarkets or bus stations. Other services include live real-estate listings, daily job-search updates, and notifications of free cultural events. Apps also enable citizens to suggest actions to improve the city, participate in yes/no votes, and freely exchange city information over social networks; and another service, "Staying Safe in Seoul", alerts citizens of emergency situations brought on by heavy rain, snow, typhoons or fires (https://www.itu.int).

Seoul's next-generation online reservation system allows citizens to search for, book and pay for public services instantly. The one-stop, integrated reservation system lists over 150 services under categories such as education, infrastructure, cultural tourism, commodities and medical treatment. Seoul Metropolitan Government and its affiliates offer over 30,000 public services, from education courses to reserving public sporting or cultural facilities, using gymnasiums or registering at day-care centers. Of these 30,000-plus services, 17,000 offer online reservation.

Smart Mobility

The most crucial aspect is the mobility choices the city already has, or would aim at creating to become efficient. Mobility options begin with primarily planning and forecasting the needs of a growing population and businesses and then incorporated into the city design (horizontal or vertical), and the planning of its transportation routes and mediums. Smart choices of mobility veer around the creation of new development patterns and implementing transportation improvements in order to maximise opportunities for multi-modal transportation (Shah et al., 2015). In the case of Seoul, South Korea, their mobility choice marked a shift from surface roads that enhanced 'automobility' in favour of enhanced public amenities and quality of urban living through mass transportation systems. 74 lineal km of road lanes were built to accommodate a BRT network. Many regular bus routes were reconfigured to better feed into the city's extensive subway system, and an integrated fare and transfer system between buses and the metro rail network established (Pucher et al., 2005). Bus-operating speeds increased from an average of 11 to more than 21 kph. According to a study by Seoul Development Institute (2005), bus reliability improved, because of regular traffic flows; dedicated lanes reduced accidents by 27% one year and ridership improved by 60% for BRT services; introduction of real-time traffic information system and in-vehicle navigation aids was introduced to guide traffic flows and alert motorists; one-way arterial couplets and curbside parking was substantially curtailed to help expedite traffic flows; and an innovative license plate scheme that required the motorists to leave their cars at home, once every ten days helped ease the mobility issues of Seoul. Good information systems help enable the overall city's system to function to its optimum, with the least spillages of stressed and scarce resources. Agile systems evolve smartly, based on memory and experience.

A number of studies which examined the promise of ITS (Intelligent Transport Systems) (e.g. the Information Technology and Innovation Foundation (ITIF) study), have repeatedly identified three nations as global leaders in smart mobility: South Korea, Japan and Singapore. These countries have laid powerful policies to successful Smart Mobility (Dia, H. 2015) such as; National ITS (Intelligent Transport Systems) Vision- From the outset, these governments articulated and owned a clear vision for ITS and linked it to national Information Technology policies and long-term strategies for improving road safety and the quality of life for their citizens; Commitment to Funding-As a percentage of GDP, South Korea and Japan each invested more than twice as much in ITS than the United States; Partnership and Collaboration-The public and private sectors in these leading nations played an important role in co-developing platforms that enabled government, industry, academic and professional associations to collaborate on development of ITS at both local and national levels; Planning for Deployment-A big portion of the funding available to these countries was allocated to supporting ITS technology development, test-beds and proof of concept demonstrations as a precursor to wide-spread deployment. This approach also helped to inform and educate their citizens about the tangible benefits of smart technologies in transport; Standardisation-Leading countries developed national ITS architectures which provided the basis for interoperable ITS applications and assisted in the delivery of consistent, cohesive and cost-efficient services to citizens. For example, establishing common standards for electronic toll collection (ETC) in Japan encouraged high market penetration and uptake of on-board devices in more than 70 percent of vehicles. In Singapore, the single national standard of ETC also facilitated the implementation of a city-wide congestion charging scheme from early 1998; Innovation and Competition-This policy principle recognises the role of the private sector in developing and making available to governments and citizens innovations and technologies to improve their lives. Examples include alignment between the transport and telecommunications industries, where ITS was recognised as being inseparable from wireless technologies (for cooperative mobility applications) and high speed networks (for video transmission).

In summary, the policy principles highlighted here apply both to nations embarking on new smart infrastructure projects, as well as developed economies looking to make the most of existing assets. But to be successful, these systems must be viewed as part of a holistic sustainable mobility vision. To drive change, governments must also move beyond a project-by-project view, spur close coordination between government agencies, and provide clarity about effective engagement between the public and private sectors. (Dia, 2015)

Smart Living

Citizens in cities that are considered smart are sure to make better use of the natural resources and environment conscious (Shah et al., 2015). The city management takes effort to provide better options to the citizens in commuting and recreating, and enjoying a better quality of life compared to other city residents that lack them (Shapiro, 2006). These cities score higher on providing cultural opportunities, public safety, health and housing, educational facilities to all segments of the population, touristic attractiveness and social cohesion. Automobile dependency is low hence citizens do not indulge much in sedentary lifestyles. Due to mass transport facilities, the statistics on road accidents is lesser.

Smart cities are focused on being a desirable place to live, work and spend time in. Concerns that influence the quality of life in the region include culture, health, safety and tourist attraction. Key questions include how to ensure the city remains liveable for children and how to effectively address people living independently for longer.

As part of the European project City-zen, a Serious Game is being developed by UK Partner Clicks and Links that is to be deployed in the city of Grenoble as well as Amsterdam Nieuw-West. The goal is to engage youths in these cities to save energy by raising awareness and changing behaviour through the familiar medium of gaming. Gaming tends to have a bit of a negative reputation, but consider the opportunity of leveraging the following traits for a positive objective, such as saving energy. Games are addictive, fun and engaging which means that they can be great tools for education (Amsterdam 2015).

The *geWoonboot* was built in 2007 by the DeltaWonen housing corporation, as a prototype of a self-sufficient floating home. In the Netherlands, as a water-rich country and with a view to climate change, the houseboat is a beautiful example of the possibilities in sustainable living at the time it was built. Situated on the creative wharf, it is currently an inspiring meeting place for all kinds of organisations and companies (Amsterdam 2015). The water used to flush the toilet is rainwater filtered by the moss sedum bed on the roof. The sewage is transported to a helophyte filter system and a floating bamboo and reed garden at the back of the boat. Once purified, the water is pumped back into the boat. Solar panels provide the boat with electricity and a solar boiler heats it. In addition, a heat exchanger takes warmth from the surface water of the river.

Gaasperdam a neighborhood in Amsterdam is moving towards a process of innovative urban renewal, Bottom-up, using their energy of residents and the qualities already available in the area. Local people's energy is an available commodity; residents have declared that they are aware of their local environment and want to work towards a better living space for the community (Amsterdam 2015).

A package of services is developed with the help of residents, for their benefit, concerning more comfort, high-speed internet, insights in energy consumption and lower energy costs. In short, a modernisation of local people's living space, without extra costs.

Cooperation is a key in 'Smart Living in Gaasperdam'. By cooperating, energy is created in several fields of interest:

1. By popular demand: the development of novel services for residents, together with residents and companies.
2. Residents become aware and are taking the director's seat.
3. The area will become more attractive.
4. Reduction of CO2.
5. The development of new profit models

Smart People

The characteristics of people in 'smart' include their level of qualification, affinity to lifelong learning, composition of social and ethnic plurality of the city, flexibility, creativity, cosmopolitanism/open mindedness, and propensity to participate in public life (Giffinger, Fertner et al, 2012). The type and composition of the city's populace strengthens the city's ability to attract investment and to make it a thriving hub of knowledge and culture. In the day to day workings of city life, people are empowered with information and communication technologies to add choices and efficiency to their lives (Shah et.al, 2015). A city is nothing without its residents and visitors. A smart society is a vital aspect of a smart city – it's all about social and human capital. Linked to this is the development of residents themselves, but also the level of creativity and the quality of social interaction between people.

'Change by Us' is a new platform in Amsterdam. The platform started in New York and is known by the name of Change by Us NYC. It allows innovative partnership in the city between citizens, city leaders, welfare organizations, housing corporations and government. When ideas are made visible it becomes very clear for everyone what kinds of things are going on in the neighbourhood. Anyone is welcome to join in to help convert an idea into a feasible project. This way residents, businesses, government and welfare organizations work together on a finer neighbourhood and a liveable city.

OTHER SMART CITY INITIATIVES

1. **GIFT City-Gujarat, India:** Gujarat International Finance Tec-City (GIFT) is an upcoming Smart city at Gandhi Nagar in Gujarat. It covers 886 acres (359 Ha) of area. It provides high quality physical infrastructure (electricity, water, gas, district cooling, roads, telecoms and broadband), special economic zone (SEZ), international education zone, an international techno park, shopping

malls, stock exchange and service units. The Gujarat Urban. Development Company Limited (GUDCOL) and Infrastructure Leasing & Financial Services (IL&FS) have established a Joint Venture Company. "Gujarat International Finance Tee-City Company Limited" (GIFTCL) for development of the project. The estimated cost of the project is USD 10,500 Million. It will be a globally benchmarked smart city having an ultimate built up area of 85,000,000 sq ft (7,900,000 sq.m), in which state of art connectivity, communication, technology, security and other services have been planned. It is envisaged to become a model of smart and sustainable city. It will provide a safe and clean environment, and ecological integrity through low energy consumption, smart energy grids, and piped gas, internet gateway, District Cooling System, Pneumatic waste management and intelligent/ bionic. systems. Recycling and reduction in wastes will help to protect the natural environment. The city has been planned by East China Architectural Design & Research Institute (ECADI) and Fairwood Consultants India. (Jain, 2015)

Ahmadabad, Airport, Gandhinagar and other cities are connected with each other with a mix of multi-modal transport systems (MRTS/LRTS/BRTS, etc.).Besides Bus Rapid System (BRT), modal split has been used for Walk-to-work, Metro and electric personal rapid transit systems are planned for intra- mobility.

In GIFT City, Gujarat, some of the smart features are Perennial Water Front ensured through construction of three barrages on river Sabarmati, Proposed Landscaped promenade at the river bank along GIFT, Underground cabling for power distribution within GIFT, Gas Insulated Switchgears (GIS) substation for Sub-transmission and Distribution within GIFT city, Compact substation (CSS) in buildings, Automated Waste Collection System (AWS) through chute system, Waste sucked through pipes at a speed of 90 km/hr, Waste Treatment through Plasma Technology, Envisaged Modal Split of 10:90 between private and public transport. (GIFT City 2015)

2. **Singapore:** Singapore stands out as a beacon of economic vitality. With nearly 5 million people sharing an island smaller than New York City, Singapore faces a continuing challenge in managing the impacts of its high population density, especially traffic congestion. As a result of heavy investment in its public transportation infrastructure, Singapore has created one of the most modern, affordable and heavily used public transport networks in the world, where the number of people using the trains is 1.6 million and 3 million people riding the bus on any given day.

High population density on a small island required Singapore to increase the use of its public; transport and keep traffic congestion at bay. Thus the Singapore

Land Transport Authority (LTA) along with IBM has created a simplified system which streamlines fare payment for riders, giving, them a choice of cards which they can also use for multiple purposes. It enables riders to use a single card of their choice to pay for all modes of travel—bus, rail and taxis—as well as car parking and vehicle congestion charges, thereby making it more convenient for riders to pay their fares. This is known as Symphony for e-Payment (SeP), it includes, ability to accommodate multiple fare card issuers, increasing convenience for commuters and lowering costs. It helps in reduction in revenue leakage by 80 percent from "lost" transactions due to systems issues and 2 percent reduction in the overall life cycle cost of the fare processing system while doubling its performance capacity to 20 million fare transactions per day. (Source: IBM)

This system gathers insights from the 20 million trip-related transactions generated each day—information on where, when and how riders are travelling and translate them into a more convenient and affordable travel experience for Singapore's citizens by optimizing routes, schedules and fares.

3. **Masdar City:** One of the world's first zero-carbon cities, Masdar City, is under construction in Abu Dhabi, United Arab Emirates. Inspired from an ancient walled city the project design utilises modern technologies in its daily operations. Construction of the city began in 2008 with Abu Dhabi Future Energy Company being primarily responsible for the project. The 6 million square metre projects are powered entirely with renewable energy, including solar and wind power. Zero-carbon cities ensure that the city is entirely carbon-free and thereby create optimal living conditions that greatly help keep adverse pollution effects in check. The core factors behind the creation and smooth functioning of the city are environmental sustainability and clean technology. The city is designed to be the home for 50,000 inhabitants and as of now it has occupancy of 4000 and above. In the next five years, the population size is estimated to grow to 10,000.The mixed-use, high-density city will also include a university, innovation centre, company head-quarters and several economic zones.
4. **Songdo-The Digital City:** Songdo, also known as ubiquitous City or U-City, located on reclaimed land near the Yellow Sea, is a $35 project that started in 2005 in South Korea. Cisco is responsible for all technological operations and is the mastermind behind operations in Songdo. While the city's information systems are all inter-connected, there are sensors on everything that are responsible for the smooth operations. Each home in the city has a telepresence system with video conference facilities so as to help deliver education, health care and government services. Offices and schools are also linked to

the network. Songdo is expected to be completed by now and would be home to 65,000 people and 30,000 workers.

5. **Rio de Janeiro's:** In Rio de Janeiro's operation center, wall of screen keep officials updated with live information with the help of 560 cameras, provide information about weather forecasting and also smart map that can analyze 60 layers of data from and around the city. (Gopalakrishnan, A. 2014)
6. **Bangalore:** In Bangalore, pilot project is started with Bangalore police to run kiosk in mall, where FIRs can be electronically filled. Secondly parking can also be made smarter with app in your phone that tells where are the empty slots in particular place, help in reducing driving as well as congestion. (Gopalakrishnan, A. 2014)
7. **Zhengzhou, China:** In China, the Zhengzhou Yuntong bus company launched the project that the cities across china are now rushing to track. It installed GPS tracking system on every bus and also launched an application that allow parents, public, government, administration to track their movements which appreciably increases security. Secondly government has also launched a pilot project that monitors migrant and unemployed residents and will equipped ever home with panic button which is connected to nearby police station, hospital, fire station for immediate support. (Krishnan, A. 2014)
8. **Malta:** In Malta, integrating both water and power systems, will be able to identify water leaks and electricity losses in the grid. 250,000 interactive meters will monitor electricity usage in real time, set variable rates, and reward customers who consume less. (Bristol City Council 2015)
9. **Bordeaux, France:** In Bordeaux, France, there is LED-based street-lighting that dim in response to changing environmental conditions. Embedded sensors measure the light, temperature, intensity of current and detect motion. Light are connected to the internet enabling remote control and maintenance. (Bristol City Council 2015).
10. **Copenhagen, Denmark:** In Copenhagen, Denmark, there are hybrid e-bikes that also function as mobile sensing units. It captures energy dissipated while cycling and braking and saves it for when you need a bit of a boost. It also maps pollution levels, traffic congestion, and road conditions in real-time. It uses apps in mobile to feed pollution data etc. into city dashboard. (Bristol City Council, 2015).
11. **London:** In London and Stockholm, there is congestion charging schemes with number plate recognition systems. London feeds traffic flow data into London Congestion Analysis Project turning it into valuable journey time information. (Bristol City Council, 2015).

12. **San Francisco:** In San Francisco, smart parking meters and parking wireless sensors are installed in pilot areas across the City in 2010 feeding into data hub. Real-time parking data to be available, both on SFpark.org and as open source data feed. It allows demand-responsive pricing. (Bristol City Council, 2015)

LAVASA CASE STUDY: SMART INITIATIVES

Lavasa is a private hill city of about 100 Square Kilometers developed by HCC (Hindustan Construction Company). It is located in the western region of India in the picturesque landscape of the Sahayadri Mountains. A three hour drive from Mumbai and an hour drive from Pune, the city is one fifth of the size of Greater Mumbai. The master plan is based on the principles of New Urbanism that brings together all the components essential to daily life in a more organized manner. A unique architectural design that is sensitively integrated with natural environment, enhancing the scenic beauty of the area and over 70% of the area that is to be greenified through detailed landscaping and re-forestation. Lavasa aims to provide a perfect work - life balance with a unique combination of technology and infrastructure advancements. The city will have a 365 day economy with a host of non polluting industries being the main economic driver; these include R&D and training centres, IT and biotech industry, KPOs (Knowledge Process Outsourcing) and those related to art, fashion and animation. It has many firsts to its credit - technology leadership, e-Governance; the first Indian city developed using Geographical Information System (GIS), use of Bio mimicry as a science in town planning and use of innovative techniques like hydro seeding in environment management. It is planned for a permanent population of around 0.3 Million residents and a tourist inflow that is envisaged at 2 million per annum. With all urban advantages in a natural setting, it is envisioned to be a more livable city of the future where residents can live, work, learn and play in harmony with nature.

Lavasa is one of the first Smart city of India, privately developed. According to The Wall Street Journal, among the digital experiences, Lavasa homes offers touch panel automation, occupancy-based lighting, door sensors, motion sensors, beam detectors and on-call transport services, assets tracking automatically through RFID tags, all residential & buildings are connecting by secure IP, networking technologies to help centralize management operations, logistic supply centers boasting of roads, airport, rail, dry dock and operation centre and smart transportation and JIT delivery (Madakam, S. 2013).

The project is striving towards industrialization where the building components such as windows, doors have been standardized and making the required materials such as bricks, crusher and RMC plant on the site itself. The Mugaon town in Lavasa city uses locally available materials like stone, bamboo and sustainable technologies like hollow fly ash-concrete blocks, solar panels, green roofs, micro wind turbines and solar heaters.

Lavasa is designed as five urban towns in the valleys and slopes of the seven surrounding hills. A sustainable environment has been created to make use of the latest technology as well as efficient use of time, material, space and other resources. The structures have design recommendations which try to act like the natural environment like, the pavements allow the water to permeate back into the ground, the roof vaporizes the water falling in it, etc. The roads were inspired by the local anthills which are stable even during heavy rain falls. This is done by channelizing the water and slows them. The spaces are connected well with walkways which are formal and informal. The planted trees would maintain the soil quality, store water throughout the dry season and also provide a canopy to control evaporation. (Nallathiga, R. 2015).

Six Dimensions of Lavasa City

The city of Lavasa is planned on six dimensions to address the issues of Urbanization, pollution and lack of infrastructure. These six dimensions are; Smart Economy - Innovation and Competitiveness, Smart Mobility- Transport and Infrastructure, Smart Environment - Sustainability and Resources, Smart People - Creativity and Social Capital, Smart Living - Quality of Life and Culture and Smart Governance - Empowerment and Participation

Smart Practises at Lavasa

Lavasa has developed an Environment Management Plan (EMP) which addresses environment concerns at several levels. Topsoil management, tree transplantation, mass plantation, hydro seeding, soil, water and air pollution control and monitoring systems are few of the initiatives taken for the project. Artificial water holes, wildlife corridors and releasing fish in dam water to maintain a healthy biodiversity are also practised.

- **Hydro-Seeding:** It is a planting process that uses slurry of seed and mulch. It helps in mass plantation, prevents soil erosion and facilitates quick vegetation.

- **Soil Bioengineering:** To protect the slope during monsoon and to preserve the green cover, living plant materials were used to prevent it from getting washed away.
- **Water Conservation:** Trenches are dug on the sides of roads for the rain water to percolate into the soil. Mass plantations are done on the site in order to raise the water table of the region.
- **Water Treatment:** UV and Ozonisation is used for sewage water treatment.
- **Air Treatment:** Water is sprinkled while carrying out construction to settles the dust particles and dust masks are used (Nallathiga, R. 2015).

The site is hilly and very steep at some places. Multiple roofed houses at different levels with integrating courts have been proposed rather than large structures with bulky roofs as they would blend into the natural environment. The privacy of the houses have also been maintained by locating them at different angles. The houses look out to green spaces and not to other structure (Nallathiga, R. 2015).

OBSERVATIONS

Replacement of the actual city infrastructures is often unrealistic in terms of cost and time. However, with recent advances in technology, we can infuse our existing infrastructures with new intelligence. By this, we mean digitizing and connecting our systems, so they can sense, analyze and integrate data, and respond intelligently to the needs. This will make smart cities resilient enough to the risks that they face.

Various trends such as new revenue models, collaboration models and sharing economy are often the results of factors like stimulation of innovation, entrepreneurship, productivity and international appeal. The public and private sectors play an important role in co-developing platforms that enables government, industry, academic and professional associations to collaborate on development of infrastructure at both local and national levels. The city shall have a 365 day economy with a host of non polluting industries being the main economic driver; these include R&D and training centres, IT and biotech industry, KPOs and those related to art, fashion and animation. Also it shall have an Environment Management Plan which addresses ecology concerns such as topsoil management, tree transplantation and mass plantation to processes of hydro seeding, soil, water, and air pollution control and monitoring system. These factors are very important in order to manage continuing pressures for development and growth while reducing their vulnerability to a range of potential crisis.

Innovative practices such as floating garden where the water used to flush the toilet is rainwater filtered by the moss sedum bed on the roof and river water heating where solar panels provide the boat with electricity and a solar boiler heats it. In addition, a heat exchanger takes warmth from the surface water of the river. The trenching should help with ensuring infiltration and the groundcover with minimizing runoff, but storm water management, water efficiency and reuse would be a welcome addition to the sustainability measures, many of which like natural bio filtration could probably be easily integrated into the nature walks. Such practices can prepare cities very well for the problems related to climate change. To include social infrastructure, ensure the affordability of housing units and offer training opportunities somewhat eases the awkward, unnatural separation of classes and make cities resilient against social disturbances.

Allowing innovative partnership in the city between citizens, city leaders, welfare organizations, housing corporations and government, where ideas are made visible and becomes very clear for everyone what kind of things are going on in the neighbourhood. Anyone is welcome to join in to help convert an idea into a feasible project. This way residents, businesses, government and welfare organizations work together on a finer neighbourhood and a liveable city. This makes cities inclusive and sensitive to the diverse needs of their population.

RECOMMENDATIONS

From the detailed study of various smart cities around the world including Lavasa, GIFT city and Bangalore from India, following recommendations are derived for the development of smart and sustainable cities.

- To facilitate citizens to screen, assess, and actively participate in the decisions on public policy-making and government expenditure.
- To make use of Mobile Web technology and mobile applications to provide citizens with unique services over mobile devices.
- To develop online reservation system that allows citizens to search for, book and pay for public services instantly.
- To combine diverse technologies to reduce environmental impact due to excessive energy consumption and offer better lives to citizens.
- To develop a clear vision for Intelligent Transport System (ITS) and link it to national policies and long term strategies for improving road safety and quality of life.

- To forge public private partnerships within the nation and view the investments as creating a platform through which the private sector can develop value-added products and services.
- To develop national ITS architectures which provide the basis for interoperable ITS applications and assist in the delivery of consistent, cohesive and cost-efficient services to citizens.
- To move from a political or jurisdiction-based system of allocating transport investment to one that uses performance and cost-benefit analysis as the basis for investment decisions.
- To engage youths in these cities to save energy by raising awareness and changing behaviour through the familiar medium of gaming.
- To allow innovative partnership in the city between citizens, city leaders, welfare organisations, housing corporations and government.
- To develop a 365 day economy with a host of non polluting industries being the main economic driver; these include R&D and training centres, IT and biotech industry, KPOs and those related to art, fashion and animation.
- Homes shall offer touch panel automation, occupancy-based lighting, door sensors, motion sensors, beam detectors and on-call transport services.
- The smart city shall include assets tracking automatically through RFID tags, all residential & buildings connected by secure IP, networking technologies to help centralize management operations, logistic supply centres boasting of roads, airport, rail, dry dock and operation centre and smart transportation and JIT delivery.
- To use locally available materials like stone, bamboo and sustainable technologies like hollow fly ash-concrete blocks, solar panels, green roofs, micro wind turbines and solar heaters.
- To develop an Environment Management Plan which addresses ecology concerns such as topsoil management, tree transplantation and mass plantation to processes of hydro seeding, soil, water, and air pollution control and monitoring system.
- The structures shall have design recommendations which try to act like the natural environment like, the pavements allow the water to permeate back into the ground, the roof vaporises the water falling in it, etc.
- **Hydro-Seeding:** It is a planting process that uses slurry of seed and mulch. It helps in mass plantation, prevents soil erosion and facilitates quick vegetation.
- **Soil Bioengineering:** To protect the slope during monsoon and to preserve the green cover, living plant materials were used to prevent it from getting washed away.

- **Water Conservation:** Trenches are dug on the sides of roads for the rain water to percolate into the soil. Mass plantations are done on the site in order to raise the water table of the region.
- **Water Treatment:** UV and Ozonisation is used for sewage water treatment.
- **Air Treatment:** Water is sprinkled while carrying out construction to settles the dust particles and dust masks are used.

REFERENCES

Adams, G., & Gerard, D. (2000). Smart growth and transportation: Opportunities and challenges for Austin. *Institute of Transportation Engineers Journal*, *70*(11), 30–34.

Ahuja, D., & Tatsutani, M. (2009). Sustainable energy for developing countries, *SAPIENS. Surveys, 2*(1).

Branchi, P. E., Fernández-Valdivielso, C., & Matias, I. R. (2014). Analysis Matrix for Smart Cities. *Future Internet*, 6, 61-75. http://www.mdpi.com/1999-5903/6/1/61

Bristol City Council. (2015). Retrieved from http://www.bristol.gov.uk/

Cavada, M., Christopher, D. F. R., & Hunt, D. V. L. (2014). Smart cities: Contradicting definitions and unclear measures. *Conference Proceedings Papers*. Retrieved from http://sciforum.net/conference/wsf-4/paper/2454/download/pdf

Census of India. (2011). *Population Census-2011*. Government of India.

Cervero, R. (2000). Urban reclamation and regeneration in Seoul, Republic of Korea. In S. Lehmann (Ed.), Low carbon cities: Transforming urban systems (pp. 224-234). New York: Routledge, Taylor and Francis Group.

DeAngelis, S. (2015). Smart Cities: Big Data and Resiliency. *Wired Innovation Insights*. Retrieved from http://insights.wired.com/profiles/blogs/smart-cities-big-data-and-resiliency#ixzz3nD44Qjkw

Dia, H. (2015). *10 Powerful Policies to Successful Smart Mobility*. Retrieved from http://www.sustainablecitiescollective.com/

Feiock, R. C. (2002). A Quasi-market framework for local economic development competition. *Journal of Urban Affairs*.

Giffinger, F., Kramer, M., & Pichler, M. (2012). European smart model. Universities of Vienna, Universities of Delft, Universities of Lubijana. Retrieved from http://www.smart-cities.eu/model_1.html

GIFT City. (2015). Retrieved from http://giftgujarat.in/

Gopalakrishnan, A. (2014). *Art of Smart: Can technology transform urban living in India?* (pp. 25–28). India Today

Hawkins, C., Ye, L., Mandpe, S., & Meyer, P. B. (2005). What is "smart growth?" - Really? *Journal of Planning Literature*, *19*(3), 301–315. doi:10.1177/0885412204271668

Hollands, R. G. (2008). Will the real smart city please stand up? Intelligent, progressive or entrepreneurial. *City*, 12.

Jain, A. (2015). *Smart Cities: Vision and Action*. New Delhi: Discovery Publishing House Pvt. Ltd.

Jawaid, M.F., & Khan, S.A. (2015, March). Evaluating the need for smart cities in India. *International Journal of Advance Research in Science And Engineering*, 4(SI 1).

Kang, Y., & Zang, L. (2014). Comparative Study of Smart Cities in Europe and China 2014. *Proceedings of 4th International Conference on Logistics, Informatics and Service Science*. Commercial Press China and Springer-Verlag Berlin Heidelberg.

Keshava, S.R. (2014). Indian Urbanization under Globalization; Development or Delusion! *International Journal in Multidisciplinary and Academic Research*, 2(1).

Krishnan, A. (2014). Art of Smart: Can technology transform urban living in India? *India Today*, p. 30.

Lavasa. (2015). Retrieved from http://www.lavasa.com/

Lee, J.H., Hancock, M.G., & Hu, M.C. (2013). *Towards an effective framework for building smart cities: Lessons from Seoul and San Francisco. Technological Forecasting and Social Change.*

Madakam, S., & Ramaswamy, R. (2013). The State of Art: Smart Cities in India: A Literature Review Report. *International Journal of Innovative Research and Development*, *2*(12).

Nallathiga, R., Tewari, K., Saboo, A., & Varghese, S. (2015). Evolution of Satellite township development in Pune: A Case Study. *Evolution; International Journal of Organic Evolution.*

NIUA. (2012). *Report on Urban Infrastructure and Services, Summary and Recommendations*. National Institute of Urban Affairs.

Parasuraman, E.N. (n. d.). Blog: *So, what is the smartness quotient of your city?* Schneider Electric.

Pucher, J., Park, H., Kim, M. H., & Song, J. (2005). Public transport reforms in Seoul: Innovation motivated by funding crisis. *Journal of Public Transportation*, *8*(5), 41–62. doi:10.5038/2375-0901.8.5.3

Shah, M. N., Nagargoje, S., Shah, C. (2015). Assessment of Ahmedabad (India) and Shanghai (China) on Smart City Parameters applying the Boyd Cohen Smart City Wheel. *Proceedings of CRIOCM 2015*.

Shapiro, J. M. (2006, May). Smart Cities: Quality of life, productivity, and the growth effects of human capital. *The Review of Economics and Statistics*, 88(2), 324-335.

The Smart City Wien Initiative Report. (2014, July). Owner and publisher: Vienna City Administration.

Wankhade, K. (2012). *JNNURM: An Opportunity for Environmentally Sustainable Urbanisation.* IGC Working Paper.

Chapter 10
Smart Resilience:
Resilience Dimension of Smart Cities

Jing Pan
JCM International Engineering Management Institute, China

ABSTRACT

Resilient cities are characterized by enough capacity to respond to internal and external changes and uncertainties. Crisis arises from unprepared residents and inadequate capacity of a community to adapt to hazards. It leads to pollution and wastes due to the recovery and reconstruction activities. Empirical data show that unavailability of accurate building information and poorly informed public contribute to most of the crisis. Building resilience is a fundamental component of community resilience. It determines the resilient capacity of a community. The chapter presents a theoretical framework of Building Resilience Information Modeling (BRIM), including a dynamic real-time hazard mapping module and a scaled building vulnerability indicator module, and how it supports multiple stakeholders to collaborate in decision making. It presents not only a solution to community resilience as a use case of smart cities, but also an innovative way of integrating social and environmental aspects in built environments in terms of collective intelligence oriented to anticipating hazards so as to resolve them proactively and effectively.

DOI: 10.4018/978-1-5225-0302-6.ch010

INTRODUCTION

The number and scale of natural disasters have drastically increased in recent decades. Number of natural disasters took place in the 1990s was four times that of the 1950s (Smith & Petley, 2009). The losses from natural disasters averaged $65 billion a year in the 1990s, which was over seven times since the 1960s (Ferris et al., 2013). Such increased natural disasters have a significant impact on built facilities. Take meteorological disasters for example, they cost over $50 billion damage in public and private facilities, and affected me than 6 million facility occupants (Ferris et al., 2013).

To recover from disasters, a tremendous amount of energies and resources are consumed for managing debris in the disaster-stricken area, and to rebuild physically damaged utilities, infrastructures, and properties. In some cases, disasters leads to environmental pollutions that take hundreds of years to recover. Moreover, Natural disasters contributes to only a portion of crisis to human societies.

Currently, crisis is considered of disruptive nature, and practitioners and scholars globally have made efforts in predicting the likelihood and severity of such disasters. Under such paradigm, crisis happens when a community suffers from losses of an uncertainty. Such losses happen to communities that are not prepared or not adequately prepared for such uncertainties. Various reasons lead to the unpreparedness in a community, such as failure to predict the uncertainty, including the strength, location, and time of such uncertainties. In an investigation of the American public, 45% of Americans being investigated never think an emergency will happen to them (The Council for Excellence in Government, 2006).

The chapter presents innovative ways to build adaptive infrastructures and to increase the capacity of the public to adapt to change and uncertainties. It demands a platform to manage facility information that meets the need of different user groups including facility occupants, facility management team, emergency management officials and partners, and the local community. A specific information model, the Building Resilience Information Modeling (BRIM), is developed for the purpose. It is determined as a three-module system to provide and update facility information and to generate resilience indicators for facility vulnerability monitoring and response.

BRIM aims to achieve resilience by creating and using knowledge and information from individuals in a community. It integrates the social aspect in the form of collected intelligence, and the physical resilience of the built environment. BRIM provides a smart solution for community resilience by providing easy access to acknowledge and to update building information, and by engaging building residents in getting prepared and engaged in the process of the real-time hazard mapping module. It encourages decision making on individual levels, and aims to increase community resilience capacity through educating individuals and communities in a proactive approach.

BACKGROUND

Crisis rises from unpreparedness against disasters. As one of the well-studied field in the disaster cycle, preparedness. To solve the problem of low efficiency, efforts of harvesting crowd wisdoms is gradually changing the situation of solely relying on official information for decision making.

Crowd wisdom capturing is enabled by the technology leap in implementing smart buildings. Such public participation changes of disaster management paradigm, from reactive to proactive – appropriate understanding of uncertainty, decrease vulnerability by building community resilience. The paradigm is changing from top down to bottom up – top-down is effective with the reactive paradigm, however, the change of paradigm calls for the change of acting process and sequence as well. It is evolving from responding to crisis to understanding the nature of uncertainty as the new norm, and decrease the vulnerability by making the community and built environments resilient to changes.

The first section introduces the concept of community resilience and building resilience. It also states how a smart building system facilitates the implementation of building resilience and how a resilient community energizes a smart city in return. This section presents the challenges in practice to enable effective knowledge transfer.

FACILITY RESILIENCE

Crisis rises from unpreparedness against disasters and uncertainties (Boin & McConnell, 2007; Saul, 2004). Disasters and uncertainties are irreducible and hard to predict (Berkes, 2007). They affect a facility in three ways: 1) the physical facility; 2) the socioeconomic activities located in the facility; and 3) the occupants in the facility. The occupants in the facility as well as the activities that take place in the facility spread the impact of disasters from a single facility to a larger community. Increasing facility resilience is the key to community resilience for it enhances the capacity of a community in terms of internal and external changes and uncertainties (Boin & McConnell, 2007; Center for Community enterprise, 2000; Kumar, 2005; Magis, 2007; Saul, 2004).

Community resilience comes from the sustainable study (Adger et al., 2005; Goodwin, 2003). Community resilience has since been extended to depict the capacity of a community to respond to disasters (Kumar, 2005; Magis, 2007). It implies the change of disaster management paradigm from seeking stability to accepting uncertainty and change as the new community norm.

For this research, facility is the basic level of community resilience building. It shares all the features a community is defined with. First it defines a geographical area and the social and institutional attributes within it (Gusfield, 1975; Chaskin et al., 2001). Secondly, it defines the relationships between people who reside in the area, including shared interests, beliefs, and access to resources (Carroll & Lee, 1990; Chaskin et al. 2001). In a facility, the vulnerability of the built environment is integrated with the vulnerability of resident groups. it is the basic level of community resilience building, and it is therefore essential to build resilience on facility levels to reach the overall resilience on community level.

Facility resilience is therefore to build the capacity on a facility level to respond to potential hazards. Moreover, a resilient facility leads to a resilient community. It goes one step further in disaster management – building up the capacity to disasters and uncertainties through experience – Not simply respond to disasters, but intentionally develop capacity to absorb effects of disasters (Berkes & Seixas, 2005). This capacity of a facility and its occupants includes adapted/evolved facilities (Berkes, 2007; Goodwin, 2003), actively engaged community resident groups (Magis, 2007), and stronger bonding among occupant groups (Costello & Johns, 2006).

CHALLENGES OF COMMUNITY RESILIENCE

The number and scale of natural disasters have drastically increased in recent decades. Number of natural disasters took place in the 1990s was four times that of the 1950s (Smith & Petley, 2009). The losses from natural disasters averaged $65 billion a year in the 1990s, which was over seven times since the 1960s (Ferris et al., 2013). Such increased natural disasters have a significant impact on communities and cities. Take meteorological disasters for example, they cost over $50 billion damage in public and private facilities, and affected more than 6 million facility occupants (Ferris et al., 2013). During a crisis, the functionality of the individual and of the community during such extreme situations, and their ability to adapt and restore the losses, are significant factors associated with coping with crisis (Cohen et al., 2013). Currently, a tremendous amount of resource is needed to repair/replace these facilities, which worsen facility management with congestion, loss of productivity, extra consumption of energy, and air pollution issues. Moreover, meteorological disasters, and natural disasters as well, is just a tiny portion of the facility vulnerability problem the human society is facing. It generally calls attention to rethink the disaster management practice.

Key challenges facing many communities to build community resilience are:

Responsive Culture of Disaster Management

Disaster management is under the reactive paradigm. It is characterized by perceiving disasters as disruptive events. A tremendous amount of resource is needed to repair/replace these facilities, which worsen facility management with congestion, loss of productivity, extra consumption of energy, and air pollution issues. Moreover, disaster preparedness is now perceived as a public function. It is many times used to evaluate the effectiveness of the public agencies (Shaw & IEDM, meteorological disasters, and natural disasters as well, is just a tiny portion of the facility vulnerability problem the human society is facing. It generally calls attention to rethink the disaster management practice.

Paradigm in Which Disaster and Crisis Are Considered as Disruptive Happenings Rather Than a Norm

Disaster management is under the reactive paradigm. It is characterized by perceiving disasters as disruptive events. A tremendous amount of resource is needed to repair/replace these facilities, which worsen facility management with congestion, loss of productivity, extra consumption of energy, and air pollution issues. Moreover, disaster preparedness is now perceived as a public function. It is many times used to evaluate the effectiveness of the public agencies (Shaw & IEDM, meteorological disasters, and natural disasters as well, is just a tiny portion of the facility vulnerability problem the human society is facing. It generally calls attention to rethink the disaster management practice.

The other important aspect of facility resilience is the ability of its occupants to learn that changes is unpredictable but constant, and to build the capacity through responding to disasters (Saul, 2004). Engaging the public can make a substantial contribution to improved community outcomes through the sharing of information, identifying problems, and crafting possible solutions collectively (Walters et al., 2000). Research in community shows that such active engagement could result in value change regarding change and uncertainties to achieve sustained facility resilience (Schiderman, 2010; Trujillo et al., 2000). How to engage the public is therefore one of the major challenges in building disaster resilience (Nakagawa & Shaw, 2004; Yang & Callahan, 2007).

Lack of Benchmarking on Resilience of Building/Facility Level

Among the components of community resilience, building facilities generate and host an array of socioeconomic activities. Facility resilience is therefore directly correlated to the psychological and economic resilience of a community (Norris et al., 2002). It is therefore important to investigate facility resilience building by its existing issues and seeking solutions to such issues. However, none research was retrievable on facility resilience at the time of this manuscript was prepared.

Lack of Consolidated Information Retrieval Mechanism to Enable Emergent Decision Making

Crisis results from unprepared residents and inadequate capacity of a facility to adapt to hazards (Mayunga, 2007; McLoughlin, 1985). During a crisis, the response and recovery team relies largely on intuition and is experience-based (Lam et al., 2009). Many public buildings have paper-based design documents, which can be digitalized with professional efforts (Haron et al., 2010). Some facility mangers do not have access to such documents, and if these documents are ever needed, the facility management team has to describe or draft drawings for the need of emergency response and mitigation (Rozenfeld et al., 2009).

Availability of building information is therefore essential to build adaptive facilities. Previous research shows that 2D drawings are completely outdated in four years on average (Taneja et al., 2010). Updating information manually by integrating service history with the existing engineering data is costly (CRCCI, 2007). A dynamic and intuitive facility information system is necessary to increase the preparedness of a facility by enabling in-time access to accurate building information.

As the challenges and issues identified in specific, a series of strategies are formed and presented in the following section.

STRATEGY #1: SMART BUILDING TO BRIDGE THE GAP

Facility management involves the efforts of a facility owner contribute to comforts of facility occupants. However, in a typical construction project today, limited information about the operation of the building are deliverable to facility management teams (Rauschert et al., 2002). The information the facility management team really needs comes in paper form (or in portable document format), written operation and maintenance (O&M) manuals, warrantees, etc. The relevant data then has to be manually extracted from these sources, formatted, and typed into the owner's Facility Management System (FMS) so that the information about the equipment, spare parts, and tasks needed to maintain the building can begin. In some cases, this process takes so long that necessary preventative maintenance tasks are not performed, equipment fails prematurely, and warranties are voided. However, emerging information technologies have provided solutions to such challenges.

Building Information Modeling (BIM), as one of the latest software technology, holds the potential in addressing FM problems with its visualization, data distribution, and analysis capabilities. Whilst some researchers explored BIM applications in design and construction (Khanzode et al., 2008; Khemlani, 2009; Babic et al., 2010), BIM also has gained increasing recognition in its capability for facility management (Eastman et al. 2008). Representative works include but not limited to building information maintenance (Hwang and Liu, 2010), maintainability checking (Dehlin and Olofsson, 2008), energy analysis (Cho et al., 2010), and sustainability (Arayici et al., 2011).

BIM has been tested as knowledge transfer among stakeholders and user groups. Knowledge management is an essential task of facility resilience that can be significantly benefited by the adoption of BIM. BIM provides a unified data repository of all building components, allowing an infinite range of data to be attached to the building components (Sabol, 2008). Information required by daily facility management tasks can be efficiently collected, stored, and distributed through a network, which also lays a solid foundation for future extension about occupant learning and knowledge transferring.

A number of studies on the adoption and usage of information technologies to facilitate knowledge transfer have been conducted in business management (Davenport & Short, 2003; Sabherwal & Sabherwal, 2005; Sher & Lee, 2004). Little research is retrievable in the context of facility management. The objective of the proposed BRIM is therefore to provide a theoretical framework of the facility resilience-oriented information system, and the associated disaster resilience building on a facility level.

Becerik-Gerber et al. (2012) conducted a persona and expert interview and summarized a list of FM application areas that can be potentially benefited from BIM, including locating facility components in real-time, facilitating real-time data access, creating and updating digital assets, emergency management, personal training and development, controlling and monitoring energy, checking maintainability, space management, and so on. The potential application areas demonstrated promising and strong capability of BIM in real time locating facility components, communicating with and training facility occupants, and conducting spatial and semantic analysis to support decision making under extreme events.

Despite the potential and demonstrated benefit, BIM is not widely adopted in FM. According to the survey by Becerik-Berber et al. (2012), only 42% respondents who are FM professionals indicated that BIM was currently used in the operation and maintenance stage, and only 40% of the non-BIM users indicated they have the plan to adopt BIM within 2 years.

To build adaptive infrastructures and to increase capacity in facility occupants, an information modeling framework is necessary in order to collect and manage facility information that meet needs of different user groups including facility occupants, facility management team, emergency management officials and partners, and the local community where the facility resides. A specific information model, the Facility Resilience Information Modeling (BRIM), is developed for facility resilience building. It is determined as a three-module system to provide and update facility information, and to generate resilience indicators for facility vulnerability monitoring and response.

STRATEGY #2: COLLECTED INTELLIGENCE TO BRIDGING THE GAP IN A SMART BUILDING SYSTEM

Complexity and urgency keep on increasing related to the community resilience building, especially in a building, a system of physical and social systems (Gilbert, 2010). During a crisis and its aftermath, the process no longer remains on any single function line. An effective response activity includes the real-situation mapping, first aids, evacuation and rescuing, and planning of further recovery efforts (FEMA, 2006). Such activities are most efficiently performed by a network of public agencies and private contractors. The network of public and private forces requires additional efforts from the policy makers to readjust the bids for each agent within the network based on the dynamically changing situation (Engelbart, 1963). Such enormous amount of work for updating the bids and analyzing the situation partly, compared with the limited intelligence during a regional crisis from the public sector, party explains the perceived ineffectiveness of the government agencies in terms of disaster response and recovery, as shown from a series of research on the factors accounting for the disaster preparedness.

Collected intelligence is to optimize solutions to complex situations by enabling community level intelligence base. Unpreparedness in the traditional approach of disaster management is in many cases caused by the limited intelligence of the public agencies that are responsible to evaluate the situation and to manage the disaster response activities in the urgent timeframe and complex situations within a community during a crisis (Shaw and IEDM, 2010). An system is therefore needed so to harvest the crowd intelligence so to meet the needs of dynamic understanding of the resilience level within a building, as well as during a crisis.

The alternated reality game system is therefore designed to capture and integrate intelligence across different resident groups within a community. The goal of such game system is to approach "community-intellect" system, for the purpose of building resilience, the game is to collect human intelligence, and building shared identities via virtual bonding, so to meet the needs of intelligence level to enable best practices in building resilience.

This chapter presents a critical innovation point that integrates collected intelligence into building and community resilience. After introducing the concept of collective intelligence and its usability in the built environment, the needs for such a change in academia and in practice are reviewed. The rest of this section then presents an innovative mechanism in knowledge-based resilience building in the built environment.

Understanding of uncertainties as normal.

STRATEGY #3: BUILDING RESILIENCE INFORMATION MODELING

Strategy #3.1: Record Risks

BRIM integrates facility management information systems including geospatial information, engineering data, architectural data, maintenance and service history, asset inventory, space assignment, and occupant dynamics, together with emergency management systems including meteorological information, warning, evacuation planning, hazardous materials handling, and business continuity. It enables automated information update that links archival facility information with real-time facility management related information by connecting existing paper-based facility documents with real-time service records.

BRIM contains three modules – central repository for facility information, real-time hazard mapping, and resilience indication portal. The central repository serves is the essential component in BRIM. It is the one place for managing all relevant information for the existing facility. Such information can be categorized into the three main types of facility management information, which include the information relevant to the normal operation of a facility and the comfort of its occupants; emergency management information which is typically needed for hazard-related activities, such as preparedness, mitigation, response, and recovery, that improves facility resilience by increasing the adaptivity of a facility and its occupants; and construction management information which include both the historical construction documents and the remodeling and repairing documents involving professional contractors. Such information structure is shown in Figure 1. The central repository defines the building in digital format. It is preliminarily necessary for the second and third modules to operate. It breaks a facility and its occupants down into controllable categories, with geographical information to mapping service requests, and functional attributes to link through different modules.

Strategy #3.2: Public Participation and Acknowledging the Real Time Hazards

Maintenance, repairs, and change of space assignment all contribute to the dynamic nature of facility information (Leite et al., 2011). The second module in BRIM seeks to build up the reference system to enable automated facility information updates within BRIM. Assets will be tagged with remote sensing-able signs to allow auto-tracking. Space assignment, maintenance and service data will be digitalized to enable automated supply into the central repository of BRIM. Such information will be integrated with the as-built 3D models using the proposed 4-D referencing system.

A 4D data referencing system is necessary to allow real-time hazard mapping in facility resilience building. The 4D referencing system is constructed by adding the temporal dimension to the geometric 3-parameter facility model. Figure 2 shows the second module and its interface. Such 4D referencing systems allow standardized

Figure 1. Information integration in BRIM

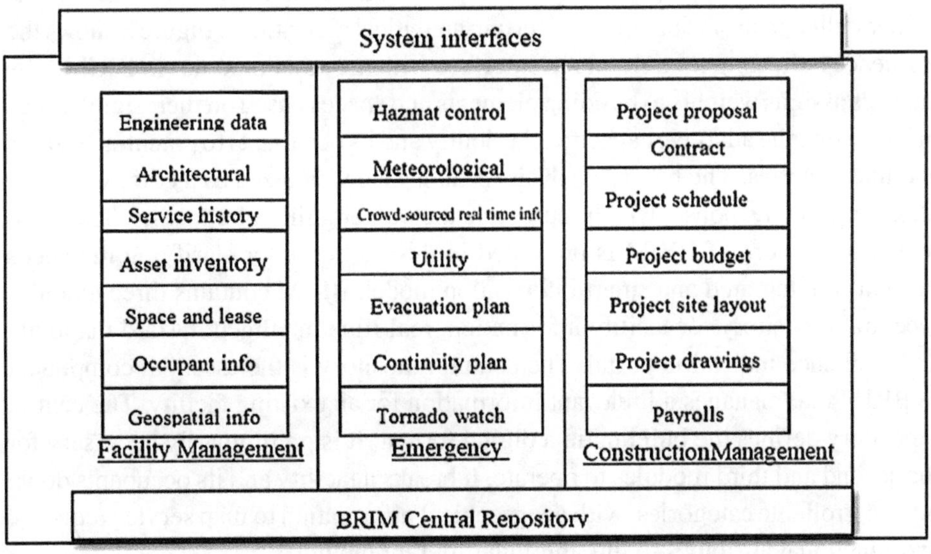

Figure 2. Real-time mapping module in BRIM

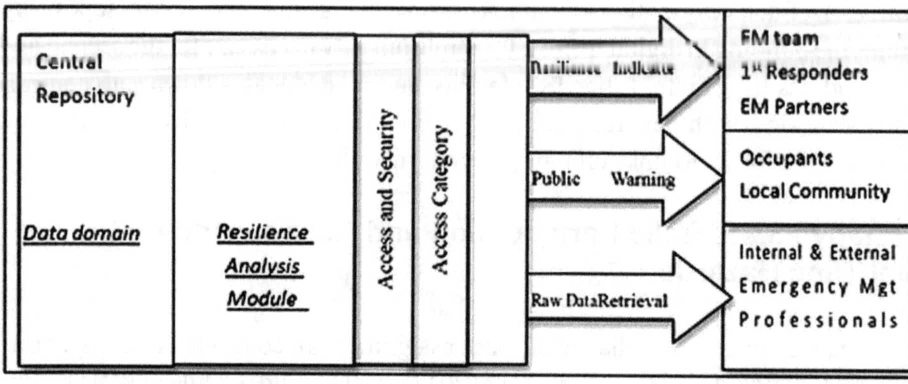

data inputs from multiple users, in this case, the facility management officials as well as the occupants. The real-time mapping module also enables an dynamic 4D information model that provides raw data for decision making during a likely crisis, or for further processing in the third module to generate an standardized resilience indicator (RI), which is easy to communicate and effective especially during a crisis when time and communication devices are limited.

Strategy #3.3: Resilience Building and Education

As the first two modules accomplish integration of existing facility information with real-time service mapping, the third module of BRIM assigns such raw data into the vulnerability analysis domain with certain analytical algorithms. Figure 3 shows the resilience indicating module and its interface with the other modules in BRIM. The analysis assigns weights to building elements and spaces based on their significance in increasing the adaptive capacity of a facility and its occupants to potential hazards and uncertainties. The hazard analysis result is also visualized as a color-coded RI for emergency response, which represents the vulnerability of a facility. The theoretical framework of BRIM is presented in this section after clarifying the needs of such an integrated and smart information model. BRIM contains three modules – central repository for facility information, real-time intelligent hazard mapping, and resilience indication portal. The central repository is the essential component in BRIM that manages all relevant information for an existing facility. The central repository defines the building in a digital format. It is preliminarily necessary for the second and third modules to operate. It breaks a facility and its occupants down into controllable categories, with geographical information to map service requests, and functional attributes to link through different modules.

Figure 3. Resilience indicating module in BRIM

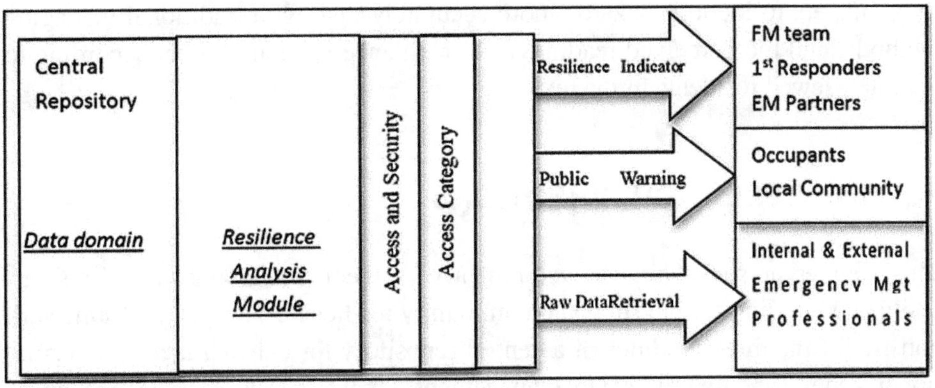

The second module in BRIM seeks to build up the reference system to enable automated facility information updates within BRIM. Assets will be tagged using sensing technology to allow auto-tracking. Space assignment, maintenance and service data will be digitalized to enable automated supply into the central repository of BRIM. Such information will be integrated with the as-built 3D models using a 4-D referencing system.

As the first two modules accomplish integration of existing facility information with real-time service mapping, the third module of BRIM assigns such raw data for vulnerability analysis. Weights will be assigned to building elements and spaces based on their significance in increasing the adaptive capacity of a facility and its occupants to potential hazards and uncertainties. The hazard analysis result is also visualized as a color-coded RI for emergency response, which represents the vulnerability of a facility.

CONCLUSION AND SIGNIFICANCE

BRIM integrates facility management information systems including geospatial information, engineering data, architectural data, maintenance and service history, asset inventory, space assignment, and occupant dynamics, together with emergency management systems including meteorological information, warning, evacuation planning, hazardous materials handling, and business continuity. It enables automated information update that links archival facility information with real-time facility management related information by connecting existing paper-based facility documents with real-time service records. BRIM was implemented and tested in a research building for the validation of the system. The results indicate that the real-time hazard mapping and scaled vulnerability indicator modules enable facility managers to identify hazards more accurately than with traditional managing methods, and the perceived readiness and resilience level are higher according to the interviewed residents in the test.

FUTURE RESEARCH DIRECTIONS

This chapter presents initial development of a theoretical framework for smart resilience building for building and community resilience. The current framework consists of the three modules of a central repository for existing facility information in general and for emergency management, a real-time hazard/service mapping module for automated 4D facility information modeling, and a resilience analysis module to generate a comprehensive analysis of the vulnerability of the facility with a color-coded RI as the final product. The future work will complement the development of a systematic tool to support for linking existing facility information with real-time service/hazard information collected from multiple users including both facility management officials and facility occupants, and carrying out analysis of realistic hazard analysis and resilience indicating algorithm. The ultimate goal is to develop a systematic approach to facility resilience information model design.

Resilience mapping as the further product as the results of the collected intelligence harvested through the BRIM.

Resilience is far beyond disaster resilience. Sustainability, in the future, such as energy performance management tasks and information, can be integrated into the same system as well. Using the same idea, the idea of smart buildings can makee cities smarter and more resilient.

REFERENCES

Berkes, F., & Ross, H. (2013). Community resilience: Toward an integrated approach. *Society & Natural Resources*, *26*(1), 5–20. doi:10.1080/08941920.2012.736605

Ferris, E., Petz, D., & Stark, C. (2013). The year of recurring disasters: a review of natural disasters in 2012. Retrieved from http://www.brookings.edu/~/media/research/files/reports/2013/03/natural%20disasters%20review/brookings_review_natural_disasters_2012.pdf

Lam, P. T., Wong, F. W., & Tse, K. T. (2009). Effectiveness of ICT for construction information exchange among multidisciplinary project teams. *Journal of Computing in Civil Engineering*, *24*(4), 365–376. doi:10.1061/(ASCE)CP.1943-5487.0000038

Magis, K. (2010). Community resilience: An indicator of social sustainability. *Society & Natural Resources*, *23*(5), 401–416. doi:10.1080/08941920903305674

Nakagawa, Y., & Shaw, R. (2004). Social capital: A missing link to disaster recovery. *International Journal of Mass Emergencies and Disasters*, *22*, 5–34.

Norris, F. H., Stevens, S. P., Pfefferbaum, B., Wyche, K., & Pfefferbaum, R. L. (2008). Community resilience as a metaphor, theory, set of capacities, and strategy for disaster readiness. *American Journal of Community Psychology*, *41*, 127–150. PMID:18157631

Rozenfeld, O., Sacks, R., & Rosenfeld, Y. (2009). 'CHASTE': Construction hazard assessment with spatial and temporal exposure. *Construction Management and Economics*, *27*(7), 625–638. doi:10.1080/01446190903002771

Saul, J. L. J. (2004). Facilitating family and community resilience in response to major disaster. Living beyond loss: Death in the family.

Schilderman, T. (2010). Putting people at the centre of reconstruction. In M. Lyons, T. Schilderman, & C. Boano (Eds.), Building back better: delivering people-centered housing reconstruction at scale (pp. 7-38). Warwickshire, UK: Practical Action Publishing Ltd. doi:10.3362/9781780440064.002

Taneja, S., Akcamete, A., Akinci, B., Garrett, J., Soibelman, L., & East, E. W. (2010). Analysis of three indoor localization technologies to support facility management field activities.*Proceedings of the International Conference on Computing in Civil and Building Engineering*, Nottingham, UK.

The Council for Excellence in Government. (2006). Are we ready? Introducing the Public Readiness Index: a survey-based tool to measure the preparedness of individuals, families and communities. Retrieved from http://www.fema.gov/media-library-data/20130726-1910-25045-5489/public_readiness_index.pdf

Trujillo, M., Prdpmez, A., & Hernandez, C. (2000). *Risk-mapping and local capacities: Lessons from Mexico and Central America.* Oxfam Working Papers, Oxfam, Great Britain.

Yang, K., & Callahan, K. (2007). Citizen involvement efforts and bureaucratic responsiveness: Participatory values, stakeholder pressures, and administrative practicality. *Public Administration Review*, *67*(2), 249–264. doi:10.1111/j.1540-6210.2007.00711.x

Chapter 11
Total Quality Management in Smart City Development

Yomi Babatunde
University of the Witwatersrand, South Africa

ABSTRACT

Wastes and pollution are associated with an uncoordinated urbanization trend. This exploratory study investigates total quality management's (TQM) role in smart city development. It delineates smart cities and its conjoined relationship with digital cities. It submits to the complementary relationship between TQM and quality function deployment (QFD) and highlights stakeholder engagement as central to smart city development strategy, underpinned by social theory. It distils that stakeholders and built environment professionals need to work collaboratively to maximize the benefit of smart cities as being an innovation value chain, leading to the use of IT-enabled platform such as building information modelling (BIM). This culminates in the design of an integrative framework with ICT (focusing on BIM) and TQM serving as the information architecture and the ideological premise respectively. It then presents the close-loop (front-end) and open-loop (back-end) approaches to smart city development, discusses future research directions and concludes with implications.

DOI: 10.4018/978-1-5225-0302-6.ch011

INTRODUCTION

Urbanization rate has exceeded 50% in all countries globally and over 70% in various European and Asian countries (Kourtit, Nijkamp, & Arribas, 2012). Smart cities are nurtured by information and communication technology (ICT) (Allwinkle and Cruickshank (2011). Consequently, it is possible to define a smart city based on its ICT, which has been concluded as being a meta-factor in smart city initiatives due to its overarching influence on other factors (Chourabi et al., 2012). The construction industry is infamous for its waste generation, low ICT adoption, and mounting stakeholders' dissatisfaction. Total Quality Management (TQM), being an organization-wide approach, is poised to address this dissatisfaction attributable to construction industry's fragmentation along its supply or value chain (Conti, 2010) to derive the maximum economic benefit (Garcia-Bernal & Garcia-Casarejos, 2014). Conversely, as of date, no study has investigated the role of TQM in developing smart cities. Hence, the aim of this chapter is to investigate the role of TQM in smart city development. Smart cities contribute to sustainable developments through their networked infrastructures (i.e. ICT) and, by so doing, minimize urban waste and pollution (Schaffers et al., 2011). This can also be supported by an IT-enabled construction environment's "ability to develop visual representation of the building and, as a result, manage the construction process by visualizing the process and the product at any time" (Chartered Institute of Building [CIOB], 2014, p. 137). This allows project participants to increase productivity and reduce waste by identifying potential problems earlier and resolving such a problem collaboratively.

While there are socio-cultural, economic, and environmental benefits to smart city development, conceptualizing and implementing a smart city can and do pose challenges to stakeholders (including the clients) and built environment professionals. As such, the term "smart cities" often connotes developments with high ICT component, whereas the conceptualization of real smart cities requires a holistic view of changing cultures and behaviors in addition to providing enabling technology (Allwinkle and Cruickshank, 2011; Deakin & Al Waer, 2011a, b). Smart city development strategies can either be at the national or local level (Angelidou, 2014) while smart city approaches can either be front-end (i.e. business models precede information architecture) or back-end (i.e. information architecture precedes business models) (Kuk & Janssen, 2011). This underscores the need for stakeholder engagement as supported by increasing realization of the role of social theory in the built environment (Goss, 1988; Hillier, 2008). On the flip side, conflicting technical and behavioral variables are a common feature of stakeholder engagement and which can be constructive or destructive in nature (Leung & Yu, 2014). This underscores the use of IT-enabled collaborative construction project environment, unlike the adversarial traditional (i.e. face-to-face) project environment (Xu, Shen,

Fan, Li, & Fan, 2012). As an IT-enabled platform, building information modelling (BIM) plays an important role in information system to supporting a collaborative and integrated construction approach, where all stakeholders edit and/or retrieve information from commonly shared models (Merschbrock & Munkvold, 2012). BIM can, thus, serve as the information architecture to operationalize smart cities' innovation value chain construct (Cosgrave, Aarbuthnot & Tryfonas, 2013). This value chain involves stakeholders (i.e. the non-built environment professionals with vested interests in smart cities) and built environment professionals (i.e. the construction practitioners explicating smart cities).

TQM's role both in public administration (Tomaževič, Seljak & Aristovnik, 2015) and in service organizations (Calabrese & Corbò, 2015) is brought to bear considering BIM's need for changes to well-established processes, organizational roles, contractual practices, and collaborative arrangements. This ensures that a consideration for the well-being of people is ingrained in the strategic decision for smart city development. Consequently, problems and challenges related to urban waste and pollution are addressed from project inception by creating the necessary harmony between the physical city and social city (Hirtle, Richter, Srinivas, & Firth, 2010; Vaughan, 2007). This systemic thinking can, in turn, help to create better urban environmental quality and human well-being for a more sustainable urban development (Panagopoulos, Duque, & Dan, in press). TQM emerged in parallel with environmental consciousness, which means that it is possible (and logical) to construe total quality environmental management (TQEM) (Borri & Boccaletti, 1995). Quality function deployment (QFD) has also emerged in parallel with TQM (Akao & Mazur, 2003) making it possible for the two concepts being able to also work in a conjoined manner so much so that there is the concept of total QFD (TQFD) (Devadasan, Kathiravan, & Thirunavukkarasu, 2006). Consequently, similar to TQEM, QFD can also be applied to environmentally conscious design to derive "QFD for environment" (QFDE) (Masui, Sakao, Kobayashi, &Inaba, 2003). But while there are many benefits of QFD as there are of TQM, including environmental decision-making (Chan & Wu, 2002; Mehrjerdi, 2010), QFD's benefits have been related to the more frequently explored project improvement benefit (Carnevalli & Miguel, 2008).

If TQM and QFD complement each other (as typified by TQFD) and the two concepts have environmental consciousness embedded in them (as also typified by TQEM and QFDE), a plausible research statement thus become how TQM can be adopted to improve smart city development. The foregoing research statement holds because there is no existing study that has investigated the role of TQM in smart city development, despite environmental concern being part of TQM as supported by Ho's (2010) integrated lean TQM framework. Following, the three objectives of this chapter include to: (1) design an integrative framework for ICT and TQM in smart city development for maximum value proposition between stakeholders and built environment professionals; (2) model the front-end and back-end approaches to smart city development for better conceptualization of the current and future environmental impacts; and (3) theorize the importance of TQM in smart city development, based on the preceding two objectives. Fulfilling these three objectives will help to achieve the aim of this study, which is to investigate the role of TQM in smart city development. The nature of the research statement lends this study to being mainly exploratory. Unlike a confirmatory research that gathers data to test a *priori* alternative hypothesis, an exploratory research gathers data to determine whether any interesting *posteriori* hypothesis might be generated (Jaeger and Halliday, 1998). As such, the research methodology adopted is literature reviews (Callahan, 2014 provides critical insights on literature reviews). An integrative literature review has been adopted being a distinctive form of research methodology that generates new knowledge about the topic reviewed (Torraco, 2005).

BACKGROUND

Cities' energy consumption and CO_2 emission generation have been estimated at 75% and 80% respectively. This has led to the European Commission proposing a new city model called "the smart city" (Lazaroiu & Roscia, 2012). The genealogy of the term smart city is subject to debates. It can, plausibly, be construed as having been in existence as a practice before its being treated as a theoretical construct. On the other hand, a plausible proxy for its theoretical construct is its having emerged following Brundtland's (1987) report on sustainable development. Some other authors (Albino, Berandi, & Dangelico, 2015) have opined that the term was first used in the 1990s when the California Institute for Smart Communities focused on the significance of new ICT for modern infrastructures within cities. One of the pioneering studies on Singapore as being a smart city also emerged in the late 1990s (Mahizhnan, 1999). As it has turned out, the original governance-oriented approach to smart cities, which emphasizes the role of social capital and relations in urban development has, however over the years, been diffused with the "urban labelling" phenomenon (Albino et al., 2015). Hence, discussions on smart cities must give credence to Hollands' (2008) rhetorical question *will the real smart city please stand up?* The study asserts that, beyond the prevailing self-labelling and self-congratulatory positions, progressive smart cities begin with the people on mind (as against IT) and balances power between the uses of IT at all levels and economic growth with sustainability (Hollands, 2008).

Immediate subsequent scholarly debates also shifted attention to distinguishing between intelligent cities and smart cities and challenged the unspoken assumption of the corporate branding needs of marketing campaigns (Deakin & Al Waer, 2011a, b). Smart cities have also been viewed as not being mutually exclusive of digital cities as the latter is a fundamental framework for the former, which is also more application and service-oriented (Li, Shan, Shao, Zhou, & Yao, 2013). Despite the delineation, there has yet to be a globally-accepted definition of smart cities. This has led to the submission that smart city's evolution and conceptualization will continue to depend on the local context factors (Neirotti, De Marco, Cagliano, Mangano, &Scorrano, 2014). The reason for the submission can be ascribed to the pervasive application of the term to hard domains (e.g. buildings and energy grids) and soft domains (e.g. education and policy) (Albino et al., 2015). The category of soft domains (i.e. "bandwidth of the mind") is, nonetheless, most important as it is the people, not the hardware and software, who make the IT work and work for the good of other people (Mahizhnan, 1999). It then follows that smart cities can be defined in terms of the six main domains including: (1) natural resources and energy, (2) transport and mobility, (3) buildings, (4) living, (5) government, and (6) economy and people, against which compliance index can then be computed (Neirotti et al., 2014). The characteristics of smart cities, on the other hand, include smart economy, smart mobility, smart environment, smart people, smart living, and smart governance (Albino et al., 2015; Lazaroiu & Roscia, 2012; Lombardi, Giordano, Farouh, & Youssef, 2012; Piro, Cianci, Grieco, Boggia, & Camarda 2014). This study, nevertheless, agrees with the pragmatic definition of smart cities as being an urban environment which, supported by pervasive ICT systems, is able to offer advanced and innovative services to citizens in order to improve the overall quality of their life (Piro et al., 2014). This definition resonates with the more universal and overarching conclusion that descriptions of smart cities now include qualities of people and communities as well as ICTs (Albino et al., 2015).

URBANIZATION CRISIS AND SMART CITY DEVELOPMENT

Urbanization involves the two-way process of multiplication of points of concentration (i.e. physical) and increase in size of individual concentration (i.e. social) until urban maturity is reached (Babatunde, 2014). From the middle Ages in Europe to the Industrial Revolution in 1800s, rural-urban divide has been serving as a catalyst for rural-urban migratory pattern, as people search for better economic opportunities and social benefits that urban areas offer (Washburn & Sindhu, 2010). What has, perhaps, been unprecedented is the notion of "urban hell" (resulting from a massive migration to cities often without jobs thereby promoting the ills of urbanization) associated with the urbanized societies emerging from the beginning of the 19th

century. The concentration of resources and facilities in cities has sparked the overwhelming rural-urban migration and the unanticipated challenges for governance in cities and the beginning of the clamor for smart city initiatives and associated concepts (Perera, Zaslavsky, Christen, & Georgakopoulos, 2014).

Making cities smart can mitigate the problems associated with rapid urbanization (Chourabi et al., 2012). These problems include scarcity of resources, inadequate and deteriorating infrastructure, energy shortages and price instability, global environmental weirding and human health concerns, and demand for better economic opportunities and social benefits (Washburn & Sindhu, 2010). The problems have seen to the urgent call for city's smartness, while non-clarity about what constitutes the smartness has, conversely, seen to the "self-congratulatory" nature that cities make of being smart. And this dichotomy seems to be degenerating the situation that it has become clear that the neo-liberal pathway to "smart(er)" cities through gradual transitioning from an intelligent city to a smart city should replace the neo-conservative and insufficiently progressive image-building and city marketing campaigns (Allwinkle and Cruickshank, 2011). This realization has also necessitated the need for emerging cities to move from being intelligent to being smart (Deakin & Al Waer, 2011a, b), and, possibly, even progressively smarter in order to be truly smart cities of the future (Batty et al. 2012).

Smart city is an urban innovation that is aimed at resolving increasing urbanization and its associated crisis (Goodspeed, 2015). This crisis can also (and should) be addressed from the onset through sustainable construction and waste minimization. Sustainable construction involves "creating construction items using best-practice clean and resource-efficient techniques from the extraction of the raw materials to the demolition and disposal of its components" (Ofori, 2000 cited in Yates, 2013, p.283). A little wonder that while sustainable construction can be promoted through fiscal incentives or penalties as well as regulations (Pitt, Tucker, Riley, & Longden, 2009), its main seven principles include reducing resource consumption, reusing resources, using recyclable resources, protecting nature, eliminating toxics, applying life-cycle costing, and focusing on quality (Kibert, 2008 cited in Yates, 2013, p.283). Beyond being a sociological construct, TQM is also a process construct constituting an integral approach to waste minimization in process industries (Petek & Glavič, 1996) to impacting on manufacturing operations management through its role in developing synergies between operational performance and environmental excellence (Hanna & Newman, 1995). Construction activities span processes and services along the entire supply chain to requiring TQM (Oakland & Marosszeky, 2006) and there exists frameworks for building TQM into construction activities (Low & Teo, 2004; Babatunde & Low, 2015). Against the backgrounds of urbanization, smart city, and TQM, subsequent discussions will culminate in the integrative framework for smart city development, the analysis of the strategic considerations for the different approaches to smart city development, and the modeling of the different approaches.

Built Environment Social Theory in Smart City Development

Smart city development strategy can either be at the national or local level, with each strategy having its unique advantages and disadvantages for new and existing cities. This underscores that irrespective of the level adopted, stakeholder engagement should be employed to better understand and respond to local characteristics, priorities, as well as the needs of the cities, in addition to global market forces and available technology (Angelidou, 2014). The identified role of stakeholder engagement in the creation of smart cities gives credence to the importance of social theory in the built environment; in particular, design-level theories (e.g. "space of space" or spatiality as supported by Hillier's (2008) space syntax paradigm). Space syntax investigates the relationship between human societies and space from the perspective of a general theory of the structure of inhabited space in all its diverse forms (Bafna, 2003). Space syntax asserts that human societies use space as a key and necessary resource in organizing themselves and, by so doing so, the space of inhabitation is configured (Bafna, 2003). Space syntax has taken a unique approach to the concept of intelligibility (i.e. the degree of correlation between connectivity and integration values), defining it as the global structure of an environment from a reading of its local properties (Bafna, 2003). Space syntax relates to automated route guidance systems, which takes cues from human-generated verbal directions on the notion that granularity of spatial information precedes complex directions and eventual wayfinding for people (Hirtle et al., 2010).

The social theory has also been extended to a unified theory of the built environment as being a complex social-ecological system, where multiple-related metabolisms interact at different scales (Moffat & Kohler, 2008). The resultant built environment model allows changes in social, economic, and environmental conditions to be assessed and used for improved understanding of the impact of different managerial and social policies at the macro and micro levels (Moffat & Kohler, 2008). In addition, social theory in the built environment has given rise to the user-centered theory, which is built around the two concepts including: (1) building user's experience, and (2) user-building relationship (Vischer, 2008). The user-centered theory underpins an approach capable of changing conventional ways of planning, constructing, thinking about and occupying buildings and space while also serving as users' feedback into the supply chain (Vischer, 2008). Access to a transparent system for assessing the built environment through the users' experience will generate information that can ultimately affect buildings' economic values. And this corroborates the systems approach view of smart cities as being an innovation value chain (i.e. bottom up Living Lab and top down Innovation District), where users co-create with developers and researchers such that multiple stakeholders are involved (Cosgrave et al., 2013).

Information Technology in Smart City Stakeholder Engagement

With the role of social theory acknowledged, the involvement of multiple stakeholders can help to generate information for addressing some of the identified challenges associated with smart cities. These include interrelations of components and performance indicators (Lombardi et al., 2012), spatial modelling issues affecting accurate simulations for future scenarios (Schmitt, 2013), impact on energy consumption and eventual sustainability (Khansari, Mostashari & Mansouri, 2014), and overall monitoring of initiatives and indices (Marsal-Llacuna, Colomer-Llinàs & Meléndez-Frigola, 2015). Among built environment professionals, IT plays an important role in achieving real-time and transparent information. Unlike the, often, adversarial traditional (i.e. face-to-face) project environment, an IT-led project environment is more collaborative and, thus, essential for the sustainable development of construction organizations in the new century (Xue et al., 2012). This can also be supported by the more effective use of IT system to enhance group decision in stakeholder involvement while implementing value management methodology in construction briefing (Luo, Shen, Fan, & Xue, 2011).

Still in relation to an IT-enabled collaborative project working environment, building information modelling (BIM) has been receiving attention. BIM enables interoperability beyond the technical level to encompass value proposition (i.e. business level) in the built environment (Grilo & Jardim-Goncalves, 2010). And while the project benefits of BIM within the built environment must be considered with respect to BIM implementation challenges (Bryde, Broquetas & Volm, 2013), there are other BIM-specific inter-related causal requirements to fulfil in order to achieve the project business outcome (Fox, 2014). The unfolding and inevitable changing roles of built environment professionals arising from BIM increasing adoption has led to factors for successful collaboration through BIM being surmised as "POWER": Product information sharing, Organizational roles synergy, Work processes coordination, Environment for teamwork, and Reference data collection (Sebastian, 2011). It can then be plausibly construed that BIM, being an integral part of IT in construction, is central to the development of smart cities. In other words, BIM can help to gather and disseminate information from the multiple stakeholders and diverse built environment professionals. There is still a misconception about BIM and its definition, which can be attributed to its still being relatively new; however, it has been surmised that as BIM becomes a mainstream methodology, the formal definition will be less of an issue as professionals start to understand the inherent attributes and characteristics (Aranda-Mena, Crawford, & Froese, 2009).

Total Quality Management for Smart City Administration

Similar to BIM's premise on integration, TQM is the quality-centered management approach of an organization, based on the participation of all its members, which aims at long-term success through customer satisfaction and benefits to all members of the organization and the society as a whole (Harrington, Voehl & Wiggin, 2012). The need for TQM in the construction industry relates to the construction industry's infamous waste generation, low ICT adoption, and mounting stakeholder dissatisfaction. And while the fragmented nature of the construction industry constitutes a major impediment to managing quality along its supply chain, TQM, Lean System, Six Sigma, and Just-In-Time are management techniques that are increasingly being applied (Dahlgaard & Dahlgaard-Park, 2006; Dubey & Singh, 2015).

But being an organization-wide approach, TQM most addresses the fragmentation of the construction industry along the supply chain due to its organization-wide quality focus and systems thinking for value generation (Conti, 2010). TQM, thus, has an impact on customer (or stakeholder) satisfaction, profitability, and productivity (Tanninen, Puumalainen & Sandström, 2010), where improved quality precedes increased productivity (Nebl & Schroeder, 2011). With these benefits, TQM is also increasingly being adapted in public administration (Dobbs, 1994; Matei & Lazar, 2011; Pollit, 2006; Swiss, 1992; Tomaževič et al., 2015). The economic importance of the construction industry (Gang & Low, 2011) has also, in a similar vein, influenced the industry's sustained clamor for TQM implementation (Garcia-Bernal & Garcia-Casarejos, 2014; Harrington et al., 2012).

It is then possible to conclude that ICT (e.g. BIM) and TQM can work in a conjoined fashion to facilitate smart city development strategies (national or local level). This, in turn, must give credence to stakeholder engagement in order to better explicate the systems approach for innovation in the built environment value chain. This helps to achieve the expected or desired value proposition across the built environment's project phases. Figure 1 presents the integrative framework of how ICT can serve as the information architecture linking the stakeholders (i.e. non-built environment professionals) and the built environment professionals. The framework is premised on stakeholder engagement being the basis for smart cities' development strategies, which, in turn, influence ICT. Being a meta-factor, ICT then serves as (or supports) the information architecture for the built environment professionals' project management process groups (Project Management Institute [PMI], 2013) and/or stages of project development (CIOB, 2014). TQM spans through and moderates these social and technical processes.

Figure 1. Integrative framework

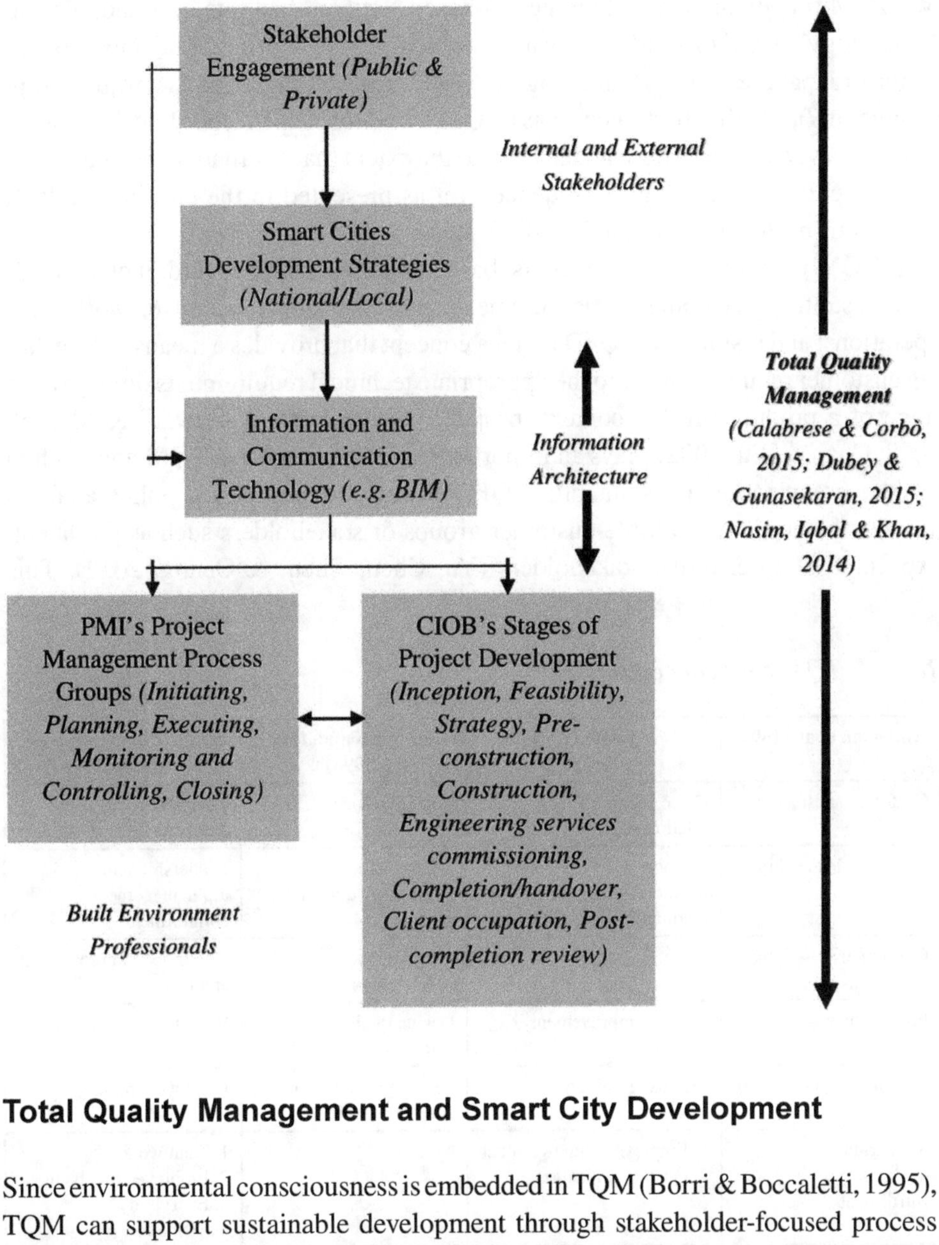

Total Quality Management and Smart City Development

Since environmental consciousness is embedded in TQM (Borri & Boccaletti, 1995), TQM can support sustainable development through stakeholder-focused process management (e.g. Isaksson, 2006) and an integrated management system that is quality-, environment-, as well as health and safety-focused (e.g. Ho, 2010). Notably, since smart city development is premised on and about sustainable development. TQM practices first emerged in Japan before spreading to the US, where it became an established concept (Akao & Mazur, 2003). As such, criteria established in the two TQM frameworks including the US-based National Institute of Standards and Technology's (NIST) Baldrige Award (www.nist.gov/baldrige/) and Japan-based Union of Japanese Scientists and Engineers' (JUSE) Deming Prize (www.juse.or.jp/deming_en/), ideally, best encompass TQM's core concepts. The Baldrige Award has, moreover, become more widespread to the extent that alternative terms adopted by different authors still give credence to it as presented in the ensuing Table 1, adapted from Bou-Llusar et al.'s (2009) study.

If TQM practices or core concepts, based on the Baldrige Award, include leadership; strategy; customers; measurement, analysis, and knowledge; workforce; operations; and results; then QFD being a concept that provides a means of translating customer requirements into the appropriate technical requirements through each stage of a product [in this context, project] development (Carnevalli & Miguel, 2008; Chan & Wu, 2002) plays an important role. Premised on a TQM approach to achieve organizational sustainability, QFD can also help in public policy analysis to meet the needs of multiple-customer groups or stakeholders such as the latent, expectant, and definitive stakeholders (Yu, Chen, Chen, & Chang, 2012). This

Table 1. TQM core concepts

Anderson et al. (1994)	Powell (1995)	Tummala and Tang (1996)	Sila and Ebrahimpour (2002)
Customer satisfaction	Close customer relationship	Customer focus	Customer focus and satisfaction
Visionary leadership	1. Committed leadership 2. Adoption and communication of TQM	1. Leadership 2. Strategic quality planning	Leadership and top management commitment
Continuous improvement	None	Continuous improvement	Continuous improvement and innovation
Process management	Process improvement	Design quality, speed and prevention	Process management
Internal cooperation	Increased training	People participation and partnership	Employee training
3. Learning 4. Employee fulfillment	1. Employee empowerment 2. Open organization	None	1. Teamwork 2. Employee involvement 3. Everybody's participation
External cooperation	1. Benchmarking 2. Closer supplier relationship	None	None
None	Measurement	Fact-based management	Quality information and performance measurement

becomes a viable tool to deal with the multiple stakeholders that are involved in and also impacted by smart city development. The scenario of multiple stakeholders constitutes challenges to smart city innovation (Nam & Pardo, 2011; Naphade, Banavar, Harrison, Paraszczak, & Morris, 2011) so much so that a strategy needs to be developed. A strategy is based on the three key principles of: creating a unique and valuable position involving a different set of activities, making trade-offs in competing activities, and creating fit among the different set of activities (Porter, 2000). In addition to supporting QFD, TQM is also a strategy that supports organizational performance (Terziovski & Samson, 2000). The support can be attributed to its being linked to organizational characteristics, notably degree of innovation (Hoang, Igel & Laosirihongthong, 2010).

TQM processes and outcomes are congruent with motivation, learning, and change in social systems (Hackman & Wagerman, 1995). Once TQM policy is in place, it serves as a platform to then implement a natural environmental quality planning program, such as a more holistic environmentally friendly strategic TQM (Madu, Kuei, & Winokur, 1995). Here again the relationship between TQM, as an all-encompassing technical and social construct, as well as IT, as an enabling meta-factor binding smart city development chain together, comes to the fore; more so, since it has been established that IT in TQM has a significantly lower impact on supplier quality assurance (Ang, Davies, & Finlay, 2001). Smart city development is, in the least, a large complex system involving multiple stakeholders with conflicting needs that QFD, as a part of TQM, purposes to address.

SMART CITY DEVELOPMENT STRATEGIES

While ICT is the meta-factor impacting on all the other factors of a smart city, digital connectivity is not to be confused with social capital or smart city. This is because people and the human capital sides of smartness precede IT, which cannot by itself automatically transform and improve cities (Allwinkle and Cruickshank, 2011). It is this realization that underpins analyzing smart city context through, for example, the holistic smart city intellectual capital perspective (Dameri & Ricciardi, 2015). Ten dimensions have also been identified to deconstruct smart city through the social innovation cross-disciplinary perspective, which also provides a continuum from the mainstream to an emergent and critical transnational version of the smart city (Calzada & Cobo, 2015). In brief, Calzada & Cobo's (2015) ten dimensions include: (1) who (social and digital divide); (2) how (individualism vs collectivism); (3) systems (socio vs technical); (4) governance (master planning vs emergent plan and top down vs bottom up); (5) information (overload vs scarcity); (6) what (social networking vs social capital); (7) space (context collapse – fragmented vs contextualized); (8) design (ambient commons – infrastructure-centered vs people-centered); (9) socio-political processes (control & normative vs free & emergent); and (10) political economy (profitable vs unprofitable).

The human versus IT decision suggests that smart cities can either be achieved by first putting the infrastructure in place and then striving for the business benefits (back-end approach) or letting the business benefits drive the infrastructure or information architecture that undergirds smart cities (front-end approach). The benefits of TQM such as satisfaction, profitability, and productivity are also central to any decision making and, consequently, instrumental to the business model adopted. A business model is derived from the main mission and strategy of a public organization and contains the rationale and elements required to fulfil the mission successfully, including the rationale to generate value (Kuk & Janssen, 2011). It then follows that implicit in development strategies for a smart city is the choice of either putting business model before information architecture (i.e. front-end approach) or vice versa (i.e. back-end approach). The front-end approach has the objective of ensuring demand-driven service delivery, while the objective of the back-end approach is achieving efficiency by processing information quickly at low cost (Kuk & Janssen, 2011). It would then appear, hypothetically, that the private entities would prefer the front-end approach while the public entities would opt for the back-end approach.

And maybe there is the possibility to, somehow, have a hybrid approach. After all, it is important that governments begin to adopt an approach that transcends solely adopting either of the two approaches in favor of developing both business models and supporting information architectures simultaneously (Kuk & Janssen, 2011). This can only be possible when public and private sector actors coordinate their activities and resources on an open innovation platform (e.g. through BIM) aligned to their developmental stage as well as their embedded cultural and social capabilities (Lee, Hancock & Hu, 2014). This possibility can be depicted on Figure 1 (presented earlier) by visualizing two bi-directional relationships: one between smart city development strategies and stakeholder engagement and another between smart city development strategies and ICT. This is where TQM as an organizational-wide approach and QFD as a technique for capturing the voice of the multiple stakeholders can play another important role as discussed earlier. While this is a possible scenario, it can still be construed that conclusion in favor of the front-end approach and back-end approach to smart cities is facilitated by decision matrices (e.g. Branchi, Fernández-Valdivielso & Matias, 2014) based on smart city indicators (e.g. Lazaroiu & Roscia, 2012; Lombardi et al., 2012). How the approach preferred is explicated between the public and private sector actors is beyond the scope of this study.

Conceptualizing a smart city must strategically align the three main dimensions including technology, people, and institutions such that there is integration of infrastructures and technology-mediated services, social learning for strengthening human infrastructure, and governance for institutional improvement and citizen engagement (Nam & Pardo, 2012). These dimensions for smart city conceptualization must align with technology's role as creating forward-looking visions for the city of the future that introduce a whole new level of knowledge management and innovation capabilities in the urban context to either necessitating technology push or demand pull for smart city solutions (Angelidou, 2015). And while a smart city strategy is easier to implement in a case of transforming existing cities, the same cannot be said for transforming existing services of such cities toward smart services to the extent that a smart solution is required (Letaifa, 2015). A smart city development strategy must respond to in-migration from the new millennials and highly-educated in-migrants since smart city's growth arises from population redistribution within the state as against population growth at the state level (Winters, 2011). This accords well with the finding that bigger (and smart) cities attract more skilled workers such that human capital accumulates more quickly in such urban areas (Glaeser & Resseger, 2010). This explains such a city's vibrancy and new business venturing as well as the need for public policies to respond to such (Gross & Campbell, 2015). Figure 1 presented earlier brings these multi-faceted issues together and relates them to the different phases of a construction project, including a smart city development and underscores the role of TQM as a discrete concept or a conjoined concept with QFD, for example.

SOLUTIONS AND RECOMMENDATIONS

Smart cities not only minimize urban waste, they are also important for minimizing urban pollution. The social-ecological system of an urban space development supports that multiple-related metabolisms interact at different scales (Moffat & Kohler, 2008). This means that there is a relationship between activities and urban spaces. As such, cities are both physical (a collection of buildings linked by space) and social (a complex system of human activity linked by interaction). It is out of this seeming segregation between the physical city and social city and, as a result, need for configuration that space syntax has emerged (Vaughan, 2007), albeit debates about its inconsistencies (Hillier & Penn, 2004; Ratti, 2004a, b). This precedes the concept of spatial granularity, defined as a function of the harmony existing between the physical city and social city. Granularity is responsible for spatial information being complicated and "tricky" to the extent that directions and eventual wayfinding for people become difficult (Hirtle et al., 2010). Granularity also contributes to an urban space being aggressive and unnatural for people due to pollution (mainly, air and noise). Air being among the most important environmental factors in the relationship between environmental qualities in inhabited areas and population health and eventual welfare.

Green spaces (in theory and in practice) contribute to air quality and this needs to be monitored for pollution. A multi-disciplinary approach for attaining a holistic understanding of urban environmental quality and human well-being in sustainable urban development has also been underscored (Panagopoulos et al., 2015). In the absence of such a harmony, it is difficult to sustain urbanization. Sustainable urbanization has been operationalized as being both SMART (specific, measurable, achievable, relevant, and time-bound) and possessing CLASS (cultural-acceptability, livability, affordability, sociability, and scalability) (Babatunde, 2014). The conjoined concept subsequently referred to as SMART CLASS views urbanization as a biological cell that can self-renew, subject to its development strategies. As discussed earlier under the section on urban crisis and smart city development, spaces have, erstwhile, been categorized as being rural or urban. Subsequent increasing (and unabated) urbanization now necessitates urbanized societies to move from being intelligent to being smart cities in order to address the growing urbanization crisis. This pressure, coupled with the applications or lack of the other considerations discussed in the preceding sections, has nurtured the "self-congratulatory" neo-conservative view of smart cities and the neo-liberal pathway to "smart(er)" cities at different extents.

It is then possible to have the two scenarios of smart city development strategies as being close-loop (hypothetically, based on a local strategy and/or front-end approach) or open-loop (hypothetically, based on national strategy and/or back-end approach) as presented in Figures 2 and 3 respectively. Figure 2 depicts the more prevalent neo-conservative view of smart cities in which urban hell has necessitated immediate strategies for developing smart cities and smarter cities, as the former seemed failing to adequately alleviate the pains of urbanization. The more entrepreneurial outlook of smart city development as typified by Figure 2 does not align with smart cities being a complex and dynamic organism (Zubizarreta, Seravalli, & Arrizabalaga, in press). In particular, the socio-ecological systems are becoming blighted areas (characterized by pollution and waste) rather than integrated green spaces of the intelligent and/or smart cities (Pickett et al., 2011; Yli-Pelkonen & Niemela, 2005). Conversely, the smart and smarter cities of Figure 3 harmonize social city (i.e. people) and physical city (i.e. activities) as is now being increasingly advocated (Batty et al. 2012; Chourabi et al., 2012; Yigitcanlar & Teriman, 2015).

Holistically conceptualized, smart cities can be more inclusive; thus, contribute to sustainable development. In the sphere of public administration, stakeholder engagement should precede development strategies for smart cities. At the level of the built environment professionals, a collaborative working environment should

Figure 2. Close-loop Smart city development strategy, adapted from Babatunde (2014)

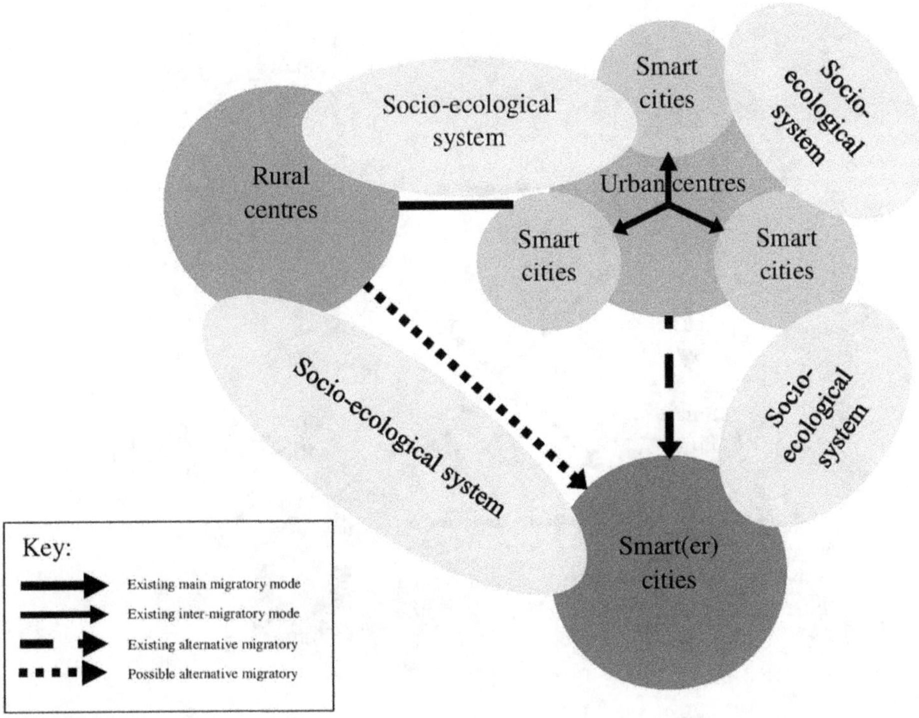

also be developed. Results of stakeholder engagement and collaborative working environment should determine the information architecture for smart cities. (This is the back-end approach.) TQM, as a management technique premised on organization-wide approach to ensuring quality, will then ensure consistent innovation and organizational performance among stakeholders, built environment professionals as well as the ICT binding them together as presented in Figure 1. Figures 2 and 3 contextualize the impacts of the close-loop and open-loop smart city development strategies. The two figures (i.e., Figures 2 and 3) have been modelled to serve as bases for discussions among stakeholders and built environment professionals for the necessary immediate, yet holistic, actions to make the now mostly intelligent cities to be smart cities and progressively smarter cities.

FUTURE RESEARCH DIRECTIONS

In the now digitalized world, the smartness of cities should be an integral part of any building development process. Having said, this should be addressed from the

Figure 3. Open-loop Smart city development strategy, adapted from Babatunde (2014)

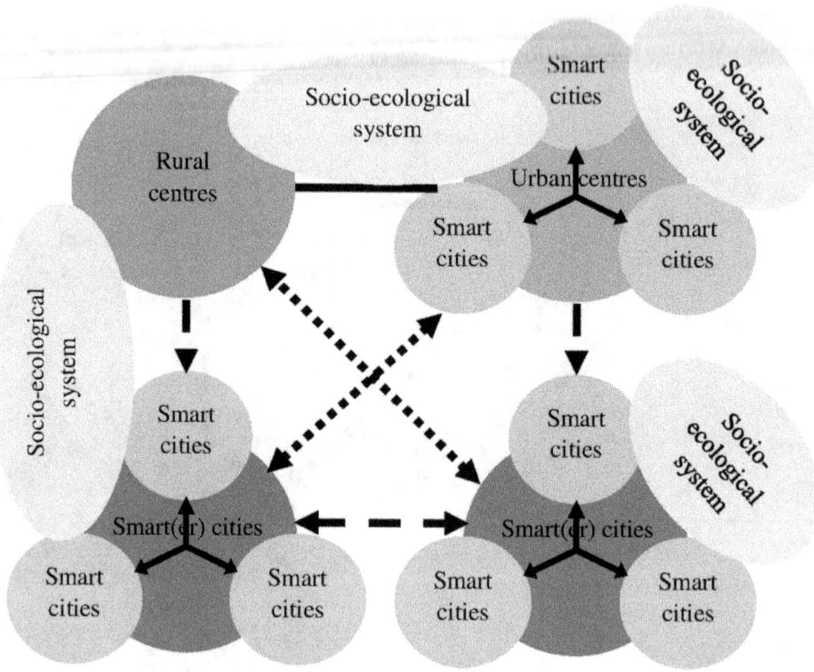

master planning phase such that subsequent developments can then readily fit in (i.e. open-loop approach). However, this can only be achieved for green field projects. Most of the existing cities are now built up, meaning that strategic choices need to be made on the infrastructures that should be provided to make them smart (or smarter). Transforming existing services of transformed cities to become smart is a bigger challenge than transforming such cities in the first instance (Letaifa, 2015). This is a case where the front-end approach to smart city development will tend to dominate. As such, in view of the now more than 50% urbanization rate globally, future research directions should focus on how to address the prevailing urbanization crisis in a way that is sustainable, socially, economically, environmentally, and culturally. The focus should be on "transforming existing cities and their services" to become smart cities rather than taking the entrepreneurial outlook of "branding new cities" as being smart. The branding approach is exclusive and tend to undermine the more collaborative and consensus-building aspirations of the networking paradigm (Allwinkle & Cruickshank, 2011) such as that of BIM and TQM.

This chapter has attempted to apply TQM to the development of smart cities. This has been done at the conceptual level to better appreciate the linkage of the different considerations for smart cities. In reality, TQM is a system theory thinking that pervades all activities of any progressive economy. TQM precedes stakeholder engagement, ICT (e.g. BIM), and the operations of the built environment professionals. TQM also applies to the activities of the public and private sector players. However, the challenges that could be encountered when implementing TQM and BIM as applicable to the development of smart cities are beyond the scope of this study. Future research projects should look into the challenges and validate, empirically, if TQM can indeed be applied to smart cities. To corroborate, cities now operate in a dynamic and competitive world to the extent that they have to exploit their indigenous strength in order to be distinct ("centrifugal") while, nonetheless, learning from each other ("centripetal") (Kourtit et al., 2012). And while geometry and spatial issues have changed over the centuries, they are still the foundations for urban design decisions and will continue to be underlying properties of existing and future smart cities (Schmitt, 2013). Smart city development is about sustaining the environment by minimizing life cycle waste; as such, premised on sustainable construction (Yates, 2013) as well as strategic TQM (Madu et al., 1995; Petek & Glavič, 1996).

CONCLUSION

The aim of this chapter was to investigate the role of TQM in smart city development. The first objective of this chapter was to design an integrative framework for ICT and TQM in smart city development for maximum value proposition between stakeholders and built environment professionals. Figure 1 fulfils this objective. The second objective was to model the front-end and back-end approaches to smart city development for better conceptualization of the current and future environmental impacts. Figures 2 and 3 fulfil this objective. The last objective of this study was to theorize the importance of TQM in smart city development. First, this study has, hypothetically, put it that private entities would prefer the front-end approach of smart city development (Figure 2) while public entities would prefer the back-end approach of smart city development (Figure 3). A plausible theory thus become that the preferred approach is influenced by TQM maturity of the public or private entities (e.g. as being fad, fashion or fit (Wiele, William & Dale, 2000)) and stakeholder type (e.g. as being latent, expectant, or definitive (Yu et al., 2012)). This study has also discussed possible decision matrices and smart city indicators that can be used to validate these two propositions both in theory and in practice. This is another area for a future larger research project. It should be noted that construction industry's waste generation and associated pollution can be attributed to its low ICT adoption and TQM adoption. Properly-conceived smart cities can mitigate waste generation and pollution. This chapter has brought to the fore some considerations to contribute to the emerging research on smart cities as a solution for reducing urban waste and pollution. There are misconceptions on the concept of smart cities as Washburn & Sindhu (2009) have rightly noted among the top management personnel. TQM, on the other hand, has been around longer, making it a more familiar concept among the top management personnel as supported by Porter (2000). This finding is a managerial implication of this study. In addition, an attempt has been made in this study to relate TQM to smart city development, albeit more conceptually. However, this suffices considering the exploratory nature of this study. Being a technique used to capture the diverse needs of the multiple stakeholders, QFD complements TQM (Devadasan et al., 2006). Environmental consciousness is embedded in TQM (Borri & Boccaletti, 1995; Ho, 2010; Isaksson, 2006) while QFD can also be applied to environmentally conscious design (Masui et al., 2003). These findings have added to the existing conceptual understanding of smart city development strategies. With the increasing trend towards smart city development, there are definitely policy implications that TQM can help to achieve. In closing, there are indeed more questions than answers when it comes to applying TQM to smart city development. This study, more than being conclusive, has opened up different areas for subsequent discussions on smart city development and TQM ranging from short-term to long-term considerations.

REFERENCES

Akao, Y., & Mazur, G. H. (2003). The leading edge in QFD: Past, present and future. *International Journal of Quality & Reliability Management*, *20*(1), 20–35. doi:10.1108/02656710310453791

Albino, V., Berardi, U., & Dangelico, R. M. (2015). Smart Cities: Definitions, Dimensions, Performance, and Initiatives. *Journal of Urban Technology*, *22*(1), 3–21. doi:10.1080/10630732.2014.942092

Allwinkle, S., & Cruickshank, P. (2011). Creating Smart-er Cities: An Overview. *Journal of Urban Technology*, *18*(2), 1–16. doi:10.1080/10630732.2011.601103

Ang, C.-L., Davies, M., & Finlay, P. N. (2001). An empirical study of the use of information technology to support total quality management. *Total Quality Management*, *12*(2), 145–157. doi:10.1080/09544120120011370

Angelidou, M. (2014). Smart city policies: A spatial approach. *Cities (London, England)*, *41*, 3–11. doi:10.1016/j.cities.2014.06.007

Angelidou, M. (2015). A conjucture of four forces. *Cities (London, England)*, *47*, 95–106. doi:10.1016/j.cities.2015.05.004

Aranda-Mena, G., Crawford, J., Chevez, A., & Froese, T. (2009). Building information modelling demystified: Does it make business sense to adopt BIM? *International Journal of Managing Projects in Business*, *2*(3), 419–434. doi:10.1108/17538370910971063

Babatunde, O. K. (2014). Sustaining Urbanization in a Developing Economy: A Case Study of Nigeria. In H. J. Kauffman (Ed.), *Urbanization: Global Trends, Roles of Climate Change and Effects on Biodiversity* (pp. 75–143). New York: Nova Science Publishers, Inc.

Babatunde, Y., & Sui Pheng, L. (2015). TQM implementation through ISO 9001: Findings from Chinese construction firms in Nigeria. *The TQM Journal*, *27*(6), 671–682. doi:10.1108/TQM-06-2015-0076

Bafna, S. (2003). Space Syntax: A Brief Introduction to Its Logic and Analytical Techniques. *Environment and Behavior*, *35*(1), 17–29. doi:10.1177/0013916502238863

Batty, M., Axhausen, K. W., Giannotti, F., Pozdnoukhov, A., Bazzani, A., Wachowicz, M., & Portugali, Y. et al. (2012). Smart cities of the future. *The European Physical Journal. Special Topics*, *214*(1), 481–518. doi:10.1140/epjst/e2012-01703-3

Borri, F., & Boccaletti, G. (1995). From total quality management to total quality environmental management. *The TQM Magazine*, *7*(5), 38–42. doi:10.1108/09544789510098614

Branchi, P. E., Fernández-Valdivielso, C., & Matias, I. R. (2014). Analysis Matrix for Smart Cities. *Future Internet*, *6*(1), 61–75. doi:10.3390/fi6010061

Brundtland, G. H. (1987). *Report of the World Commission on Environment and Development: Our Common Future*. Oslo: United Nations Publications.

Bryde, D., Broquetas, M., & Volm, J. M. (2013). The project benefits of Building Information Modelling (BIM). *International Journal of Project Management*, *31*(7), 971–980. doi:10.1016/j.ijproman.2012.12.001

Calabrese, A., & Corbò, M. (2015). Design and blueprinting for total quality management implementation in service organizations. *Total Quality Management & Business Excellence*, *26*(7-8), 719–732. doi:10.1080/14783363.2014.881970

Callahan, J. L. (2014). Writing Literature Reviews: A Reprise and Update. *Human Resource Development Review*, *13*(3), 271–275. doi:10.1177/1534484314536705

Calzada, I., & Cobo, C. (2015). Unplugging: Deconstructing the Smart City. *Journal of Urban Technology*, *22*(1), 23–43. doi:10.1080/10630732.2014.971535

Carnevalli, J. A., & Miguel, P. C. (2008). Review, analysis and classification of the literature on QFD – Types of research, difficulties and benefits. *International Journal of Production Economics*, *114*(2), 737–754. doi:10.1016/j.ijpe.2008.03.006

Chan, L.-K., & Wu, M.-L. (2002). Quality function deployment: A literature review. *European Journal of Operational Research*, *143*(3), 463–497. doi:10.1016/S0377-2217(02)00178-9

Chartered Institute of Building. (2014). *Code of Practice for Project Management for Construction and Development* (5th ed.). Oxford, UK: Wiley-Blackwell.

Chourabi, H., Gil-Garcia, J. R., Pardo, T. A., Nam, T., Mellouli, S., Scholl, H. J., . . . Nahon, K. (2012). Understanding smart cities: An integrative framework. *Proceedings of the45th Hawaii International Conference onSystem Science (HICSS)* (pp. 2289-2297). Washington DC. NY: IEEE Computer Society.

Conti, T. (2010). Systems thinking in quality management. *The TQM Journal*, *22*(4), 352–368. doi:10.1108/17542731011053280

Cosgrave, E., Arbuthnot, K., & Tryfonas, T. (2013). Living Labs, Innovation Districts and Innovation Market Places: A Systems Approach for Smart Cities. *Procedia Computer Science*, *16*, 668–677. doi:10.1016/j.procs.2013.01.070

Dahlgaard, J. J., & Dahlgaard-Park, S. M. (2006). Lean production, six sigma quality, TQM, and company culture. *The TQM Magazine*, *18*(3), 263–281. doi:10.1108/09544780610659998

Dameri, R. P., & Ricciardi, F. (2015). Smart city intellectual capital: An emerging view of territorial systems innovation management. *Journal of Intellectual Capital*, *16*(4), 860–887. doi:10.1108/JIC-02-2015-0018

Deakin, M., & Al Waer, H. (2011a). From intelligent to smart cities. *Intelligent Buildings International*, *3*(1), 133–139. doi:10.1080/17508975.2011.586673

Devadasan, S. R., Kathiravan, N., & Thirunavukkarasu, V. (2006). Theory and practice of total quality function deployment: A perspective from traditional pump-manufacturing environment. *The TQM Magazine*, *18*(2), 143–161. doi:10.1108/09544780610647865

Dobbs, M. F. (1994). Continuous Improvement as Continuous Implementation: Implementing TQM in the City of Santa Ana. *Public Productivity & Management Review*, *18*(1), 89–100. doi:10.2307/3380697

Dubey, R., & Gunasekaran, A. (2015). Exploring soft TQM dimensions and their impact on firm performance: Some exploratory empirical results. *International Journal of Production Research, 53*(2), 371–382. doi:10.1080/00207543.2014.933909

Dubey, R., & Singh, T. (2015). Understanding complex relationship among JIT, lean behaviour, TQM and their antecedents using interpretive structural modelling and fuzzy MICMAC analysis. *The TQM Journal, 27*(1), 42–62. doi:10.1108/TQM-09-2013-0108

Fox, S. (2014). Getting real about BIM: Critical realist descriptions as an alternative to the naïve framing and multiple fallacies of hype. *International Journal of Managing Projects in Business, 7*(3), 405–422. doi:10.1108/IJMPB-12-2013-0073

Garcia-Bernal, J., & Garcia-Casarejos, N. (2014). Economic analysis of TQM adoption in the construction sector. *Total Quality Management & Business Excellence, 25*(3), 209–221. doi:10.1080/14783363.2012.728848

Giang, D. T. H., & Sui Pheng, L. (2011). Role of construction in economic development: Review of key concepts in the past 40 years. *Habitat International, 35*(1), 118–125. doi:10.1016/j.habitatint.2010.06.003

Glaeser, E. L., & Resseger, M. G. (2010). The complementarity between cities and skills. *Journal of Regional Science, 50*(1), 221–244. doi:10.1111/j.1467-9787.2009.00635.x

Goodspeed, R. (2015). Smart cities: Moving beyond urban cybernetics to tackle wicked problems. *Revue Economique et Sociale, 8*(1), 79–92.

Goss, J. (1988). The Built Environment and Social Theory: Towards an Architectural Geography. *The Professional Geographer, 40*(4), 392–403. doi:10.1111/j.0033-0124.1988.00392.x

Grilo, A., & Jardim-Goncalves, R. (2010). Value proposition on interoperability of BIM and collaborative working environments. *Automation in Construction, 19*(5), 522–530. doi:10.1016/j.autcon.2009.11.003

Gross, S., & Campbell, N. (2015). Central city vibrancy and new business venturing. *Journal of Entrepreneurship and Public Policy, 4*(2), 257–271. doi:10.1108/JEPP-01-2014-0002

Hackman, R., & Wageman, R. (1995). Total Quality Management: Empirical, Conceptual, and Practical Issues. *Administrative Science Quarterly, 40*(2), 309–342. doi:10.2307/2393640

Hanna, M. D., & Newman, W. R. (1995). Operations and environment an expanded focus for TQM. *International Journal of Quality & Reliability Management, 12*(5), 38–53. doi:10.1108/02656719510089984

Harrington, H. J., Voehl, F., & Wiggin, H. (2012). Applying TQM to the construction industry. *The TQM Journal, 24*(4), 352–362. doi:10.1108/17542731211247373

Hillier, B. (2008). Space and spatiality: What the built environment needs from social theory. *Building Research and Information, 36*(3), 216–230. doi:10.1080/09613210801928073

Hillier, B., & Penn, A. (2004). Rejoinder to Carlo Ratti. *Environment and Planning. B, Planning & Design, 31*(4), 501–511. doi:10.1068/b3019a

Hirtle, S., Richter, K.-F., Srinivas, S., & Firth, R. (2010). This is the tricky part: When directions become difficult. *Journal of Spatial Information Science, 1*, 53–73.

Ho, S. K. M. (2010). Integrated lean TQM model for sustainable development. *The TQM Journal, 22*(6), 583–593. doi:10.1108/17542731011085294

Hoang, D. T., Igel, B., & Laosirihongthong, T. (2010). Total quality management (TQM) strategy and organizational characteristics: Evidence from recent WTO member. *Total Quality Management & Business Excellence, 21*(9), 931–951. doi: 10.1080/14783363.2010.487680

Hollands, R.G. (2008). Will the real smart city please stand up? *City: analysis of urban trends, culture, theory, policy, action*, 12(3), 303-320.

Isaksson, R. (2006). Total quality management for sustainable development – Process based system models. *Business Process Management Journal, 12*(5), 632–645. doi:10.1108/14637150610691046

Jaeger, R. G., & Halliday, T. R. (1998). On Confirmatory versus Exploratory Research. *Herpetologica, 54*(Suppl.), S64–S66.

Khansari, N., Mostashari, A., & Mansouri, M. (2014). Conceptual Modelling of the Impact of Smart Cities on Household Energy Consumption. *Procedia Computer Science, 28*, 81–86. doi:10.1016/j.procs.2014.03.011

Kourtit, K., Nijkamp, P., & Arribas, D. (2012). Smart cities in perspective – a comparative European study by means of self-organizing maps. *Innovation: The Journal of Social Science Research, 25*(2), 229–246.

Kuk, G., & Janssen, M. (2011). The Business Models and Information Architectures of Smart Cities. *Journal of Urban Technology*, *18*(2), 39–52. doi:10.1080/10630732.2011.601109

Lazaroiu, G. C., & Roscia, M. (2012). Definition methodology for the smart cities model. *Energy*, *47*(1), 326–332. doi:10.1016/j.energy.2012.09.028

Lee, J. H., Hancock, M. G., & Hu, M.-C. (2014). Towards an effective framework for building smart cities: Lessons from Seoul and San Francisco. *Technological Forecasting and Social Change*, *89*, 80–99. doi:10.1016/j.techfore.2013.08.033

Lertaifa, S. B. (2015). How to strategize smart cities: Revealing the SMART model. *Journal of Business Research*, *68*(7), 1414–1419. doi:10.1016/j.jbusres.2015.01.024

Leung, M.-Y., & Yu, J. (2014). Value methodology in public engagement for construction development projects. *Built Environment Project and Asset Management*, *4*(1), 55–70. doi:10.1108/BEPAM-05-2012-0033

Li, D., Shan, J., Shao, Z., Zhou, X., & Yao, Y. (2013). Geomatics for smart cities – concept key techniques, and applications. *Geo-spatial. Information Science*, *16*(1), 13–24.

Lombardi, P., Giordano, S., Farouh, H., & Yousef, W. (2012). Modelling the smart city performance. *Innovation – The European Journal of Social Science Research*, 25(2), 137-149.

Low, S. P., & Teo, J. A. (2004). Implementing Total Quality Management in Construction Firms. *Journal of Management Engineering*, *20*(1), 8–15. doi:10.1061/(ASCE)0742-597X(2004)20:1(8)

Luo, X., Shen, G. Q., Fan, S., & Xue, X. (2011). A group decision support system for implementing value management methodology in construction briefing. *International Journal of Project Management*, *29*(8), 1003–1017. doi:10.1016/j.ijproman.2010.11.003

Madu, C. N., Kuei, C.-H., & Winokur, D. (1995). A strategic total quality management (STQM) approach. *Futures*, *27*(8), 839–856. doi:10.1016/0016-3287(95)00048-2

Mahizhnan, A. (1999). Smart cities: The Singapore case. *Cities (London, England)*, *16*(1), 13–18. doi:10.1016/S0264-2751(98)00050-X

Marsal-Llacuna, M.-L., Colomer-Llinàs, J., & Meléndez-Frigola, J. (2015). Lessons in urban monitoring taken from sustainable and livable cities to better address the Smart Cities initiative. *Technological Forecasting and Social Change*, *90*, 611–622. doi:10.1016/j.techfore.2014.01.012

Masui, K., Sakao, T., Kobayashi, M., & Inaba, A. (2003). Applying Quality Function Deployment to environmentally conscious design. *International Journal of Quality & Reliability Management*, *20*(1), 90–106. doi:10.1108/02656710310453836

Matei, L., & Lazar, C.-G. (2011). Quality Management and the Reform of Public Administration in Several States in South-Eastern Europe. Comparative Analysis. *Theoretical and Applied Economics*, *4*(557), 65–98.

Mehrjerdi, Y. Z. (2010). Applications and extensions of quality function deployment. *Assembly Automation*, *30*(4), 388–403. doi:10.1108/01445151011075843

Merschbrock, C., & Munkvold, B. E. (2012). A Research Review on Building Information Modeling in Construction – An Area Ripe for IS Research. *Communications of the Association for Information Systems*, *31*(10), 207–228.

Moffat, S., & Kohler, N. (2008). Conceptualizing the built environment as a social-ecological system. *Building Research and Information*, *36*(3), 248–268. doi:10.1080/09613210801928131

Nam, T., & Pardo, T. A. (2011). Smart City as Urban Innovation: Focusing on Management, Policy, and Context.*Proceedings of the 5th International Conference on Theory and Practice of Electronic Governance (ICEGOV),*Tallinn, Estonia (pp. 185-194). ACM. doi:10.1145/2072069.2072100

Nam, T., & Pardo, T. A. (2012). Conceptualizing Smart City with Dimensions of Technology, People, and Institutions.*Proceedings of the 12th Annual International Conference on Digital Government Research,*College Park, MD (pp.282-291).

Naphade, M., Banavar, G., Harrison, C., Paraszczak, J., & Morris, R. (2011). Smarter Cities and Their Innovation Challenges. *Computer*, *44*(6), 32–39. doi:10.1109/MC.2011.187

Nasim, K., Iqbal, M. Z., & Khan, I. A. (2014). Antecedents of TQM implementation capability: A review with conceptual. *Total Quality Management & Business Excellence*, *25*(12), 1395–1409. doi:10.1080/14783363.2013.807682

Nebl, T., & Schroeder, A.-K. (2011). Understanding the interdependencies of quality problems and productivity. *The TQM Journal*, *23*(5), 480–495. doi:10.1108/17542731111157590

Neirotti, P., De Marco, A., Cagliano, A. C., Mangano, G., & Scorrano, F. (2014). Current trends in Smart City initiatives: Some stylised facts. *Cities (London, England)*, *38*, 25–36. doi:10.1016/j.cities.2013.12.010

Oakland, J., & Marosszeky, M. (2006). *Total Quality in the Construction Supply Chain*. Burlington, MA: Butterworth-Heinemann (An Imprint of Elsevier).

Panagopoulos, T., Duque, J. A. G., & Dan, M. B. (in press). Urban planning with respect to environmental quality and human well-being. *Environmental Pollution*. Retrieved from http://www.sciencedirect.com/science/article/pii/S0269749115003711

Perera, C., Zaslavsky, A., Christen, P., & Georgakopoulos, D. (2014). Sensing as a service model for smart cities supported by Internet of Things. *Transactions on Emerging Telecommunications Technologies*, *25*(1), 81–93. doi:10.1002/ett.2704

Petek, J., & Glavič, P. (1996). An integral approach to waste minimization in process industries. *Resources, Conservation and Recycling*, *17*(3), 169–188. doi:10.1016/0921-3449(96)01151-2

Pickett, S. T. A., Cadenasso, M. L., Grove, J. M., Boone, C. G., Groffman, P. M., Irwin, E., & Warren, P. et al. (2011). Urban ecological systems: Scientific foundations and a decade of progress. *Journal of Environmental Management*, *92*(3), 331–362. doi:10.1016/j.jenvman.2010.08.022 PMID:20965643

Piro, G., Cianci, I., Grieco, L. A., Boggia, G., & Camarda, P. (2014). Information centric services in Smart Cities. *Journal of Systems and Software*, *88*, 169–188. doi:10.1016/j.jss.2013.10.029

Pitt, M., Tucker, M., Riley, M., & Longden, J. (2009). Towards sustainable construction: Promotion and best practices. *Construction Innovation*, *9*(2), 1471–4175. doi:10.1108/14714170910950830

Pollit, C. (2006). Changing European states, changing public administration: Anti-statist reforms and new administrative directions: Public administration in the United Kingdom. In E. Otenyo & N. S. Lind (Eds.), *Comparative Public Administration: The Essential Readings* (pp. 777–792). Oxford, UK: Jai Press (an imprint of Elsevier). doi:10.1016/S0732-1317(06)15036-8

Porter, M. E. (2000). What is Strategy? *Harvard Business Review*, 2000, 4134.

Project Management Institute. (2013). *A Guide to the Project Management Body of Knowledge (PMBOK Guide)* (5th ed.). USA: Project Management Institute.

Ratti, C. (2004a). Space syntax: Some inconsistencies. *Environment and Planning. B, Planning & Design*, *31*(4), 487–499. doi:10.1068/b3019

Ratti, C. (2004b). Rejoinder to Hillier and Penn. *Environment and Planning. B, Planning & Design*, *31*(4), 513–516. doi:10.1068/b3019b

Schaffers, H., Komninos, N., Pallot, M., Trousse, B., Nilsson, M., & Oliveira, A. (2011). Smart Cities and the Future Internet: Towards Cooperation Frameworks for Open Innovation. *Future Internet Assembly, 6656*, 431–446. doi:10.1007/978-3-642-20898-0_31

Schmitt, G. (2013). Spatial modelling issues in future smart cities. *Geo-spatial. Information Science, 16*(1), 7–12.

Sebastian, R. (2011). Changing roles of the clients, architects and contractors through BIM. *Engineering, Construction, and Architectural Management, 18*(2), 176–187. doi:10.1108/09699981111111148

Swiss, J. E. (1992). Adapting Total Quality Management (TQM) to Government. *Public Administration Review, 52*(4), 356–362. doi:10.2307/3110395

Tanninen, K., Puumalainen, K., & Sandström, J. (2010). The power of TQM: Analysis of its effects on profitability, productivity and customer satisfaction. *Total Quality Management & Business Excellence, 21*(2), 171–184. doi:10.1080/14783360903549949

Terziovski, M., & Samson, D. (2000). The effect of company size on the relationship between TQM strategy and organizational performance. *The TQM Magazine, 12*(2), 144–149. doi:10.1108/09544780010318406

Tomaževič, N., Seljak, J., & Aristovnik, A. (2015). *TQM in public administration organizations: an application of data envelopment analysis in the police service.* Total Quality Management & Business Excellence; doi:10.1080/14783363.2015.1007861

Torraco, R. J. (2005). Writing Integrative Literature Reviews: Guidelines and Examples. *Human Resource Development Review, 4*(3), 356–367. doi:10.1177/1534484305278283

Vaughan, L. (2007). The spatial syntax of urban segregation. *Progress in Planning, 27*(3), 205–294. doi:10.1016/j.progress.2007.03.001

Vischer, J. C. (2008). Towards a user-centred theory of the built environment. *Building Research and Information, 36*(3), 231–240. doi:10.1080/09613210801936472

Washburn, D., & Sindhu, U. (2010). Helping CIOs Understand "Smart City" Initiatives. *Growth*, 2010, 17.

Wiele, A., van der, Williams, A.R.T., & Dale, V.G. (2000). Total quality management: Is it a fad, fashion, or fit. *Quality Management Journal, 7*(2), 65–79.

Winters, J. V. (2011). Why are smart cities growing? Who moves and who stays. *Journal of Regional Science, 51*(2), 253–270. doi:10.1111/j.1467-9787.2010.00693.x

Xu, X., Shen, Q., Fan, H., Li, H., & Fan, S. (2012). IT supported collaborative work in A/E/C projects: A ten-year review. *Automation in Construction, 21*, 1–9. doi:10.1016/j.autcon.2011.05.016

Yates, J. K. (2013). Sustainable methods for waste minimization in construction. *Construction Innovation, 13*(3), 281–301. doi:10.1108/CI-Nov-2011-0054

Yigitcanlar, T., & Teriman, S. (2015). Rethinking sustainable urban development: Towards an integrated planning and development process. *International Journal of Environmental Science and Technology, 12*(1), 341–352. doi:10.1007/s13762-013-0491-x

Yli-Pelkonen, V., & Niemelä, J. (2005). Linking ecological and social systems in cities: Urban planning in Finland as a case. *Biodiversity and Conservation, 14*(8), 1947–1967. doi:10.1007/s10531-004-2124-7

Yu, C.-H., Chen, C.-K., Chen, W.-H., & Chang, H.-C. (2012). Developing a revised QFD technique to meet the needs of multiple-customer groups: A case of public policy analysis. *Total Quality Management & Business Excellence, 23*(11-12), 1413–1431. doi:10.1080/14783363.2012.704266

Zubizarreta, I., Seravalli, A., & Arrizabalaga, S. (in press). Smart City Concept: What It Is and What It Should Be. *Journal of Urban Planning and Development*. Retrieved from http://ascelibrary.org/doi/abs/10.1061/(ASCE)UP.1943-5444.0000282

KEY TERMS AND DEFINITIONS

Building Information Modeling (BIM): A collaborative working platform that allows sharing of real-time and future project information.

Built Environment: The physical and social spaces created by construction and, by extension, economic activities of the affected spaces.

Information and Communication Technology (ICT): Technological infrastructures serving as the backbone for information gathering and sharing.

Quality Function Deployment (QFD): A technique for capturing multiple groups' or stakeholders' requirements to better meet their (conflicting) needs.

Smart Cities: Sustainable cities supporting social, economic, environmental and cultural activities of its current and future inhabitants through its ICT.

Social Theory: This acknowledges and tends to minimize diversities in the thoughts and actions of different actors in a given environment.

Space Syntax: A concept of spatiality, which acknowledges the notion of "space of space". It purposes to harmonize the physical city and the social city.

Stakeholders: These are non-built environment professionals, who are either affected by or have vested interests in built environment activities.

Total Quality Management (TQM): A systems-thinking or organization-wide approach to ensure productivity and overall quality in activities involving people.

Urbanization: The chain process of rural-urban migration, urban densification, and urban transformation. It could represent a positive or (more commonly) a negative phenomenon.

Compilation of References

2010. *Erste Verordnung zur Durchführung des BundesImmissionsschutzgesetzes - Verordnung über kleine und mittlere Feuerungsanlagen - 1.BImSchV.* Saarbrücken: Bundesministeriums der Justiz und für Verbraucherschutz in Zusammenarbeit mit der juris GmbH.

Abo-Qudais, S., & Abu-Qdais, H. (2005). Perceptions and attitudes of individuals exposed to traffic noise in working places. *Building and Environment, 40*(6), 778–787. doi:10.1016/j.buildenv.2004.08.013

ACEA. (2015). *Overview of purchase and tax incentives for electric vehicles in the EU in 2015.* Brussels: ACEA - European Automobile Manufacturers Association.

Achillas, C., Vlachokostas, C., Moussiopoulos, N., Banias, G., Kafetzopoulos, G., & Karagiannidis, A. (2011). Social acceptance for the development of a waste-to-energy plant in an urban area. *Resources, Conservation and Recycling, 55*(9-10), 857–863. doi:10.1016/j.resconrec.2011.04.012

Achterberg, T. (2008). SCIP: Solving constraint integer programs. *Mathematical Programming Computation, 0*(1), 1–41. doi:10.1007/mpc.v0i1.4

Adams, G., & Gerard, D. (2000). Smart growth and transportation: Opportunities and challenges for Austin. *Institute of Transportation Engineers Journal, 70*(11), 30–34.

Adams, M., Cox, T., Moore, G., Croxford, B., Refaee, M., & Sharples, S. (2006). Sustainable Soundscapes: Noise Policy and the Urban Experience. *Urban Studies (Edinburgh, Scotland), 43*(13), 2385–2398. doi:10.1080/00420980600972504

Adesanya, D. A., & Raheem, A. A. (2009). Development of corn cob ash blended cement. *Construction & Building Materials, 23*(1), 347–352. doi:10.1016/j.conbuildmat.2007.11.013

African Development Bank (2011). Infrastructure and Growth in Zimbabwe: An Action Plan for Sustained Strong Economic Growth.

Agarwal, R., & Prasad, J. (1998). A conceptual and operational definition of personal innovativeness in the domain of information technology. *Information Systems Research, 9*(2), 204–215. doi:10.1287/isre.9.2.204

Agudelo-Vera, C. M., Leduc, W. R. W., Mels, A. R., & Rijnaarts, H. H. M. (2012). Harvesting urban resources towards more resilient cities. *Resources, Conservation and Recycling*, *64*, 3–12. doi:10.1016/j.resconrec.2012.01.014

Ahuja, D., & Tatsutani, M. (2009). Sustainable energy for developing countries, *SAPIENS. Surveys, 2*(1).

Ajzen, I., & Fishbein, M. (1980). *Understanding Attitudes and Predicting Behavior*. Englewood Cliffs, NJ: Prentice-Hall Inc.

Akao, Y., & Mazur, G. H. (2003). The leading edge in QFD: Past, present and future. *International Journal of Quality & Reliability Management*, *20*(1), 20–35. doi:10.1108/02656710310453791

Akram, T., Menon, S. A., & Obaid, H. (2009). Production of low cost self compacting concrete using bagasse ash. *Construction & Building Materials*, *23*(2), 703–712. doi:10.1016/j.conbuildmat.2008.02.012

Alami, A. H. (2010). Experiments on olive husk-addition to masonry clay bricks on their mechanical properties, and their application and manufacturability as an insulating material. *Advanced Materials Research*, *83-86*, 874–880. doi:10.4028/www.scientific.net/AMR.83-86.874

Alam, M. R., Reaz, M. B. I., & Ali, M. A. M. (2012). A review of smart homes—Past, present, and future. *IEEE Transactions on* Systems, Man, and Cybernetics, Part C: Applications and Reviews, *42*(6), 1190–1203.

Albino, V., Berardi, U., & Dangelico, R. M. (2015). Smart Cities: Definitions, Dimensions, Performance, and Initiatives. *Journal of Urban Technology*, *22*(1), 3–21. doi:10.1080/10630732.2014.942092

Algin, H. M., & Turgut, P. (2008). Cotton and limestone powder wastes as brick material. *Construction & Building Materials*, *22*(6), 1078–1080. doi:10.1016/j.conbuildmat.2007.03.006

Ali, S. A., & Tamura, A. (2003). Road traffic noise levels, restrictions and annoyance in Greater Cairo, Egypt. *Applied Acoustics*, *64*(8), 815–823. doi:10.1016/S0003-682X(03)00031-8

Aliyu, A. S., Dada, J. O., & Adam, I. K. (2015). Current status and future prospects of renewable energy in Nigeria. *Renewable & Sustainable Energy Reviews*, *48*, 336–346. doi:10.1016/j.rser.2015.03.098

Al-Juruf, R. S., Ahmed, F. A., & Alam, I. A. (1988). Development of heat insulating materials using date palm leaves. *Thermal insulation*, 11, 158-68.

Allwinkle, S., & Cruickshank, P. (2011). Creating Smart-er Cities: An Overview. *Journal of Urban Technology*, *18*(2), 1–16. doi:10.1080/10630732.2011.601103

American Coal Ash Association. (2013). Coal Combustion Product (CCP) Production & Use Survey.

American Society for Testing and Materials International. (n. d.). *ASTM C* 1314-12. *Standard test method for compressive strength of masonry prisms.*

American Society for Testing and Materials International. (n. d.). ASTM C-1072, *Standard test methods for measurement of masonry flexural bond strength.*

American Society for Testing and Materials International. (n. d.). ASTM D7582 – 15, *Standard Test Methods for Proximate Analysis of Coal and Coke by Macro Thermo-gravimetric Analysis.*

Amin, N. (2011). Use of Bagasse Ash in Concrete and Its Impact on the Strength and Chloride Resistivity. *Journal of Materials in Civil Engineering*, *23*(5), 717–720. doi:10.1061/(ASCE)MT.1943-5533.0000227

Amrilphale, S. S., & Patel, M. (1987). Utilization of red mud, fly ash for manufacturing bricks with pyrophyllite. *Silic. Ind*, *52*(3-4), 31–35.

Andayesh, M., & Sadeghpour, F. (2014). The time dimension in site layout planning. *Automation in Construction*, *44*, 129–139. doi:10.1016/j.autcon.2014.03.021

Anderberg, S. (1998). Industrial metabolism and the linkages between economics, ethics and the environment. *Ecological Economics*, *24*(2-3), 311–320. doi:10.1016/S0921-8009(97)00151-1

Ang, C.-L., Davies, M., & Finlay, P. N. (2001). An empirical study of the use of information technology to support total quality management. *Total Quality Management*, *12*(2), 145–157. doi:10.1080/09544120120011370

Angelidou, M. (2014). Smart city policies: A spatial approach. *Cities (London, England)*, *41*, 3–11. doi:10.1016/j.cities.2014.06.007

Angelidou, M. (2015). A conjucture of four forces. *Cities (London, England)*, *47*, 95–106. doi:10.1016/j.cities.2015.05.004

Ang-Olson, J., Ecola, L., & Santore, M.K. (2003, July 28-30). Characteristics and Performance of Smart Growth Transportation Systems. Proceedings of the 2nd Urban street symposium, Anaheim, California, USA.

Aniefiok, E., Igweonu, I. E., Eguzo, C. V., & Robert, B. J. (2013). Integrating Smart Grid Model in Nigeria Power Network. *International Journal of Advances in Engineering & Technology*, *6*(4), 1760.

Anthopoulos, L. G., & Tsoukalas, I. A. (2006). The Implementation Model of a Digital City. The Case Study of the Digital City of Trikala, Greece. *Journal of E-Government*, *2*(2), 91–109. doi:10.1300/J399v02n02_06

Aranda-Mena, G., Crawford, J., Chevez, A., & Froese, T. (2009). Building information modelling demystified: Does it make business sense to adopt BIM? *International Journal of Managing Projects in Business*, *2*(3), 419–434. doi:10.1108/17538370910971063

Asdrubali, F., & Buratti, C. (2005). Sound intensity investigation of the acoustics performances of high insulation ventilating windows integrated with rolling shutter boxes. *Applied Acoustics*, *66*(9), 1088–1101. doi:10.1016/j.apacoust.2005.02.001

Ashour, T., Wieland, H., Georg, H., Bockisch, F. J., & Wu, W. (2010). The influence of natural reinforcement fibres on insulation values of earth plaster for straw bale buildings. *Materials & Design*, *31*(10), 4676–4685. doi:10.1016/j.matdes.2010.05.026

Australia, . (2013). *Locomotive Emissions Project-Scoping Study of Potential Measures to Reduce Emissions from New and In-Service Locomotives in NSW and Australia*. Australia: NSW EPA.

Aworemi, J., Salami, A., Adewoye, J., & Ilori, M. (2008). Impact of Socio-Economic Characteristics on Formal and Formal Public Transport Demands in Kwara State, Nigeria. *African Journal of Business Management*, 2(4), 72-76.

Babatunde, O. K. (2014). Sustaining Urbanization in a Developing Economy: A Case Study of Nigeria. In H. J. Kauffman (Ed.), *Urbanization: Global Trends, Roles of Climate Change and Effects on Biodiversity* (pp. 75–143). New York: Nova Science Publishers, Inc.

Babatunde, Y., & Sui Pheng, L. (2015). TQM implementation through ISO 9001: Findings from Chinese construction firms in Nigeria. *The TQM Journal*, *27*(6), 671–682. doi:10.1108/TQM-06-2015-0076

Babisch, W. (2008). Road traffic noise and cardiovascular risk. *Noise & Health*, *10*(38), 27. doi:10.4103/1463-1741.39005 PMID:18270405

Bafna, S. (2003). Space Syntax: A Brief Introduction to Its Logic and Analytical Techniques. *Environment and Behavior*, *35*(1), 17–29. doi:10.1177/0013916502238863

Ballesteros, M. J., Fernández, M. D., Quintana, S., Ballesteros, J. A., & González, I. (2010). Noise emission evolution on construction sites. Measurement for controlling and assessing its impact on the people and on the environment. *Building and Environment*, *45*(3), 711–717. doi:10.1016/j.buildenv.2009.08.011

Balta-Ozkan, N., Davidson, R., Bicket, M., & Whitmarsh, L. (2013). Social barriers to the adoption of smart homes. *Energy Policy*, *63*, 363–374. doi:10.1016/j.enpol.2013.08.043

Barles, S. (2010). Society, energy and materials: The contribution of urban metabolism studies to sustainable urban development issues. *Journal of Environmental Planning and Management*, *53*(4), 439–455. doi:10.1080/09640561003703772

Barnes, C. S., Alexis, N. E., Bernstein, J. A., Cohn, J. R., Demain, J. G., Horner, E., & Phipatanakul, W. et al. (2013). Climate change and our environment: The effect on respiratory and allergic disease. *The Journal of Allergy and Clinical Immunology: In Practice*, *1*(2), 137–141. doi:10.1016/j.jaip.2012.07.002 PMID:23687635

Barron, R. F. (2002). *Industrial Noise Control and Acoustics*. CRC Press. doi:10.1201/9780203910085

Bartram, L., Rodgers, J., & Woodbury, R. (2011). Smart homes or smart occupants? supporting aware living in the home. In *Human-Computer Interaction–INTERACT 2011* (pp. 52–64). Springer Berlin Heidelberg. doi:10.1007/978-3-642-23771-3_5

Batty, M., Axhausen, K. W., Giannotti, F., Pozdnoukhov, A., Bazzani, A., Wachowicz, M., & Portugali, Y. et al. (2012). Smart cities of the future. *The European Physical Journal. Special Topics*, *214*(1), 481–518. doi:10.1140/epjst/e2012-01703-3

Baykasoglu, A., Gullu, H., Canakci, H., & Ozbakir, L. (2008). Prediction of compressive and tensile strength of limestone via genetic programming. *Expert Systems with Applications*, *35*(1-2), 111–123. doi:10.1016/j.eswa.2007.06.006

Belojević, G., Öhrström, E., & Rylander, R. (1992). Effects of noise on mental performance with regard to subjective noise sensitivity. *International Archives of Occupational and Environmental Health*, *64*(4), 293–301. doi:10.1007/BF00378288 PMID:1468799

Berg, R. E., & Stork, D. G. (2005). *The Physics of Sound*. Pearson Prentice Hall.

Berkes, F., & Ross, H. (2013). Community resilience: Toward an integrated approach. *Society & Natural Resources*, *26*(1), 5–20. doi:10.1080/08941920.2012.736605

Black, D. A., Black, J. A., Issarayangyun, T., & Samuels, S. E. (2007). Aircraft noise exposure and resident's stress and hypertension: A public health perspective for airport environmental management. *Journal of Air Transport Management*, *13*(5), 264–276. doi:10.1016/j.jairtraman.2007.04.003

Blumendorf, M. (2013). Building sustainable smart homes.*Proceedings of the 1st International Conference on Information and Communication Technologies for Sustainability* (pp. 151-158).

Bogota European Commission. (2011a). *Roadmap to a Single European Transport Area – Towards a competitive and resource efficient transport system* (White Paper). European Commission, Brussels.

Borri, F., & Boccaletti, G. (1995). From total quality management to total quality environmental management. *The TQM Magazine*, *7*(5), 38–42. doi:10.1108/09544789510098614

Bouyoucos, G. J. (1962). Hydrometer method improved for making particle size analysis of soils. *Agronomy Journal*, *54*(5), 464–465. doi:10.2134/agronj1962.00021962005400050028x

Branchi, P. E., Fernández-Valdivielso, C., & Matias, I. R. (2014). Analysis Matrix for Smart Cities. *Future Internet*, 6, 61-75. http://www.mdpi.com/1999-5903/6/1/61

Branchi, P. E., Fernández-Valdivielso, C., & Matias, I. R. (2014). Analysis Matrix for Smart Cities. *Future Internet*, *6*(1), 61–75. doi:10.3390/fi6010061

Brick Development Association. (2015). Brick manufacturing rises to meet housebuilding challenge. Retrieved from http://www.brick.org.uk/2015/05/brick-manufacturing-rises-to-meet-housebuilding-challenge/

Bristol City Council. (2015). Retrieved from http://www.bristol.gov.uk/

British Standard. BS 874: Part 2. (1986). Methods for determining thermal insulating properties. London.

British Standards, B. S. I. (2009). *British Standards: Code of Practice for Noise and Vibration Control on Construction and Open Sites - Part 1: Noise, BS 5228*. British Standards Institution.

Brown, A. L., & Lam, K. C. (1987). Urban noise surveys. *Applied Acoustics*, *20*(1), 23–39. doi:10.1016/0003-682X(87)90081-8

Brundtland, G. H. (1987). *Report of the World Commission on Environment and Development: Our Common Future*. Oslo: United Nations Publications.

Bryde, D., Broquetas, M., & Volm, J. M. (2013). The project benefits of Building Information Modelling (BIM). *International Journal of Project Management*, *31*(7), 971–980. doi:10.1016/j.ijproman.2012.12.001

Bundesministerium für Ernährung und Landwirtschaft. (2014). Nationale Politikstrategie Bioökonomie. National Policy Strategy on Bioeconomy - Renewable resources and biotechnological processes as a basis for food, industry and energy.

Bundesministerium für Umwelt (BMU). (2002). Technical Instructions on Air Quality Control – TA Luft.

Bundesministerium für Umwelt Naturschutz und Reaktorsicherheit. (2012). German Resource Efficiency Programme, Deutsches Ressourceneffizienz- programm (ProgRess), 124.

Bundesregierung. Law on Life-Cycle Management, Kreislaufwirtschaftsgesetz, Gesetz zur Förderung der Kreislaufwirtschaft und Sicherung der umweltverträglichen Bewirtschaftung von Abfällen (KrWG) (2012).

Bundesregierung. The Commercial Waste Ordinance, Verordnung über die Entsorgung von gewerblichen Siedlungsabfällen und von bestimmten Bau- und Abbruchabfällen (Gewerbeabfallverordnung - GewAbfV) (2002).

Buratti, C. (2002). Indoor noise reduction index with open window. *Applied Acoustics*, *63*(4), 431–451. doi:10.1016/S0003-682X(01)00040-8

Bureau of Indian Standard. IS 1350-1. (1984). *Methods of Test for Coal and Coke*, Part I: Proximate Analysis [PCD 7: Solid Mineral Fuels]

Bureau of Indian Standard. IS 2185 Part-I. (1979). *Specifications for concrete masonry - hollow and solid concrete blocks*. BIS, New Delhi.

Bureau of Indian Standard. IS 3495 Part-I. (1992). *Methods of tests of burnt clay building bricks – determination of compressive strength*. BIS, New Delhi.

Bureau of Indian Standard. IS 3495 Part-II. (1992). *Methods of tests of burnt clay building bricks – determination of water absorption*. BIS, New Delhi.

Bureau of Indian Standard. IS 3495 Part-III. (1992). *Methods of tests of burnt clay building bricks – determination of efflorescence.* BIS, New Delhi.

Bureau of Indian Standard. IS 456. (2000). Plain and reinforced concrete – code of practice. BIS, New Delhi.

Bureau of Indian Standard. IS 4860. (1996). Specifications for acid resistant bricks. BIS, New Delhi.

Burges, C. (1998). A tutorial on support vector machines for pattern recognition. In *Data Mining and Knowledge Discovery*. Boston: Kluwer Academic Publishers.

Busch-Vishniac, I. J., West, J. E., Barnhill, C., Hunter, T., Orellana, D., & Chivukula, R. (2005). Noise levels in Johns Hopkins Hospital. *The Journal of the Acoustical Society of America, 118*(6), 3629–3645. doi:10.1121/1.2118327 PMID:16419808

Byrne, D. C., & Reeves, E. R. (2008). Analysis of Nonstandard Noise Dosimeter Microphone Positions. *Journal of Occupational and Environmental Hygiene*, *5*(3), 197–209. doi:10.1080/15459620701879438 PMID:18213533

Calabrese, A., & Corbò, M. (2015). Design and blueprinting for total quality management implementation in service organizations. *Total Quality Management & Business Excellence*, *26*(7-8), 719–732. doi:10.1080/14783363.2014.881970

Callahan, J. L. (2014). Writing Literature Reviews: A Reprise and Update. *Human Resource Development Review*, *13*(3), 271–275. doi:10.1177/1534484314536705

Calzada, I., & Cobo, C. (2015). Unplugging: Deconstructing the Smart City. *Journal of Urban Technology*, *22*(1), 23–43. doi:10.1080/10630732.2014.971535

Caragliu, A., Del Bo, C., & Nijkamp, P. (2011). Smart cities in Europe. *Journal of Urban Technology*, *18*(2), 65–82. doi:10.1080/10630732.2011.601117

Carnevalli, J. A., & Miguel, P. C. (2008). Review, analysis and classification of the literature on QFD – Types of research, difficulties and benefits. *International Journal of Production Economics*, *114*(2), 737–754. doi:10.1016/j.ijpe.2008.03.006

Cavada, M., Christopher, D. F. R., & Hunt, D. V. L. (2014). Smart cities: Contradicting definitions and unclear measures. *Conference Proceedings Papers*. Retrieved from http://sciforum.net/conference/wsf-4/paper/2454/download/pdf

Census of India. (2011). *Population Census-2011*. Government of India.

Central Vehicle Registry. (2012). Vehicle statistics given verbally.

Cervero, R. (2000). Urban reclamation and regeneration in Seoul, Republic of Korea. In S. Lehmann (Ed.), Low carbon cities: Transforming urban systems (pp. 224-234). New York: Routledge, Taylor and Francis Group.

Chan, L.-K., & Wu, M.-L. (2002). Quality function deployment: A literature review. *European Journal of Operational Research*, *143*(3), 463–497. doi:10.1016/S0377-2217(02)00178-9

Chan, M., Campo, E., Estève, D., & Fourniols, J. Y. (2009). Smart homes—current features and future perspectives. *Maturitas*, *64*(2), 90–97. doi:10.1016/j.maturitas.2009.07.014 PMID:19729255

Chan, M., Estève, D., Escriba, C., & Campo, E. (2008). A review of smart homes—Present state and future challenges. *Computer Methods and Programs in Biomedicine*, *91*(1), 55–81. doi:10.1016/j.cmpb.2008.02.001 PMID:18367286

Chartered Institute of Building. (2014). *Code of Practice for Project Management for Construction and Development* (5th ed.). Oxford, UK: Wiley-Blackwell.

Cherkassky, V., & Mulier, F. (1998). *Learning from data: Concepts, theory, and methods*. New York: Wiley.

Chiang, K. Y., Chou, H., Hua, C. R., Chien, K. L., & Cheeseman, C. (2009). Lightweight bricks manufactured from water treatment sludge and rise husks. *Hazardous Materials*, *171*(1-3), 76–82. doi:10.1016/j.jhazmat.2009.05.144 PMID:19596512

Chourabi, H., Gil-Garcia, J. R., Pardo, T. A., Nam, T., Mellouli, S., Scholl, H. J., . . . Nahon, K. (2012). Understanding smart cities: An integrative framework. *Proceedings of the45th Hawaii International Conference onSystem Science (HICSS)* (pp. 2289-2297). Washington DC. NY: IEEE Computer Society.

Christopher, K., & Leman, P. L. (1994). *Re-Thinking HOV - High Occupancy Vehicle Facilities and the Public Interest*. California: California Department of Transportation.

Chusilp, N., Jaturapitakkul, C., & Krattikomol, K. (2009). Utilization of bagasse ash as a pozzolanic material in concrete. *Construction & Building Materials*, *23*(11), 3352–3358. doi:10.1016/j.conbuildmat.2009.06.030

Cities, E. (2015, September). Retrieved from Energy Cities in Europe. Retrieved from http://www.energy-cities.eu/-Members-

Civil Building Thermal Design Code GB50176-93. (1993). Beijing: China Building Industry Press.

Coats, A. W., & Redfern, J. P. (1963). Thermogravimetric analysis: A review. *Analyst (London)*, *88*(1053), 906–924. doi:10.1039/an9638800906

Cody, J. (2009). *Regenerative breaking in an electric vehicle.* Maszyny Elektryczne: Zeszyty Problemowe.

Colwell, K. (2015, May 25). *Car and Driver*. Retrieved from Engine Stop-Start Systems on Nonhybrid Vehicles. Retrieved from http://www.caranddriver.com/features/engine-stop-start-systems-explained-tech-dept

Commission of the European Communities. (2000). Commission Decision on the European List of Waste (COM 2000/532/EC). *Official Journal of the European Communities*, *2000D0532*(01.01.2002). doi:(2000/532/EC)

Commission of the European Communities. (2014). 15 Final. Communication from the Commission to the European Parliament, the Council, the European economic and social committee and the committee of the regions a policy framework for climate and energy in the period from 2020 to 2030.

Concrete construction.net. (2015). Global demand for construction aggregates to exceed 48 billion metric tons in 2015. Retrieved from http://www.concreteconstruction.net/aggregates/global-demand-for-construction-aggregates-to-exceed-48-billion-metric-tons-in-2015.aspx

Consulting Australia, S. L. R. (2012). *Northern Sydney Freight Corridor, North Strathfield Rail Underpass-Noise and Vibration Assessment*. Transport for New South Wales.

Conti, T. (2010). Systems thinking in quality management. *The TQM Journal*, *22*(4), 352–368. doi:10.1108/17542731011053280

Cook, D. J., Youngblood, M., Heierman, E. O. III, Gopalratnam, K., Rao, S., Litvin, A., & Khawaja, F. (2003). Mavhome: an agent-based smart home.*Proceedings of the First IEEE International Conference on Pervasive Computing and Communications PerCom '03* (pp. 521–524) doi:10.1109/PERCOM.2003.1192783

Cooper, R. B., & Zmud, R. W. (1990). Information technology implementation research: A technological diffusion approach. *Management Science*, *36*(2), 123–139. doi:10.1287/mnsc.36.2.123

Cordeiro, G. C., Filho, R. D. T., Tavares, L. M., & Fairbairn, E. R. (2009). Ultrafine grinding of sugar cane bagasse ash for application as pozzolanic admixtures in concrete. *Cement and Concrete Research*, *39*(2), 110–115. doi:10.1016/j.cemconres.2008.11.005

Cosgrave, E., Arbuthnot, K., & Tryfonas, T. (2013). Living Labs, Innovation Districts and Innovation Market Places: A Systems Approach for Smart Cities. *Procedia Computer Science*, *16*, 668–677. doi:10.1016/j.procs.2013.01.070

Council on Hemispheric Affairs. (2009). Bus Rapid Transit and the Latin American City: Successes to Date, But Miles to Go. Retrieved from http://www.coha.org/bus-rapid-transit-and-the-latin-american-city-successes-to-date-but-miles-to-go/

Coutts, R. P. (1992). From forest to factory to fabrication. In R.N. Swamy (Ed.), *Proceedings of the 4th International Symposium Fibre Reinforced Cement and Concrete* (pp. 31-47).

Craven, N., Bewes, O. G., Fenech, B. A., & Jones, R. (2013). Responding to the Environmental Noise Directive by demonstrating the benefits of rail grinding on the Great Britain's railway network. *Proceedings of the Institution of Mechanical Engineers, Part F: Journal of Rail and Rapid Transit*. doi:10.1177/0954409713494948

Cristóbal, J., Guillén-Gosálbez, G., Jiménez, L., & Irabien, A. (2012). Optimization of global and local pollution control in electricity production from coal burning. *Applied Energy*, *92*, 369–378. doi:10.1016/j.apenergy.2011.11.028

Crocker, M. J. (2007). *Handbook of Noise and Vibration Control*. John Wiley & Sons. doi:10.1002/9780470209707

Crum, M. R., Premkumar, G., & Ramamurthy, K. (1996). An assessment of motor carrier adoption, use, and satisfaction with EDI. *Transportation Journal*, *35*(4), 44.

Dada, J. O. (2014). Towards understanding the benefits and challenges of Smart/Micro-Grid for electricity supply system in Nigeria. *Renewable & Sustainable Energy Reviews*, *38*, 1003–1014. doi:10.1016/j.rser.2014.07.077

Dahlgaard, J. J., & Dahlgaard-Park, S. M. (2006). Lean production, six sigma quality, TQM, and company culture. *The TQM Magazine*, *18*(3), 263–281. doi:10.1108/09544780610659998

Dameri, R. P., & Ricciardi, F. (2015). Smart city intellectual capital: An emerging view of territorial systems innovation management. *Journal of Intellectual Capital*, *16*(4), 860–887. doi:10.1108/JIC-02-2015-0018

Deakin, M., & Al Waer, H. (2011a). From intelligent to smart cities. *Intelligent Buildings International*, *3*(1), 133–139. doi:10.1080/17508975.2011.586673

DeAngelis, S. (2015). Smart Cities: Big Data and Resilency. *Wired Innovation Insights.* Retrieved from http://insights.wired.com/profiles/blogs/smart-cities-big-data-and-resiliency#ixzz3nD44Qjkw

Demir, I. (2008). Effect of organic residues addition on the technological properties of clay bricks. *Waste Management (New York, N.Y.)*, *28*(3), 622–627. doi:10.1016/j.wasman.2007.03.019 PMID:17512183

Demiris, G., Rantz, M. J., Aud, M. A., Marek, K. D., Tyrer, H. W., Skubic, M., & Hussam, A. A. (2004). Older adults' attitudes towards and perceptions of'smart home'technologies: A pilot study. *Informatics for Health & Social Care*, *29*(2), 87–94. PMID:15370989

den Boer, L., & Schroten, A. (2007). Traffic noise reduction in Europe. Report of CE Delft, Commissioned by the European Federation for Transport and Environment (T&E). Brussels.

dena, D. E.-A. (2009). *Leitfaden Energieausweis Teil 3 – Energieverbrauchsausweise, IBSN: 978-3-9812787-4-3.* Berlin: dena.

Denzin, N., & Lincoln, Y. (1994). *Handbook of Qualitative Research*. Thousand Oaks: Sage.

Department of Environment and Climate Change NSW. (2009). *Interim Construction Noise Guideline*. NSW, Australia: State of NSW and Department of Environment and Climate Change NSW.

Department of Environment and Resource Management. (2010). *State Planning Policy 5/10 Air, Noise and Hazardous Materials*. Queensland, Australia. Retrieved from http://www.ehp.qld.gov.au/noise/

Department of Environment. Climate Change and Water NSW. (2011). *NSW Road Noise Policy*. NSW, Australia. Retrieved from http://www.epa.nsw.gov.au/noise/traffic.htm

Department of Infrastructure. Energy and Resources. (2011). *State Road Noise Strategy*. Tasmania, Australia. Retrieved from http://www.transport.tas.gov.au/__data/assets/pdf_file/0016/60514/State_Road_Noise_Strategy_-_Version_1.0,_June_2011.pdf

Design Standard for Energy Efficiency of Residential Buildings in Severe Cold and Cold Zones JGJ26-2010. (2010). Beijing: China Building Industry Press.

Devadasan, S. R., Kathiravan, N., & Thirunavukkarasu, V. (2006). Theory and practice of total quality function deployment: A perspective from traditional pump-manufacturing environment. *The TQM Magazine*, *18*(2), 143–161. doi:10.1108/09544780610647865

DHL - green services, our solution for you. (2014). Berlin: DHL Headquarters.

Dia, H. (2015). *10 Powerful Policies to Successful Smart Mobility*. Retrieved from http://www.sustainablecitiescollective.com/

Ding, G. K. C. (2008). Sustainable construction—The role of environmental assessment tools. *Journal of Environmental Management*, *86*(3), 451–464. doi:10.1016/j.jenvman.2006.12.025 PMID:17289255

Dobbs, M. F. (1994). Continuous Improvement as Continuous Implementation: Implementing TQM in the City of Santa Ana. *Public Productivity & Management Review*, *18*(1), 89–100. doi:10.2307/3380697

Dominguez, E. A., & Ullmann, R. (1996). Ecological bricks made with clays and steel dust pollutant. *Applied Clay Science*, *11*(2-4), 237–249. doi:10.1016/S0169-1317(96)00020-8

Dube, D., & Chirisa, I. (2012, May). The Informal City: Assessing its Scope, Variants and Direction in Harare, Zimbabwe. *Global Advanced Research Journal of Geography and Regional Planning*, 1(1), 16-25. Retrieved from http://garj.org/garjerr/index.htm

Dubey, R., & Gunasekaran, A. (2015). Exploring soft TQM dimensions and their impact on firm performance: Some exploratory empirical results. *International Journal of Production Research*, *53*(2), 371–382. doi:10.1080/00207543.2014.933909

Dubey, R., & Singh, T. (2015). Understanding complex relationship among JIT, lean behaviour, TQM and their antecedents using interpretive structural modelling and fuzzy MICMAC analysis. *The TQM Journal*, *27*(1), 42–62. doi:10.1108/TQM-09-2013-0108

Eadie, D. T., & Santoro, M. (2006). Top-of-rail friction control for curve noise mitigation and corrugation rate reduction. *Journal of Sound and Vibration*, *293*(3–5), 747–757. doi:10.1016/j.jsv.2005.12.007

Eckl, R., & MacWilliams, A. (2009). Smart home challenges and approaches to solve them: A practical industrial perspective. In *Intelligent Interactive Assistance and Mobile Multimedia Computing* (pp. 119–130). Springer Berlin Heidelberg. doi:10.1007/978-3-642-10263-9_11

ECLAC. (2008). *Developments in public-transport systems in the cities of Latin America. Economic Commission for Latin America and the Caribbean*. ECLAC.

Ecobricks. (n. d.). Issues in the Indian Brick Sector. Retrieved from http://www.ecobrick.in/challenges_Issues_in_the_Indian_Brick_Sector.aspx

Edwards, K., & Grinter, R. (2001, 30 September–2 October). At home with ubiquitous computing: Seven challenges. *Proceedings of the 3rd International Conference on Ubiquitous Computing*, Atlanta, GA, USA (pp. 256–272).

Edwards, W. K., & Grinter, R. E. (2001). At home with ubiquitous computing: Seven challenges. In Ubicomp 2001: Ubiquitous Computing (pp. 256-272). Springer Berlin Heidelberg.

Egan, M. D. (2007). *Architectural Acoustics*. J. Ross Pub.

Eiselt, H., & Marianov, V. (2015). Location modeling for municipal solid waste facilities. *Computers & Operations Research*, *62*, 305–315. doi:10.1016/j.cor.2014.05.003

EPA VIC. (2008). NOISE CONTROL GUIDELINES. Victoria Government, Australia.

EPA. (2000). *NSW Industrial Noise Policy*. NSW, Australia: Environment Protection Authority. Retrieved from http://www.epa.nsw.gov.au/noise/industrial.htm

EPA. (2013). *Rail Infrastructure Noise Guideline*. NSW, Australia: State of NSW and Environment Protection Authority. Retrieved from http://www.epa.nsw.gov.au/noise/railnoisegl.htm

EPEA i VHE. (2008). *Ökologisches Leistungsprofil von Verfahren zur Behandlung von biogenen Reststoffen Kompass für die Entscheidungsfindung vor dem Hintergrund der geplanten Überarbeitung*.

European Commission. (2009). The EU Ecolabel for Tourist Accommodation - 2009/578/EC. Brussels.

European Commission. (2013). Mobilising Intelligent Transport Systems for EU cities. Commission staff working document SWD527, Brussels.

European Commission. (2015). Erst sechs, dann 28 - Europäische Nachbarschaftspolitik und Erweiterungsverhandlungen. Brussels: European Commission.

European Parliament and Council. Directive 2008/98/EC of the European Parliament and of the Council of 19 November 2008 on waste and repealing certain directives (Waste framework directive)., L 312/3 3–30 (2008). doi:2008/98/EC.; 32008L0098

Eurostat. (2015). Generation of waste [env_wasgen]. Retrieved from http://appsso.eurostat.ec.europa.eu/nui/show.do?dataset=env_wasgen&lang=en

Fadare, S.O., & Hay, A.M. (1990). Housing Densities, Life Style and Travel Behavior in A Third World City, The Example of Ibadan. *Nig. J. Econ And Soc. Stud.*, 32(2).

Falk, S. A., & Woods, N. F. (1973). Hospital Noise — Levels and Potential Health Hazards. *The New England Journal of Medicine*, *289*(15), 774–781. doi:10.1056/NEJM197310112891504 PMID:4728750

Faria, K. P., Gurgel, R. F., & Holanda, J. F. (2012). Recycling of sugarcane bagasse ash waste in the production of clay bricks. *Journal of Environmental Management*, *101*, 7–12. doi:10.1016/j.jenvman.2012.01.032 PMID:22387325

Feiock, R. C. (2002). A Quasi-market framework for local economic development competition. *Journal of Urban Affairs*.

Ferris, E., Petz, D., & Stark, C. (2013). The year of recurring disasters: a review of natural disasters in 2012. Retrieved from http://www.brookings.edu/~/media/research/files/reports/2013/03/natural%20disasters%20review/brookings_review_natural_disasters_2012.pdf

Fichman, R. G. (1992, December). Information technology diffusion: a review of empirical research. *Proceedings of theInternational Conference on Information Systems* (pp. 195-206).

Fichman, R. G. (1992, December) Information technology diffusion: a review of empirical research. In J. De Gross, J. Becker, & J. Elam (Eds.), *Proceedings of the 13th International Conference on Information Systems*, Dallas (pp. 195-206).

Fife, D., & Rappaport, E. (1976). Noise and hospital stay. *American Journal of Public Health*, *66*(7), 680–681. doi:10.2105/AJPH.66.7.680 PMID:937616

Fischer, C. (2007). Influencing electricity consumption via consumer feedback: a review of experience.*Proceedings of the European Council for an Energy Efficient Economy (ECEEE)* (pp. 1873-1884).

Fogari, R., Zoppi, A., Vanasia, A., Marasi, G., & Villa, G. (1994). Occupational noise exposure and blood pressure. *Journal ofHypertension*, *12*(4), 475–479. doi:10.1097/00004872-199404000-00019 PMID:8064173

Fox, S. (2014). Getting real about BIM: Critical realist descriptions as an alternative to the naïve framing and multiple fallacies of hype. *International Journal of Managing Projects in Business*, *7*(3), 405–422. doi:10.1108/IJMPB-12-2013-0073

Franssen, E. M., van Wiechen, C. M. A. G., Nagelkerke, N. J. D., & Lebret, E. (2004). Aircraft noise around a large international airport and its impact on general health and medication use. *Occupational and Environmental Medicine*, *61*(5), 405–413. doi:10.1136/oem.2002.005488 PMID:15090660

Fricke, F. (1985). Construction Site Noise. *Architectural Science Review*, *28*(1), 12–22. doi:10.1080/00038628.1985.9696566

Frost and Sullivan. (2010). A Smarter Transportation System for the 21st Century.

Ganesan, K., Rajagopal, K., & Thangavel, K. (2007). Evaluation of bagasse ash as supplementary cementitious material. *Cement and Concrete Composites*, *29*(6), 515–524. doi:10.1016/j.cemconcomp.2007.03.001

Gansterer, M. (2013). *Managing traffic and transport in urban areas. Sustainable solutions for traffic problems*. Vienna, Austria: VCÖ - Mobilität mit Zukunft.

Gan, W. Q., Davies, H. W., Koehoorn, M., & Brauer, M. (2012). Association of Long-term Exposure to Community Noise and Traffic-related Air Pollution With Coronary Heart Disease Mortality. *American Journal of Epidemiology*, *175*(9), 898–906. doi:10.1093/aje/kwr424 PMID:22491084

Garcia-Bernal, J., & Garcia-Casarejos, N. (2014). Economic analysis of TQM adoption in the construction sector. *Total Quality Management & Business Excellence*, *25*(3), 209–221. doi:10.1080/14783363.2012.728848

Gascón, D. (2015, May 15). Vehicle Traffic Monitoring Platform with Bluetooth Sensors over ZigBee. *Libellium World*. Retrieved from http://www.libelium.com/vehicle_traffic_monitoring_bluetooth_sensors_over_zigbee/#!prettyPhoto

Gehring, U., Tamburic, L., Sbihi, H., Davies, H. W., & Brauer, M. (2014). Impact of Noise and Air Pollution on Pregnancy Outcomes. *Epidemiology (Cambridge, Mass.)*, *25*(3), 351–358. doi:10.1097/EDE.0000000000000073 PMID:24595395

German Federal Ministry of Education and Research. (2011). The German Government's raw materials strategy.

Gershenson, C. (2015, May). *Smart traffic light systems as a pilot project at Universidad Nacional Autónoma de México*. Retrieved from http://green.wiwo.de/ampeln-mit-kopfchen-forscher-entwickeln-das-schlaue-verkehrslicht/

Ghavami, K. (1995). Ultimate load behavior of bamboo reinforced lightweight concrete beams. *Cement and Concrete Composites*, *17*(4), 281–288. doi:10.1016/0958-9465(95)00018-8

Ghavami, K. (2005). Bamboo as reinforcement in structural concrete elements. *Cement and Concrete Composites*, *27*(6), 637–649. doi:10.1016/j.cemconcomp.2004.06.002

Giang, D. T. H., & Sui Pheng, L. (2011). Role of construction in economic development: Review of key concepts in the past 40 years. *Habitat International*, *35*(1), 118–125. doi:10.1016/j.habitatint.2010.06.003

Giffinger, F., Kramer, M., & Pichler, M. (2012). European smart model. Universities of Vienna, Universities of Delft, Universities of Lubijana. Retrieved from http://www.smart-cities.eu/model_1.html

GIFT City. (2015). Retrieved from http://giftgujarat.in/

Gilchrist, A., Allouche, E. N., & Cowan, D. (2003). Prediction and mitigation of construction noise in an urban environment. *Canadian Journal of Civil Engineering*, *30*(4), 659–672. doi:10.1139/l03-019

Glaeser, E. L., & Resseger, M. G. (2010). The complementarity between cities and skills. *Journal of Regional Science*, *50*(1), 221–244. doi:10.1111/j.1467-9787.2009.00635.x

Goines, L., & Hagler, L. (2007). Noise pollution: A modem plague. *Southern Medical Journal*, *100*(3), 287–294. doi:10.1097/SMJ.0b013e3180318be5 PMID:17396733

Goldberg, D. E. (1989). *Genetic Algorithms in Search, Optimization, and Machine Learning* (p. 432). Massachusetts: Addison-Wesley Professional.

Golub, A.D. (2003). Welfare Analysis Of Informal Transit Services. In *Brazil And The Effects Of Regulation.* Institute Of Transportation Studies, University Of California at Berkeley.

Gong, X. Z., Akashi, Y., & Sumiyoshi, D. (2012). Optimization of passive design measures for residential buildings in different Chinese areas. *Building and Environment*, *58*, 46–57. doi:10.1016/j.buildenv.2012.06.014

Goodspeed, R. (2015). Smart cities: Moving beyond urban cybernetics to tackle wicked problems. *Revue Economique et Sociale*, *8*(1), 79–92.

Gopalakrishnan, A. (2014). *Art of Smart: Can technology transform urban living in India?* (pp. 25–28). India Today.

Goss, J. (1988). The Built Environment and Social Theory: Towards an Architectural Geography. *The Professional Geographer*, *40*(4), 392–403. doi:10.1111/j.0033-0124.1988.00392.x

Government of Zimbabwe. (1993). *Economic Structural Adjustment Programme*. Harare: Government Printers.

Gram, H. E. (1983). Durability of natural fibres in concrete. Stockholm: Swedish Cement and Concrete Research Institute. Retrieved from http://www.engineeringtoolbox.com/specific-heat-capacity-d_339.html

Grilo, A., & Jardim-Goncalves, R. (2010). Value proposition on interoperability of BIM and collaborative working environments. *Automation in Construction*, *19*(5), 522–530. doi:10.1016/j.autcon.2009.11.003

Gross, S., & Campbell, N. (2015). Central city vibrancy and new business venturing. *Journal of Entrepreneurship and Public Policy*, *4*(2), 257–271. doi:10.1108/JEPP-01-2014-0002

Gungor, V. C., Sahin, D., Kocak, T., Ergüt, S., Buccella, C., Cecati, C., & Hancke, G. P. (2012). Smart grid and smart homes: Key players and pilot projects. *Industrial Electronics Magazine, IEEE*, *6*(4), 18–34. doi:10.1109/MIE.2012.2207489

Hackman, R., & Wageman, R. (1995). Total Quality Management: Empirical, Conceptual, and Practical Issues. *Administrative Science Quarterly*, *40*(2), 309–342. doi:10.2307/2393640

Hammad, A. W. A., Rey, D., & Akbarnezhad, A. (2014). A mixed-integer nonlinear programming model for minimising construction site noise levels through site layout optimization. *Presented at the 31st International Symposium on Automation and Robotics in Construction and Mining, ISARC '14*(pp. 722–729).

Hammad, A. W. A., Akbarnezhad, A., & Rey, D. (2016). A multi-objective mixed integer nonlinear programming model for construction site layout planning to minimise noise pollution and transport costs. *Automation in Construction*, 61, 73 – 85. http://dx.doi.org/10.1016/j.autcon.2015.10.010

Hammad, A. W. A., Akbarnezhad, A., Rey, D., & Waller, S. T. (2015). A computational method for estimating travel frequencies in site layout planning. *Journal of Construction Engineering and Management*, 142(5). http://dx.doi.org/10.1061/(ASCE)CO.1943-7862.0001086

Hanna, D., Blouin-Demers, G., Wilson, D. R., & Mennill, D. J. (2011). Anthropogenic noise affects song structure in red-winged blackbirds (Agelaius phoeniceus). *The Journal of Experimental Biology*, *214*(21), 3549–3556. doi:10.1242/jeb.060194 PMID:21993783

Hanna, M. D., & Newman, W. R. (1995). Operations and environment an expanded focus for TQM. *International Journal of Quality & Reliability Management*, *12*(5), 38–53. doi:10.1108/02656719510089984

Harrington, H. J., Voehl, F., & Wiggin, H. (2012). Applying TQM to the construction industry. *The TQM Journal*, *24*(4), 352–362. doi:10.1108/17542731211247373

Harris, C. M. (1991). *Handbook of acoustical measurements and noise control*. McGraw-Hill New York.

Hassan, M. M., Ahmed, S., Rahman, K. A., & Biswas, T. (2008). Pattern of medical waste management: Existing scenario in Dhaka City, Bangladesh. *BMC Public Health*, *8*(1), 36–45. doi:10.1186/1471-2458-8-36 PMID:18221548

Hastie, T., Tibshirani, R., & Friedman, J. (2001). *The elements of statistical learning: Data mining, inference and prediction*. Berlin: Springer. doi:10.1007/978-0-387-21606-5

Hausba (2014) *Hausba Smarthomes* Retrieved fromhttp://www.hausbassmarthomes.com

Hawkins, C., Ye, L., Mandpe, S., & Meyer, P. B. (2005). What is "smart growth?" - Really? *Journal of Planning Literature*, *19*(3), 301–315. doi:10.1177/0885412204271668

Heckel, M. (May 2015). *Exportchancen, Wachstum Ohne Grenzen*. Von Smart cities entstehen in aller Welt - auch mit deutscher Hilfe. Retrieved from http://www.der-betrieb.de/content/hbnews/ abgerufen

Heiskanen, E., Hyvönen, K., Niva, M., Pantzar, M., Timonen, P., & Varjonen, J. (2007). User involvement in radical innovation: Are consumers conservative? *European Journal of Innovation Management*, *10*(4), 489–509. doi:10.1108/14601060710828790

Henry, G., & Heinke, G. W. (1996). *Environmental Science and Engineering* (2nd ed.). Englewood, NJ: Prentice-Hall.

Hideo Konishi, S.-M. (2010). *Carpooling and Congestion Pricing: HOV and HOT lanes*. Boston: Department of Economics Boston College.

Hillier, B. (2008). Space and spatiality: What the built environment needs from social theory. *Building Research and Information*, *36*(3), 216–230. doi:10.1080/09613210801928073

Hillier, B., & Penn, A. (2004). Rejoinder to Carlo Ratti. *Environment and Planning. B, Planning & Design*, *31*(4), 501–511. doi:10.1068/b3019a

Hirtle, S., Richter, K.-F., Srinivas, S., & Firth, R. (2010). This is the tricky part: When directions become difficult. *Journal of Spatial Information Science, 1*, 53–73.

Hoang, D. T., Igel, B., & Laosirihongthong, T. (2010). Total quality management (TQM) strategy and organizational characteristics: Evidence from recent WTO member. *Total Quality Management & Business Excellence, 21*(9), 931–951. doi:10.1080/14783363.2010.487680

Hollands, R. G. (2008). Will the real smart city please stand up? Intelligent, progressive or entrepreneurial. *City*, 12.

Hollands, R.G. (2008). Will the real smart city please stand up? *City: analysis of urban trends, culture, theory, policy, action*, 12(3), 303-320.

Hompel, M. t. (2015). *Cyber physical systems in logistics*. Mülheim an der Ruhr: EffizienzCluster Management GmbH.

Ho, S. K. M. (2010). Integrated lean TQM model for sustainable development. *The TQM Journal, 22*(6), 583–593. doi:10.1108/17542731011085294

IBM CPLEX Optimizer - United States. (2014, December 30). Retrieved from http://www-01.ibm.com/software/commerce/optimization/cplex-optimizer/

IBMCorporation, . (2010). *The Case for Smarter Transportation*. USA.

Industry, C. d. (2006). *Impact Assessment for Euro 6 emission limits for light duty vehicles*. Brussels: Commission of the European Communities.

International Telecommunication Union. (2010). Applications of ITU-T G.9960, In: Paper, I.T.U.T. (Ed.), ITU-T G.9961 Transceivers for Smart Grid Applications: Advanced Metering Infrastructure, Energy Management in the Home and Electric Vehicles (ITU-T, 06/2010). International Telecommunication Union Technical Paper.

Isaksson, R. (2006). Total quality management for sustainable development – Process based system models. *Business Process Management Journal, 12*(5), 632–645. doi:10.1108/14637150610691046

Iwnicki, S. (2006). *Handbook of Railway Vehicle Dynamics*. CRC Press.

Jaeger, R. G., & Halliday, T. R. (1998). On Confirmatory versus Exploratory Research. *Herpetologica, 54*(Suppl.), S64–S66.

Jahandideh, S., Jahandideh, S., Asadabadi, E. B., Askarian, M., Movahedi, M. M., Hosseini, S., & Jahandideh, M. (2009). The use of artificial neural networks and multiple linear regression to predict rate of medical waste generation. *Waste Management (New York, N.Y.), 29*(11), 2874–2879. doi:10.1016/j.wasman.2009.06.027 PMID:19643591

Jain, A. (2015). *Smart Cities: Vision and Action*. New Delhi: Discovery Publishing House Pvt. Ltd.

Japanese Industrial Standard. (1994). *JIS A 5905*. Tokyo, Japan: Fibreboards.

Jaturapitakkul, C., Kiattikomol, K., Tangchirapat, W., & Saeting, T. (2007). Evaluation of the sulfate resistance of concrete containing palm oil fuel ash. *Construction & Building Materials, 21*(7), 1399–1405. doi:10.1016/j.conbuildmat.2006.07.005

Jawaid, M.F., & Khan, S.A. (2015, March). Evaluating the need for smart cities in India. *International Journal of Advance Research in Science And Engineering*, 4(SI 1).

Jenkins, P., Smith, H., & Wang, Y. P. (2007). Planning and Housing in the Rapidly Urbanising World. London, UK: Routledge; Retrieved from http://eprints.gla.ac.uk/75688/

Jiang, Y., Lin, B. R., Zeng, J. L., & Zhu, Y. X. (2009). *Energy Efficiency of Residence* (Vol. 7). Beijing: China Building Industry Press.

Kang, Y., & Zang, L. (2014). Comparative Study of Smart Cities in Europe and China 2014. *Proceedings of 4th International Conference on Logistics, Informatics and Service Science*. Commercial Press China and Springer-Verlag Berlin Heidelberg.

Kastner, W., Neugschwandtner, G., Soucek, S., & Newman, H. M. (2005). Communication systems for building automation and control. *Proceedings of the IEEE, 93*(6), 1178–1203. doi:10.1109/JPROC.2005.849726

Katti, M., & Warren, P. S. (2004). Tits, noise and urban bioacoustics. *Trends in Ecology & Evolution, 19*(3), 109–110. doi:10.1016/j.tree.2003.12.006 PMID:16701237

Kennedy, C., Pincetl, S., & Bunje, P. (2011). The study of urban metabolism and its applications to urban planning and design. *Environmental Pollution,* 159(8-9), 1965–73. doi:10.1016/j.envpol.2010.10.022

Kenney, G. D., & Ayer, H. E. (1975). Noise Exposure and Hearing Levels of Workers in the Sheet Metal Construction Trade. *American Industrial Hygiene Association Journal, 36*(8), 626–632. doi:10.1080/0002889758507302 PMID:1227289

Keshava, S.R. (2014). Indian Urbanization under Globalization; Development or Delusion! *International Journal in Multidisciplinary and Academic Research*, 2(1).

Khadivi, M. R., & Fatemi Ghomi, S. M. T. (2012). Solid waste facilities location using of analytical network process and data envelopment analysis approaches. *Waste Management (New York, N.Y.), 32*(6), 1258–1265. doi:10.1016/j.wasman.2012.02.002 PMID:22382043

Khalafallah, A., & El-Rayes, K. (2008). Minimizing Construction-Related Security Risks during Airport Expansion Projects. *Journal of Construction Engineering and Management, 134*(1), 40–48. doi:10.1061/(ASCE)0733-9364(2008)134:1(40)

Khansari, N., Mostashari, A., & Mansouri, M. (2014). Conceptual Modelling of the Impact of Smart Cities on Household Energy Consumption. *Procedia Computer Science, 28*, 81–86. doi:10.1016/j.procs.2014.03.011

Khedari, J., Nankongnab, N., Hirunlabh, J., & Teekasap, S. (2004). New low-cost insulating particleboards from mixture of durian peel and coconut coir. *Building and Environment, 39*(1), 59–65. doi:10.1016/j.buildenv.2003.08.001

Kinder, T. (2000). A sociotechnical approach to the innovation of a network technology in the public sector – the introduction of smart homes in West Lothian. *European Journal of Innovation Management, 3*(2), 72–90. doi:10.1108/14601060010322284

King, J. L., Gurbaxani, V., Kraemer, K. L., McFarlan, F. W., Raman, K. S., & Yap, C. S. (1994). Institutional factors in information technology innovation. *Information Systems Research, 5*(2), 139–169. doi:10.1287/isre.5.2.139

King, R. P., & Davis, J. R. (2003). Community noise: Health effects and management. *International Journal of Hygiene and Environmental Health, 206*(2), 123–131. doi:10.1078/1438-4639-00202 PMID:12708234

Korjenic, A., Petranek, V., Zach, J., & Hroudova, J. (2011). Development and Performance evaluation of natural thermal insulation materials composed of renewable resources. *Energy and Building, 43*(9), 2518–2523. doi:10.1016/j.enbuild.2011.06.012

Korve, H. W., Farran, J. I., & Mansel, D. M. (1995). Integration of light rail transit into city streets. In *Transportation Research Board Conference Proceedings* (Vol. 1). Retrieved from http://trid.trb.org/view.aspx?id=457847

Kourtit, K., Nijkamp, P., & Arribas, D. (2012). Smart cities in perspective – a comparative European study by means of self-organizing maps. *Innovation: The Journal of Social Science Research, 25*(2), 229–246.

Koza, J. R. (1992). *Genetic Programming: On the Programming of Computers by Means of Natural Selection.* Cambridge, MA: MIT press.

Krishnan, A. (2014). Art of Smart: Can technology transform urban living in India? *India Today*, p. 30.

Krishpersad, M. (2006). Biodegradable fibrous thermal insulation. *Journal of the Brazilian Society of Mechanical Sciences and Engineering, 28*, 45–47.

Kuk, G., & Janssen, M. (2011). The Business Models and Information Architectures of Smart Cities. *Journal of Urban Technology, 18*(2), 39–52. doi:10.1080/10630732.2011.601109

Lam, J. C., Yang, L., & Liu, J. P. (2006). Development of passive design zones in China using bioclimatic approach. *Energy Conversion and Management, 47*(6), 746–762. doi:10.1016/j.enconman.2005.05.025

Lam, P. T., Wong, F. W., & Tse, K. T. (2009). Effectiveness of ICT for construction information exchange among multidisciplinary project teams. *Journal of Computing in Civil Engineering, 24*(4), 365–376. doi:10.1061/(ASCE)CP.1943-5487.0000038

Lan, T.-S., & Chiu, M.-C. (2008). Optimal noise control on plant using simulated annealing. *Transactions of the Canadian Society for Mechanical Engineering*, *32*(3-4), 423–437.

Lavasa. (2015). Retrieved from http://www.lavasa.com/

Lazaroiu, G. C., & Roscia, M. (2012). Definition methodology for the smart cities model. *Energy*, *47*(1), 326–332. doi:10.1016/j.energy.2012.09.028

Leccese, Z. L. (2014). *Intelligent wireless street lighting system.* Poland: Street lighting paper.

Leduc, G. (2009). *Longer and Heavier Vehicles - An overview of technical aspects.* Luxembourg: JRC Scientific and Technical Reports.

Lee, J.H., Hancock, M.G., & Hu, M.C. (2013). *Towards an effective framework for building smart cities: Lessons from Seoul and San Francisco. Technological Forecasting and Social Change.*

Lee, J. H., Hancock, M. G., & Hu, M.-C. (2014). Towards an effective framework for building smart cities: Lessons from Seoul and San Francisco. *Technological Forecasting and Social Change*, *89*, 80–99. doi:10.1016/j.techfore.2013.08.033

Lercher, P., Evans, G. W., Meis, M., & Kofler, W. W. (2002). Ambient neighbourhood noise and children's mental health. *Occupational and Environmental Medicine*, *59*(6), 380–386. doi:10.1136/oem.59.6.380 PMID:12040113

Lertaifa, S. B. (2015). How to strategize smart cities: Revealing the SMART model. *Journal of Business Research*, *68*(7), 1414–1419. doi:10.1016/j.jbusres.2015.01.024

Lertsatitthanakorn, C., Atthajariyakul, S., & Soponnonnarit, S. (2009). Techno-economical evaluation of a rice husk ash (RHA) based sand-cement block for reducing solar conduction heat gain to a building. *Construction & Building Materials*, *23*(1), 364–369. doi:10.1016/j.conbuildmat.2007.11.017

Lertsutthiwong, P., Khunthon, S., Siralertmukul, K., Noomum, K., & Chandrachang, S. (2008). New insulating particleboards prepared from mixture of solid wastes from tissue paper manufacturing and corn peel. *Bioresource Technology*, *99*(11), 4841–4845. doi:10.1016/j.biortech.2007.09.051 PMID:17977719

Leung, M.-Y., & Yu, J. (2014). Value methodology in public engagement for construction development projects. *Built Environment Project and Asset Management*, *4*(1), 55–70. doi:10.1108/BEPAM-05-2012-0033

Li, D., Shan, J., Shao, Z., Zhou, X., & Yao, Y. (2013). Geomatics for smart cities – concept key techniques, and applications. *Geo-spatial. Information Science*, *16*(1), 13–24.

Li, H., & Love, P. (1998). Site-Level Facilities Layout Using Genetic Algorithms. *Journal of Computing in Civil Engineering*, *12*(4), 227–231. doi:10.1061/(ASCE)0887-3801(1998)12:4(227)

Lima, S. A., Varum, H., Sales, A., & Neto, V. F. (2012). Analysis of the mechanical properties of compressed earth block masonry using the sugarcane bagasse ash. *Construction & Building Materials*, *35*, 829–837. doi:10.1016/j.conbuildmat.2012.04.127

Ling, I. H., & Teo, D. L. (2011). Properties of EPS RHA lightweight concrete bricks under different curing conditions. *Construction & Building Materials*, *25*(8), 3648–3655. doi:10.1016/j.conbuildmat.2011.03.061

Litman, T. (2014, November 19-21). Implementing Transport Policies and Programmes toward Realizing "Bali Vision Three Zeros - Zero Congestion, Zero Pollution, and Zero Accidents towards Next Generation Transport Systems in Asia." Paper presented at the Eighth Regional Environmentally Sustainable Transport (EST) Forum In Asia.

Li, Y., Mai, Y. W., & Ye, L. (2000). Sisal fibre and its composites: A review of recent developments. *Composites Science and Technology*, *60*(11), 2037–2055. doi:10.1016/S0266-3538(00)00101-9

Logistic cluster. (2015, May). Retrieved from http://log.logcluster.org/operational-environment/green-logistics/index.html

Lombardi, P., Giordano, S., Farouh, H., & Yousef, W. (2012). Modelling the smart city performance. *Innovation – The European Journal of Social Science Research*, 25(2), 137-149.

Losa, M., Leandri, P., & Rocchio, P. (2012). Crumb Rubber Modified Asphalt Concrete for Low Noise Surfaces. *Presented at the Asphalt Rubber Conference*, Munich. Retrieved from http://www.ra-foundation.org/crumb-rubber-modified-asphalt-concrete-for-low-noise-surfaces/

Low, S. P., & Teo, J. A. (2004). Implementing Total Quality Management in Construction Firms. *Journal of Management Engineering*, *20*(1), 8–15. doi:10.1061/(ASCE)0742-597X(2004)20:1(8)

Lund, H. (2007). Renewable energy strategies for sustainable development. *Energy*, *32*(6), 912–919. doi:10.1016/j.energy.2006.10.017

Luo, X., Shen, G. Q., Fan, S., & Xue, X. (2011). A group decision support system for implementing value management methodology in construction briefing. *International Journal of Project Management*, *29*(8), 1003–1017. doi:10.1016/j.ijproman.2010.11.003

Lusk, S. L., Kerr, M. J., & Kauffman, S. A. (1998). Use of hearing protection and perceptions of noise exposure and hearing loss among construction workers. *American Industrial Hygiene Association Journal*, *59*(7), 466–470. doi:10.1080/15428119891010217 PMID:9697294

Lyon, R. H. (2013). *Machinery Noise and Diagnostics*. Butterworth-Heinemann.

Madakam, S., & Ramaswamy, R. (2013). The State of Art: Smart Cities in India: A Literature Review Report. *International Journal of Innovative Research and Development*, *2*(12).

Madu, C. N., Kuei, C.-H., & Winokur, D. (1995). A strategic total quality management (STQM) approach. *Futures*, *27*(8), 839–856. doi:10.1016/0016-3287(95)00048-2

Magis, K. (2010). Community resilience: An indicator of social sustainability. *Society & Natural Resources*, *23*(5), 401–416. doi:10.1080/08941920903305674

MAGTiOŚ. Zarządzenie Ministra Administracji, Gospodarki Terenowej i Ochrony Środowiska z dnia 9 listopada 1982 r. w sprawie szczegółowych zasad wyznaczania granic i obszarów stref ochronnych oraz orientacyjnych wskaźników ich szerokości. (1982).

Maha, H. D., Rita, K., Samar, K., & Darine, M. (2015). Current status of dental waste management in Lebanon. *Environmental Nanotechnology. Monitoring & Management*, *4*, 1–5.

Mahizhnan, A. (1999). Smart cities: The Singapore case. *Cities (London, England)*, *16*(1), 13–18. doi:10.1016/S0264-2751(98)00050-X

Mahzuz, H. A., Ahmed, M., Ashrafuzzaman, M., Karim, R., & Ahmed, R. (2011). Performance evaluation of bamboo with mortar and concrete. *Journal of Engineering and Technology Research*, *3*(12), 342–350.

Malhotra, S. K., & Tehri, S. P. (1995). Building materials from granulated blast furnace slag – some new prospects. *Indian Journal of Engineering and Materials Sciences*, *2*, 80–82.

Malte Laub, T. M. (2015, May). *Gebäude gehen auf Sendung*. Von Genios - Fachmedien und Mittelstand. Retrieved from http://www.genios.de/presse-archiv/artikel/HB/20150515/gebaeude-gehen-auf-sendung/F75850E1-EF6D-44D2-9A4B-B2E490F7BEB7.html

Mannan, M. A., & Ganapathy, C. (2004). Concrete from an agricultural waste-oil palm shell (OPS). *Building and Environment*, *39*(4), 44–448. doi:10.1016/j.buildenv.2003.10.007

Manuel, J. (2005). Clamoring for Quiet: New Ways to Mitigate Noise. *Environmental Health Perspectives*, *113*(1), A46–A49. doi:10.1289/ehp.113-a46 PMID:15631960

Mao, Y. (2006). Study on Climate Adaptability of Human Beings to Thermal Comfort in China (Vol. 105). Xi'an: Xi'an University of Architecture and Technology.

Mark, A. A., & Russell, A. O. (2011). A comparative study of Bamboo reinforced concrete beams using different stirrup materials for rural construction. *International Journal of Civil and Structural Engineering*, *2*(1), 407–423.

Marsal-Llacuna, M.-L., Colomer-Llinàs, J., & Meléndez-Frigola, J. (2015). Lessons in urban monitoring taken from sustainable and livable cities to better address the Smart Cities initiative. *Technological Forecasting and Social Change*, *90*, 611–622. doi:10.1016/j.techfore.2014.01.012

Marsh, K. J. (1982). The concawe model for calculating the propagation of noise from open-air industrial plants. *Applied Acoustics*, *15*(6), 411–428. doi:10.1016/0003-682X(82)90003-2

Masui, K., Sakao, T., Kobayashi, M., & Inaba, A. (2003). Applying Quality Function Deployment to environmentally conscious design. *International Journal of Quality & Reliability Management*, *20*(1), 90–106. doi:10.1108/02656710310453836

Matei, L., & Lazar, C.-G. (2011). Quality Management and the Reform of Public Administration in Several States in South-Eastern Europe. Comparative Analysis. *Theoretical and Applied Economics*, *4*(557), 65–98.

Mbonu, E. S., Kennedy, C. O., Godwin, O. U., & Chinwe, G. E. (2013). E-MART: A novel Smart Home Governance System for a densely populated power challenged environment. *Proceedings of theIEEE International Conference on Emerging & Sustainable Technologies for Power & ICT in a Developing Society (NIGERCON)* (pp. 100-105). doi:10.1109/NIGERCON.2013.6715644

McKenna, E., Richardson, I., & Thomson, M. (2012). Smart meter data: Balancing consumer privacy concerns with legitimate applications. *Energy Policy*, *41*, 807–814. doi:10.1016/j.enpol.2011.11.049

McMullan, D. (1995). Scanning electron microscopy 1928–1965. *Scanning*, *17*(3), 175–185. doi:10.1002/sca.4950170309

Mehrjerdi, Y. Z. (2010). Applications and extensions of quality function deployment. *Assembly Automation*, *30*(4), 388–403. doi:10.1108/01445151011075843

Melkonyan, D. H. (2015, May). Statistical analysis of long term air pollution in North Rhine Westphalia. Duisburg, NRW, Germany.

Melkonyan, D. A. (2012). *Long-term analysis of NO, NO2 and O3 concentrations in North Rhine-Westphalia, Germany*. Essen: Elsevier.

Memon, S. A., Shaikh, M. A., & Akbar, H. (2011). Utilization of rice husk ash as viscosity modifying agent in self compacting concrete. *Construction & Building Materials*, *25*(2), 1044–1048. doi:10.1016/j.conbuildmat.2010.06.074

Merschbrock, C., & Munkvold, B. E. (2012). A Research Review on Building Information Modeling in Construction – An Area Ripe for IS Research. *Communications of the Association for Information Systems*, *31*(10), 207–228.

Milne, M., Givoni, B., & Watson, D. (1979). *Energy Conservation through Building Design*. London: McGraw-Hill.

Ministerie van Infrastructuur en Milieu. National Waste Management Plan (NWMP), Landelijk afvalbeheerplan 2009-2021 (2014). Den Haag.

Ministry of Transportation Ontario. (2015, June). *High Occupancy Vehicle (HOV) Lanes*. Retrieved from http://www.mto.gov.on.ca/english/ontario-511/hov-lanes.shtml

Ministry of Transportation, Toronto, Ontario. (2013). *Using High Occupancy Vehicle (HOV) Lanes*.

Moffat, S., & Kohler, N. (2008). Conceptualizing the built environment as a social-ecological system. *Building Research and Information*, *36*(3), 248–268. doi:10.1080/09613210801928131

Moore, G. C., & Benbasat, I. (1991). Development of an instrument to measure the perceptions of adopting an information technology innovation. *Information Systems Research*, *2*(3), 192–222. doi:10.1287/isre.2.3.192

Morrissey, J., Moore, T., & Horne, R. E. (2011). Affordable passive solar design in a temperate climate: An experiment in residential building orientation. *Renewable Energy*, *36*(2), 568–577. doi:10.1016/j.renene.2010.08.013

MUNLV. (2007). *Bauleitplanung*. Düsseldorf: Abstandserlass.

Murphy, E., & King, E. A. (2014). Noise Mitigation Approaches. In E. M. A. King (Ed.), Environmental Noise Pollution (pp. 203–245). Boston: Elsevier. Retrieved from http://www.sciencedirect.com/science/article/pii/B9780124115958000070 doi:10.1016/B978-0-12-411595-8.00007-0

Næss, P. (2001). Urban Planning and Sustainable Development. *European Planning Studies*, *9*(4), 503–524. doi:10.1080/09654310120049871

Nair, D. G., Jagadish, K. S., & Fraaij, A. (2006). Reactive pozzolanas from rice husk ash, An alternative to cement for rural housing. *Cement and Concrete Research*, *36*(6), 1062–1071. doi:10.1016/j.cemconres.2006.03.012

Naish, D. A., Tan, A. C. C., & Demirbilek, F. N. (2014). Simulating the effect of acoustic treatment types for residential balconies with road traffic noise. *Applied Acoustics*, *79*, 131–140. doi:10.1016/j.apacoust.2013.12.021

Nakagawa, Y., & Shaw, R. (2004). Social capital: A missing link to disaster recovery. *International Journal of Mass Emergencies and Disasters*, *22*, 5–34.

Nallathiga, R., Tewari, K., Saboo, A., & Varghese, S. (2015). Evolution of Satellite township development in Pune: A Case Study. *Evolution; International Journal of Organic Evolution.*

Nam, T., & Pardo, T. A. (2011). Smart City as Urban Innovation: Focusing on Management, Policy, and Context.*Proceedings of the 5th International Conference on Theory and Practice of Electronic Governance (ICEGOV),*Tallinn, Estonia (pp. 185-194). ACM. doi:10.1145/2072069.2072100

Nam, T., & Pardo, T. A. (2012). Conceptualizing Smart City with Dimensions of Technology, People, and Institutions.*Proceedings of the 12th Annual International Conference on Digital Government Research,*College Park, MD (pp.282-291).

Naphade, M., Banavar, G., Harrison, C., Paraszczak, J., & Morris, R. (2011). Smarter Cities and Their Innovation Challenges. *Computer*, *44*(6), 32–39. doi:10.1109/MC.2011.187

Nasim, K., Iqbal, M. Z., & Khan, I. A. (2014). Antecedents of TQM implementation capability: A review with conceptual. *Total Quality Management & Business Excellence*, *25*(12), 1395–1409. doi:10.1080/14783363.2013.807682

Nebl, T., & Schroeder, A.-K. (2011). Understanding the interdependencies of quality problems and productivity. *The TQM Journal*, *23*(5), 480–495. doi:10.1108/17542731111157590

Neirotti, P., De Marco, A., Cagliano, A. C., Mangano, G., & Scorrano, F. (2014). Current trends in Smart City initiatives: Some stylised facts. *Cities (London, England)*, *38*, 25–36. doi:10.1016/j.cities.2013.12.010

Neitzel, R., Seixas, N. S., Camp, J., & Yost, M. (1999). An Assessment of Occupational Noise Exposures in Four Construction Trades. *American Industrial Hygiene Association Journal*, *60*(6), 807–817. doi:10.1080/00028899908984506 PMID:10635548

Nema, P., Nema, S., & Roy, P. (2012). An overview of global climate changing in current scenario and mitigation action. *Renewable & Sustainable Energy Reviews*, *16*(4), 2329–2336. doi:10.1016/j.rser.2012.01.044

New Jersey Department of Transportation and Pennsylvania Department of Transport. (2008). Smart Transportation Guidebook: Planning and Designing Highways and Streets That Support Sustainable and Liveable Communities.

Ng, C. F.NG. (2000). Effects of building construction noise on residents: A quasi-experiment. *Journal of Environmental Psychology*, *20*(4), 375–385. doi:10.1006/jevp.2000.0177

Nguyen, T. A., & Aiello, M. (2013). Energy intelligent buildings based on user activity: A survey. *Energy and Building*, *56*, 244–25. doi:10.1016/j.enbuild.2012.09.005

Nielsen, G., & Solberg, S. (1988). Costs and effects of traffic noise abatement measures: Basis for a national programme. *Applied Acoustics*, *25*(3), 149–168. doi:10.1016/0003-682X(88)90091-6

Niser, M. (n. d.). Lee's Method. Retrieved from http://physics.niser.ac.in/labmanuals/sem1/Lee's-Method.pdf

Nissenbaum, M., Aramini, J., & Hanning, C. (2012). Effects of industrial wind turbine noise on sleep and health. *Noise & Health*, *14*(60), 237. doi:10.4103/1463-1741.102961 PMID:23117539

NIUA. (2012). *Report on Urban Infrastructure and Services, Summary and Recommendations*. National Institute of Urban Affairs.

Norris, F. H., Stevens, S. P., Pfefferbaum, B., Wyche, K., & Pfefferbaum, R. L. (2008). Community resilience as a metaphor, theory, set of capacities, and strategy for disaster readiness. *American Journal of Community Psychology*, *41*, 127–150. PMID:18157631

NTPC guidelines. (n. d.). Ch. 6 Coal Ash. Retrieved from https://www.ntpc.co.in/ash-download/1664/6/ntpc-guide-users-coal-ash-chapter-6

Nunan, J. (2010). *The complete guide to alternative home building materials & methods: including sod, compressed earth, plaster, straw, beer cans, bottles, cordwood, and*. Atlantic Publishing Group Inc.

Nurminen, T. (1995). Female noise exposure, shift work, and reproduction. *Journal of Occupational and Environmental Medicine / American College of Occupational and Environmental Medicine*, *37*(8), 945–950.

Oakland, J., & Marosszeky, M. (2006). *Total Quality in the Construction Supply Chain.* Burlington, MA: Butterworth-Heinemann (An Imprint of Elsevier).

ÖAMTC. (2014). Umweltzonen in Deutschland. Augsburg: Österreichischer Automobil-, Motorrad und Touring Club.

Obrecht, M., & Denac, M. (2013). A sustainable energy policy for Slovenia: Considering the potential of renewables and investment costs. *Journal of Renewable and Sustainable Energy, 5*(3), 032301. doi:10.1063/1.4811283

Oke, A. O., Emuoyibofarhe, J. O., & Adetunji, A. B. (2013). Development of a GSM based control system for electrical appliances. *IACSIT International Journal of Engineering and Technology, 3*(4), 443–448.

Omar, D., Nazli, S. N., & Karuppannan, S. A. L. (2012). Clinical Waste Management in District Hospitals of Tumpat, Batu Pahat and Taiping. *Procedia: Social and Behavioral Sciences, 68*, 134–145. doi:10.1016/j.sbspro.2012.12.213

Omer, A. M. (2008). Energy environment and sustainable development. *Renewable & Sustainable Energy Reviews, 12*(9), 2265–2300. doi:10.1016/j.rser.2007.05.001

Onésippe, C., Passe-Coutrin, N., Toro, F., Delvasto, S., Bilba, K., & Arsène, M. A. (2010). Sugar cane bagasse fibres reinforced cement composites, Thermal considerations. *Composites*, 41(A), 549–556.

Oniszk Popławska, A., & Kulig, A. (2014). Application of Odour Predictions to Spatial Planning, the Case of Agricultural Biogas. *Chemical Engineering Transactions, 40*, 277–282. doi:10.3303/CET1440047

Oniszk-Popławska, A. (2014). *Site visit to Kirstockach within an M-BIST project.*

Oniszk-Popławska, A. (2015). *Site visit to Trittau within an M-BIST project.*

Osada, Y. (1988). An overview of health effects on noise. *Journal of Sound and Vibration, 127*(3), 407–410. doi:10.1016/0022-460X(88)90363-X

Oyedepo, S. O. (2012). Energy and sustainable development in Nigeria: The way forward. *Energy. Sustainability Science, 2*(1), 1–17.

Paetz, A. G., Becker, B., Fichtner, W., & Schmeck, H. (2011). Shifting electricity demand with smart home technologies–an experimental study on user acceptance. *Proceedings of the30th USAEE/IAEE North American conference online proceedings* (pp. 19-20).

Paetz, A. G., Dütschke, E., & Fichtner, W. (2012). Smart homes as a means to sustainable energy consumption: A study of consumer perceptions. *Journal of Consumer Policy, 35*(1), 23–41. doi:10.1007/s10603-011-9177-2

Palmarini, N. (2012). *Smart Transportation Systems. Human Centric Solution Centers Europe.* IBM Corporation.

Panagopoulos, T., Duque, J. A. G., & Dan, M. B. (in press). Urban planning with respect to environmental quality and human well-being. *Environmental Pollution.* Retrieved from http://www.sciencedirect.com/science/article/pii/S0269749115003711

Panyakaew, S., & Fotios, S. (2008, October 22-24). Agricultural waste materials as thermal insulation for dwellings in Thailand: Preliminary results. *Proceedings of the 25th conference on Passive and Low Energy Architecture PLEA '08*, Durbin.

Papazafeiropoulou, A., Gandecha, R., & Stergioulas, L. (2005) Interpretive flexibility along the innovation decision process of the UK NHS care records service (NCRS). Insights from a local implementation case study.*Proceedings of the Thirteenth European Conference on Information Systems*Regensburg, Germany.

Parasuraman, E.N. (n. d.). Blog: *So, what is the smartness quotient of your city?* Schneider Electric.

Patrick, O., Tolulolope, O., & Sunny, O. (2013). Smart Grid Technology and Its Possible Applications to the Nigeria 330 kV Power System. *Smart Grid and Renewable Energy*, *4*(05), 391–397. doi:10.4236/sgre.2013.45045

PCUGP. Zarządzenie Prezesa Centralnego Urzędu Gospodarki Wodnej z dnia 30 maja 1967 r. w sprawie szerokości stref ochronnych ustanawianych dla ochrony powietrza atmosferycznego przed zanieczyszczeniem. (1967).

Pedersen, E., & Waye, K. P. (2007). Wind turbine noise, annoyance and self-reported health and well-being in different living environments. *Occupational and Environmental Medicine*, *64*(7), 480–486. doi:10.1136/oem.2006.031039 PMID:17332136

Penna, M., Pottstock, H., & Velasquez, N. (2005). Effect of natural and synthetic noise on evoked vocal responses in a frog of the temperate austral forest. *Animal Behaviour*, *70*(3), 639–651. doi:10.1016/j.anbehav.2004.11.022

Perera, C., Zaslavsky, A., Christen, P., & Georgakopoulos, D. (2014). Sensing as a service model for smart cities supported by Internet of Things. *Transactions on Emerging Telecommunications Technologies*, *25*(1), 81–93. doi:10.1002/ett.2704

Petek, J., & Glavič, P. (1996). An integral approach to waste minimization in process industries. *Resources, Conservation and Recycling*, *17*(3), 169–188. doi:10.1016/0921-3449(96)01151-2

Petts, J., & Gev, E. (1994). *Environmental impact assessment for waste treatment and disposal facilities.* West Sussex: John Wiley&Sons.

Pickett, S. T. A., Cadenasso, M. L., Grove, J. M., Boone, C. G., Groffman, P. M., Irwin, E., & Warren, P. et al. (2011). Urban ecological systems: Scientific foundations and a decade of progress. *Journal of Environmental Management*, *92*(3), 331–362. doi:10.1016/j.jenvman.2010.08.022 PMID:20965643

Pinch, T. J., & Bijker, W. E. (1984). The social construction of facts and artefacts: Or how the sociology of science and the sociology of technology might benefit each other. *Social Studies of Science*, *14*(3), 399–441. doi:10.1177/030631284014003004

Pinto, J., Paiva, A., Varum, H., Costa, A., Cruz, D., Pereira, S., & Agarwal, J. et al. (2011). Corn's cob as a potential ecological thermal insulation material. *Energy and Building*, *43*(8), 1985–1990. doi:10.1016/j.enbuild.2011.04.004

Piro, G., Cianci, I., Grieco, L. A., Boggia, G., & Camarda, P. (2014). Information centric services in Smart Cities. *Journal of Systems and Software*, *88*, 169–188. doi:10.1016/j.jss.2013.10.029

Pitt, M., Tucker, M., Riley, M., & Longden, J. (2009). Towards sustainable construction: Promotion and best practices. *Construction Innovation*, *9*(2), 1471–4175. doi:10.1108/14714170910950830

Pollit, C. (2006). Changing European states, changing public administration: Antistatist reforms and new administrative directions: Public administration in the United Kingdom. In E. Otenyo & N. S. Lind (Eds.), *Comparative Public Administration: The Essential Readings* (pp. 777–792). Oxford, UK: Jai Press (an imprint of Elsevier). doi:10.1016/S0732-1317(06)15036-8

Porter, M. E. (2000). What is Strategy? *Harvard Business Review*, 2000, 4134.

Premkumar, G., & Potter, M. (1995). Adoption of computer aided software engineering (CASE) technology: An innovation adoption perspective. *The Data Base for Advances in Information Systems*, *26*(2-3), 105–124. doi:10.1145/217278.217291

Premkumar, G., Ramamurthy, K., & Crum, M. (1997). Determinants of EDI adoption in the transportation industry. *European Journal of Information Systems*, *6*(2), 107–121. doi:10.1057/palgrave.ejis.3000260

Project Management Institute. (2013). *A Guide to the Project Management Body of Knowledge (PMBOK Guide)* (5th ed.). USA: Project Management Institute.

Pucher, J., Park, H., Kim, M. H., & Song, J. (2005). Public transport reforms in Seoul: Innovation motivated by funding crisis. *Journal of Public Transportation*, *8*(5), 41–62. doi:10.5038/2375-0901.8.5.3

Qiu, J., & Liu, H. M. (2011). Northeast China rural residential passive solar design exploration. In *Proceedings of International Conference of Green Building Materials and Energy-saving Construction (GBMEC 2011)*. Harbin, People's Republic of China (pp. 200-203). doi:10.4028/www.scientific.net/AMR.280.200

Quintana, G., Velasquez, J., Betancourt, S., & Ganan, P. (2009). Binderless fibre board from steam exploded babana bunch. *Industrial Crops and Products*, *29*(1), 60–66. doi:10.1016/j.indcrop.2008.04.007

Rahman, M. M., Rashid, M.H., Hossain, M. A., Hasan, M. T., & Hasan, M. K. (2011). Performance Evaluation of Bamboo Reinforced Concrete Beam. *International Journal of Engineering and Technology*, 11(4), 142-46.

Rahman, M. A. (1987). Properties of clay-sand-rice husk ash mixed bricks. *Cement Composites and Lightweight Concrete*, *9*(2), 105–108. doi:10.1016/0262-5075(87)90026-1

RailCorp. (2012). *SPC 235-RESILIENT BASEPLATES*. NSW Transport. Retrieved from http://www.asa.transport.nsw.gov.au/sites/default/files/asa/railcorp-legacy/disciplines/civil/spc-235.pdf

Rajput, D., Bhagade, S. S., Raut, S. P., Ralegaonakar, R. V., & Mandavgane, S. A. (2012). Reuse of cotton and recycle paper mill waste as building material. *Construction & Building Materials*, *32*, 470–475. doi:10.1016/j.conbuildmat.2012.02.035

Ramakrishna, G., & Sundararajan, T. (2005). Studies on the durability of natural fibres and the effect of corroded fibres on the strength of mortar. *Cement and Concrete Composites*, *27*(5), 575–582. doi:10.1016/j.cemconcomp.2004.09.008

Ratti, C. (2004a). Space syntax: Some inconsistencies. *Environment and Planning. B, Planning & Design*, *31*(4), 487–499. doi:10.1068/b3019

Ratti, C. (2004b). Rejoinder to Hillier and Penn. *Environment and Planning. B, Planning & Design*, *31*(4), 513–516. doi:10.1068/b3019b

Razavi, H., Ramezanifar, E., & Bagherzadeh, J. (2014). An economic policy for noise control in industry using genetic algorithm. *Safety Science*, *65*, 79–85. doi:10.1016/j.ssci.2013.12.010

Reddy, B. V. V., & Vyas, U. C. V. (2008). Influence of shear bond strength on compressive strength and stress-strain characteristics of masonry. *Materials and Structures*, *41*(10), 1697–1712. doi:10.1617/s11527-008-9358-x

Reinisch, C., Kofler, M. J., Iglesias, F., & Kastner, W. (2011). ThinkHome: Energy efficiency in future smart homes. *EURASIP Journal on Embedded Systems*, 2011, 18.

Rmasamy, V., & Biswas, S. (2008). Performance of Rice Husk Ash Concrete with Superplasticizers. *ICI Journal*, *4*, 27–34.

Roads and Maritime Services NSW. (2009). *Road Design Guide*. Retrieved from http://www.rms.nsw.gov.au/business-industry/partners-suppliers/aus-roads-guides/road-design.html

Roads and Maritime Services. (2015). *Noise Criteria Guideline*. NSW, Australia. Retrieved from http://www.rms.nsw.gov.au/documents/about/environment/noise-criteria-guideline-book.pdf

Rodriguez-Ubinas, E., Montero, C., Porteros, M., Vega, S., Navarro, I., Castillo-Cagigal, M., & Gutiérrez, A. et al. (2014). Passive design strategies and performance of Net Energy Plus Houses. *Energy and Building*, *83*, 10–22. doi:10.1016/j.enbuild.2014.03.074

Rogers, E. M. (2003). *Diffusion of innovations*. New York: Free Press.

Rossing, T. D., Wheeler, P. A., & Moore, F. R. (2002). *The Science of Sound*. Addison Wesley.

Roumet, C. (2015). *25 years of Energy Cities*. Brussels: The European association of local authorities in energy transition.

Rozenfeld, O., Sacks, R., & Rosenfeld, Y. (2009). 'CHASTE': Construction hazard assessment with spatial and temporal exposure. *Construction Management and Economics*, *27*(7), 625–638. doi:10.1080/01446190903002771

RTA. (2001). *RTA Environmental Noise Management Manual.* Roads and Traffic Authority of New South Wales. Retrieved from http://www.rms.nsw.gov.au/about/environment/reducing-noise/index.html

Rukzon, S., & Chindaprasirt, P. (2012). Utilization of bagasse ash in high-strength concrete. *Materials & Design*, *34*, 45–50. doi:10.1016/j.matdes.2011.07.045

Rushbrook, Ph., Chandra, C., & Gayton, S., (2000). *Starting healthcare waste management in medical institution, practical approach.* World Health Organization (WHO) Healthcare Practical Information Series No. 1.

Sales, A., & Lima, S. A. (2010). Use of Brazilian sugarcane ash in concrete as sand replacement. *Waste Management (New York, N.Y.)*, *30*(6), 1114–1122. doi:10.1016/j.wasman.2010.01.026 PMID:20163947

Sampathrajan, A., Vijayaraghavan, N. C., & Swaminathan, K. R. (1992). Mechanical and thermal properties of particle boards made from farm residues. *Bioresource Technology*, *40*(3), 249–251. doi:10.1016/0960-8524(92)90151-M

Sandberg, U. (1999, December 6-8). Abatement of Traffic, Vehicle and Tire/Road Noise - The Global Perspective. Proceedings of Internoise 99 - The 1999 International Congress On Noise Control Engineering, Fort Lauderdale, Florida, USA (Vol. 1). Retrieved from http://trid.trb.org/view.aspx?id=651529

Sata, V., Jaturapitakkul, C., & Kiattikomol, K. (2007). Influence of pozzolana from various by-product materials on mechanical properties of high-strength concrete. *Construction & Building Materials*, *21*(7), 1589–1598. doi:10.1016/j.conbuildmat.2005.09.011

Saul, J. L. J. (2004). Facilitating family and community resilience in response to major disaster. Living beyond loss: Death in the family.

Sawyer, S., & Tapia, A. (2005). The sociotechnical nature of mobile computing work: Evidence from a study of policing in the United States. *International Journal of Technology and Human Interaction*, *1*(3), 1–14. doi:10.4018/jthi.2005070101

Schaffers, H., Komninos, N., Pallot, M., Trousse, B., Nilsson, M., & Oliveira, A. (2011). Smart Cities and the Future Internet: Towards Cooperation Frameworks for Open Innovation. *Future Internet Assembly*, *6656*, 431–446. doi:10.1007/978-3-642-20898-0_31

Schilderman, T. (2010). Putting people at the centre of reconstruction. In M. Lyons, T. Schilderman, & C. Boano (Eds.), Building back better: delivering people-centered housing reconstruction at scale (pp. 7-38). Warwickshire, UK: Practical Action Publishing Ltd. doi:10.3362/9781780440064.002

Schmitt, G. (2013). Spatial modelling issues in future smart cities. *Geo-spatial. Information Science*, *16*(1), 7–12.

Scholes, W. E., & Sargent, J. W. (1971). Designing against noise from road traffic. *Applied Acoustics*, *4*(3), 203–234. doi:10.1016/0003-682X(71)90017-X

Scott, F. (2007) *Teaching Homes to be Green*, London: Green Alliance Smart Home Group Nigeria (2014) *iSmart Home Technology Integrators Limited.* Retrieved from http://www.smarthomes.com.ng

Sebastian, R. (2011). Changing roles of the clients, architects and contractors through BIM. *Engineering, Construction, and Architectural Management*, *18*(2), 176–187. doi:10.1108/09699981111111148

Sengupta, P., Saikia, N., & Borthakur, P. C. (2002). Bricks from petroleum effluent treatment plant sludge: Properties and environmental characteristics. *Journal of Environmental Engineering*, *128*(11), 1090–1094. doi:10.1061/(ASCE)0733-9372(2002)128:11(1090)

Shah, M. N., Nagargoje, S., Shah, C. (2015). Assessment of Ahmedabad (India) and Shanghai (China) on Smart City Parameters applying the Boyd Cohen Smart City Wheel. *Proceedings of CRIOCM 2015.*

Shakir, A. A., Naganathan, S., & Mustapha, N. K. (2013). Properties of bricks made using fly ash, quarry dust and billet scale. *Construction & Building Materials*, *41*, 131–138. doi:10.1016/j.conbuildmat.2012.11.077

Shapiro, J. M. (2006, May). Smart Cities: Quality of life, productivity, and the growth effects of human capital. *The Review of Economics and Statistics*, 88(2), 324-335.

Shetty, M. S. (2013). *Concrete technology – theory and practice. S.* New Delhi, India: Chand Publishing.

Skinder, B. M., Sheikh, A. Q., Ashok, K. P., & Ganai, B. A. (2013). Brick kiln emissions and its environmental impact: A review. *Journal of Ecology and the Natural Environment*, *6*(1), 1–11.

Smola, A., & Scholkopf, B. (1998). A tutorial on support vector regression. NeuroCOLT Technical Report NC-TR-98-030, Royal Holloway College, University of London, UK.

Smola, A., & Scholkopf, B. (1998). On a kernel based method for pattern recognition, regression, approximation and operator inversion. *Algorithmica*, *22*(1-2), 211–231. doi:10.1007/PL00013831

Sørensen, M., Lühdorf, P., Ketzel, M., Andersen, Z. J., Tjønneland, A., Overvad, K., & Raaschou-Nielsen, O. (2014). Combined effects of road traffic noise and ambient air pollution in relation to risk for stroke? *Environmental Research*, *133*, 49–55. doi:10.1016/j.envres.2014.05.011 PMID:24906068

Sörensson, K. (2015, June 26). Transportation situation in Malmö, County Skåne coordinated by Skånetrafiken. (C. Jenne, Interviewer).

Standards Australia. (1990).

Standards Australia. (2010). *Standards Australia 2010, Guide to noise and vibration control on construction, demolition and maintenance sites, AS2436.*

Suvimol, S., & Daungruedee, C. (2008). Bagasse ash, Effect of pozzolanic activity and application in cement use aspect. *Proceedings of the3rd ACF International Conference*.

Swiss, J. E. (1992). Adapting Total Quality Management (TQM) to Government. *Public Administration Review*, *52*(4), 356–362. doi:10.2307/3110395

Sydney Trains. (2015). Tangara. Retrieved from http://www.sydneytrains.info/about/fleet/t_sets.htm

TACA. (2015). *Belastingdienst*. Heerlen, The Netherlands: Tax and Customs Administration.

Tamburini, L., Rossi, M., & Brunelli, D. (2015). Electronic and ICT Solutions for Smart Buildings and Urban Areas. In A. Vesco & F. Ferrero (Eds.), *Handbook of Research on Social, Economic, and Environmental Sustainability in the Development of Smart Cities* (pp. 165–192). Hershey, PA, USA: IGI Global. doi:10.4018/978-1-4666-8282-5.ch009

Taneja, S., Akcamete, A., Akinci, B., Garrett, J., Soibelman, L., & East, E. W. (2010). Analysis of three indoor localization technologies to support facility management field activities.*Proceedings of the International Conference on Computing in Civil and Building Engineering*, Nottingham, UK.

Tang, P., & Venables, T. (2000). 'Smart'homes and telecare for independent living. *Journal of Telemedicine and Telecare*, *6*(1), 8–14. doi:10.1258/1357633001933871 PMID:10824384

Tanninen, K., Puumalainen, K., & Sandström, J. (2010). The power of TQM: Analysis of its effects on profitability, productivity and customer satisfaction. *Total Quality Management & Business Excellence*, *21*(2), 171–184. doi:10.1080/14783360903549949

Tashima, M.M., Da-Silva, C. R., & Akasaki, J.L. (2004). The possibility of adding RHA to the concrete. *Proceedings of the International RILEM conference on the use of recycled materials in building structures* (pp. 778-786).

Tay, J. H. (1987). Bricks manufactured from sludge. *Journal of Environmental Engineering*, *113*(2), 278–283. doi:10.1061/(ASCE)0733-9372(1987)113:2(278)

Terziovski, M., & Samson, D. (2000). The effect of company size on the relationship between TQM strategy and organizational performance. *The TQM Magazine*, *12*(2), 144–149. doi:10.1108/09544780010318406

The Australian Noise Exposure Forecast (ANEF) System. (2014). Retrieved from http://www.infrastructure.gov.au/aviation/environmental/transparent_noise/expanding/app_a.aspx

The Council for Excellence in Government. (2006). Are we ready? Introducing the Public Readiness Index: a survey-based tool to measure the preparedness of individuals, families and communities. Retrieved from http://www.fema.gov/media-library-data/20130726-1910-25045-5489/public_readiness_index.pdf

The Smart City Wien Initiative Report. (2014, July). Owner and publisher: Vienna City Administration.

The World Bank. (2015). Access to electricity (% of population). Retrieved fromhttp://data.worldbank.org/indicator/EG.ELC.ACCS.ZS?order=wbapi_data_value_2010+wbapi_data_value+wbapi_data_value-first&sort=desc

Tiwari, R., Cervero, R., & Schipper, L. (2011). Driving CO2 reduction by Integrating Transport and Urban Design strategies. *Cities (London, England)*, *28*(5), 394–405. doi:10.1016/j.cities.2011.05.005

Tojo, N., & Fischer, C. (2011). Europe as a Recycling Society. European Recycling Policies in relation to the actual. *ETC/SCP Working Paper*, (ETC/SCP working paper 2/2011).

Tomaževič, N., Seljak, J., & Aristovnik, A. (2015). *TQM in public administration organizations: an application of data envelopment analysis in the police service*. Total Quality Management & Business Excellence; doi:10.1080/14783363.2015.1007861

Tornatzky, L. G., & Klein, K. J. (1982). Innovation Characteristics and Innovation Adoption-Implementation: A Meta Analysis of Findings. *IEEE Transactions on Engineering Management*, *29*(11), 28–45. doi:10.1109/TEM.1982.6447463

Torraco, R. J. (2005). Writing Integrative Literature Reviews: Guidelines and Examples. *Human Resource Development Review*, *4*(3), 356–367. doi:10.1177/1534484305278283

Transport Research and Innovation Portal (TRIP). (2014). *Intelligent Transport Systems: Communicating Transport Research And Innovation*. European Commission.

Trujillo, M., Prdpmez, A., & Hernandez, C. (2000). *Risk-mapping and local capacities: Lessons from Mexico and Central America*. Oxfam Working Papers, Oxfam, Great Britain.

Turgut, P., & Algin, H. M. (2007). Limestone dust and wood sawdust as brick material. *Building and Environment*, *42*(9), 3399–3403. doi:10.1016/j.buildenv.2006.08.012

U.S. Department of Health and Human Services. (1988). Proposed National Strategies for the Prevention of Leading Work-related Diseases and Injuries: Noise-induced Hearing Loss (No. 89-135).

U.S. Geological Survey. (2015). Mineral Commodity Summaries, January 2015.

UNEP. (2011). *Waste and Climate Change*. Global Trends and Strategy Framework.

United Nations. (2012) *World Population Prospects: The 2012 Revision*. Retrieved from http://esa.un.org/wpp/

Vallet, M., Maurin, M., Page, M. A., Favre, B., & Pachiaudi, G. (1978). Annoyance from and habituation to road traffic noise from urban expressways. *Journal of Sound and Vibration*, *60*(3), 423–440. doi:10.1016/S0022-460X(78)80121-7

Van Audenhove, F.-J., Korniichuk, O., Daubey, L., & Pourbaix, J. (2014). The Future on Urban Mobility 2.0: Imperatives to Shape Extended Mobility Ecosystems of Tomorrow.

van Hinte, E., Peeren, C., & Jongert, J. (2007). *Superuse: Constructing New Architecture by Shortcutting Material Flows*. Rotterdam, the Netherlands: Nai Publishers.

van Kempen, E., & Babisch, W. (2012). The quantitative relationship between road traffic noise and hypertension: A meta-analysis. *Journal of Hypertension*, *30*(6), 1075–1086. doi:10.1097/HJH.0b013e328352ac54 PMID:22473017

Van Renterghem, T., & Botteldooren, D. (2009). Reducing the acoustical façade load from road traffic with green roofs. *Building and Environment*, *44*(5), 1081–1087. doi:10.1016/j.buildenv.2008.07.013

Vapnik, V. N. (1998). *Statistical learning theory*. New York: Wiley.

Vapnik, V. N. (1999). *The nature of statistical learning theory* (2nd ed.). Berlin: Springer.

Vaughan, L. (2007). The spatial syntax of urban segregation. *Progress in Planning*, *27*(3), 205–294. doi:10.1016/j.progress.2007.03.001

Vereniging van Nederlandse Gemeenten. Bedrijven en milieuzonering' uit (2009). Holland.

Verkehrsdienst, O. u. (2015, February). Freephone traffic engineering hotline since 2009. City Cologne.

Victoria EPA. (2014). Commercial and industrial noise. Retrieved from http://www.epa.vic.gov.au/your-environment/noise/commercial-and-industrial-noise#Introductiontonoisestandards

Vijayraghavan, P., & Krishnan, R. (1998). Noise in electric machines: a review. *Proceedings of theThirty-Third IAS Annual MeetingIndustry Applications Conference* (*Vol. 1*, pp. 251–258). Doi:<ALIGNMENT.qj></ALIGNMENT>10.1109/IAS.1998.732298

Vincent, E. N., & Yusuf, S. D. (2014). Integrating Renewable Energy and Smart Grid Technology into the Nigerian Electricity Grid System. *Smart Grid and Renewable Energy*, *5*(09), 220–238. doi:10.4236/sgre.2014.59021

Vischer, J. C. (2008). Towards a user-centred theory of the built environment. *Building Research and Information*, *36*(3), 231–240. doi:10.1080/09613210801936472

Wagner, A. (2010). *Informal Public Transport; Transport Policy Advisory Services: Federal Ministry for Economic Cooperation and Development Bryman, A 2001, Social Research Methods*. Oxford University Press.

Wankhade, K. (2012). *JNNURM: An Opportunity for Environmentally Sustainable Urbanisation.* IGC Working Paper.

Washburn, D., & Sindhu, U. (2010). Helping CIOs Understand "Smart City" Initiatives. *Growth*, 2010, 17.

Waste atlas 2013 report. (2013). Retrieved from http://www.atlas.d-waste.com/

WHO. (2011). *Burden of disease from environmental noise, quantification of healthy years lost in Europe*. Copenhagen, Denmark: World Health Organisation.

Wiele, A., van der, Williams, A.R.T., & Dale, V.G. (2000). Total quality management: Is it a fad, fashion, or fit. *Quality Management Journal*, *7*(2), 65–79.

Williams, D. B., & Carter, C. B. (2009). Transmission electron microscopy: a textbook for materials science (Vol. 2, pp. 559-560). Springer. doi:10.1007/978-0-387-76501-3

Wilson, C. E. (1989). *Noise control: measurement, analysis, and control of sound and vibration.* Harper & Row Limited.

Winters, J. V. (2011). Why are smart cities growing? Who moves and who stays. *Journal of Regional Science*, *51*(2), 253–270. doi:10.1111/j.1467-9787.2010.00693.x

Wit, S. d. (2015, May). *Sonnenseite - Oekologische Kommunikation*. Retrieved from http://www.sonnenseite.com/de/energie/erster-solar-radweg-effizienter-als-erwartet.html

Wolkenstein, P. (2014, September). Euro-6-Neuzulassungen - Saubere Sache. Stuttgart, Baden Wuerttemberg, Germany. Retrieved from http://www.auto-motor-und-sport.de/news/euro-6-neuzulassungen-saubere-sache-6053193.html

Wolman, A. (1965). The metabolism of cities. *Scientific American*, *3*(213). PMID:14338520

Woolgar, S. (1996). Technologies as cultural artifacts. In W. H. Dutton (Ed.), *Information and Communication Technologies: visions and realities* (pp. 87–102). Oxford: Oxford University Press.

Xiaoan, G. (2006). Railway environmental noise control in China. *Journal of Sound and Vibration*, *293*(3–5), 1078–1085. doi:10.1016/j.jsv.2005.08.058

Xu, J., Sugawara, R., Widyorini, R., Han, G., & Kawai, S. (2004). Manufacture and thermal properties of low-density binderless particleboard from kenaf core. *Journal of Wood Science*, *50*(1), 62–67. doi:10.1007/s10086-003-0522-1

Xu, W., Lo, T. Y., & Menon, S. A. (2012). Microstructure and reactivity of rice husk ash. *Construction & Building Materials*, *29*, 541–547. doi:10.1016/j.conbuildmat.2011.11.005

Xu, X., Shen, Q., Fan, H., Li, H., & Fan, S. (2012). IT supported collaborative work in A/E/C projects: A ten-year review. *Automation in Construction*, *21*, 1–9. doi:10.1016/j.autcon.2011.05.016

Yang, K., & Callahan, K. (2007). Citizen involvement efforts and bureaucratic responsiveness: Participatory values, stakeholder pressures, and administrative practicality. *Public Administration Review*, *67*(2), 249–264. doi:10.1111/j.1540-6210.2007.00711.x

Yang, L. (2003). *Building Climate Analysis and Design Strategies. Xi'an.* Xi'an University of Architecture and Technology.

Yarbrough, D. W. (2005). Apparent thermal conductivity data and related information for rice hulls and crushed pecan shells. *Thermal Conductivity*, *27*, 222–230.

Yates, J. K. (2013). Sustainable methods for waste minimization in construction. *Construction Innovation*, *13*(3), 281–301. doi:10.1108/CI-Nov-2011-0054

Yigitcanlar, T., & Teriman, S. (2015). Rethinking sustainable urban development: Towards an integrated planning and development process. *International Journal of Environmental Science and Technology*, *12*(1), 341–352. doi:10.1007/s13762-013-0491-x

Yli-Pelkonen, V., & Niemelä, J. (2005). Linking ecological and social systems in cities: Urban planning in Finland as a case. *Biodiversity and Conservation*, *14*(8), 1947–1967. doi:10.1007/s10531-004-2124-7

Yu, C.-H., Chen, C.-K., Chen, W.-H., & Chang, H.-C. (2012). Developing a revised QFD technique to meet the needs of multiple-customer groups: A case of public policy analysis. *Total Quality Management & Business Excellence*, *23*(11-12), 1413–1431. doi:10.1080/14783363.2012.704266

Zhang, Q., He, K., & Huo, H. (2012). Policy: Cleaning China's air. *Nature*, *484*(7393), 161–162. PMID:22498609

Zhao, B., Daughetry, K. E., & Zhang, L. (1992). In R. W. Piggot (Ed.), *Ingram KD, Hill R. Portland cement doped with lead compounds." Proc., 1st Int. Sump. Cement Industry Solution to Waste Management* (pp. 359–372). Calgary Alta., Canada.

Zhou, X.-, Zheng, F., Li, H.-, & Lu, C.-. (2010). An environment-friendly thermal insulation material from cotton stalk fibers. *Energy and Building*, *42*(7), 1070–1074. doi:10.1016/j.enbuild.2010.01.020

Zubizarreta, I., Seravalli, A., & Arrizabalaga, S. (in press). Smart City Concept: What It Is and What It Should Be. *Journal of Urban Planning and Development*. Retrieved from http://ascelibrary.org/doi/abs/10.1061/(ASCE)UP.1943-5444.0000282

About the Contributors

Goh Bee Hua received the NUS Overseas Graduate Scholarship to pursue her doctoral study at the University College London in 1994 and obtained her Ph.D. in Building in 1997. She had since contributed over 65 research articles to top peer-reviewed journals and several leading conferences in her field, achieving for publication performance a citations H-index of 10 which excludes self-citations. She had also published a book chapter on Green BIM in the *Handbook of Research on Building Information Modelling and Construction Informatics*, and a book on Implementing IT Business Strategy in the Construction Industry. In 2007, she had received the Outstanding Academic Achievement Award conferred by the International Federation for Information Processing's Technical Committee 8.9 on Enterprise Information Systems, IEEE Systems, Man and Cybernetics Technical Committee and Taylor & Francis Group. In 2015, she had received the American Society of Civil Engineer's Outstanding Reviewer Award for the Journal of Management in Engineering. She also holds a M.Sc. in Building and B.Sc. in Building (2nd Upper Class Hons.) from NUS, and a Diploma in Building (with Merit) from the Singapore Polytechnic.

* * *

Ali Akbarnezhad is a lecturer in the School of Civil and Environmental Engineering, University of New South Wales, Sydney, Australia. His main areas of research include sustainability assessment, operations research and construction materials. Over the past ten years, Dr Akbarnezhad has been a part of a team securing more than three million dollars competitive research funds in Singapore and Australia for fundamental research in these areas. Dr Akbarnezhad's contributions to research and innovative teaching in the field have been recognized by several awards including Japan Concrete Institute's JCI-OWICS Award, Skilled Builders Award (Building and Construction Authority of Singapore), and Vice Chancellor's Teaching Excellence Award (UNSW).

Oluwayomi Babatunde's professional career spans project management, property management, and architecture in Nigeria, Singapore and South Africa. Yomi (as preferred to be addressed) holds a PhD and MSc in project management from the National University of Singapore and M.Arch and BSc. Hons in Architecture from the Obafemi Awolowo University, Nigeria. Yomi is a registered member of the Society of Project Managers (Singapore), Project Management Institute (USA) and is a Certified International Project Manager by the American Academy of Project Management (USA) where he is also a Fellow Member. Yomi is also a registered member of the South African Facilities Management Association, where he maintains the professionalization portfolio. Yomi's research interests include international project management, total quality management, and cross-cultural management.

Elmond Bandauko is a Teaching Assistant in the Department of Rural and Urban Planning, University of Zimbabwe. His research interests include African Cities, Housing Studies, Urban Planning and Public Health, Land Management in Urban Africa. He has published several papers and book chapters in these areas. He holds a BSc (Hons) in Rural and Urban Planning from the University of Zimbabwe.

Tinashe Bobo is a student at the University of Zimbabwe studying for a Bachelor of Science Honors Degree in Rural and Urban Planning. His research interests are in areas such as urban policy, housing, informality and urban development.

Yıldırım Huseyin Dalkiliç is Assist. Prof. Dr. at Erzincan University.

Ahmed W. A. Hammad is a PhD candidate at the School of Civil and Environmental Engineering, University of New South Wales, Sydney, Australia. He holds a BE (hons) in Civil and Environmental Engineering from the University of New South Wales. Academically his research interests lay in applying operational research techniques in construction, with main focus on sustainable development.

Jagan J. is currently working as the Junior Research Fellow (JRF) for the Sponsored Research project by BRNS in the Centre for Disaster Mitigation and Management (CDMM), VIT University, Vellore. He is also pursued a M.S. (By Research) in Geotechnical Engineering in the same department. He completed his Bachelor of Engineering (B.E.) in Civil Engineering in Adhiyamaan College of Engineering, Hosur.

Christian Jenne qualified as a Mechanical Engineer after completing a three year apprenticeship in Germany and worked for many years as a Project Engineer in Germany and Switzerland in the Research & Development area. In 2007 he moved to

Ireland where he graduated with a Management Business degree from the Michael Smurfit Business School in University College Dublin. Subsequently he completed a Master's degree in Renewable Energy Systems. He is currently doing a Ph.D. with University of Duisburg-Essen under the "ATBEST" research collaboration project between Germany, Ireland, United Kingdom and Sweden investigating "Biogas storage and distribution concepts for the transportation sector".

Navneet Jha is a post-graduate student in advanced construction management from NICMAR, Pune, Maharashtra.

Urszula Koźmińska graduated from the Faculty of Architecture, Warsaw University of Technology (WUT in 2007 (MSc.Eng.Arch.). From 2010-2011, she studied design in Istituto Europeo di Design in Madrid. Currently, she is a PHD candidate at the Faculty of Architecture WUT and her research constitutes the analysis of reused materials application in urban planning and architecture. Since 2005, she has been working in architectural offices and contractors in Warsaw and Amsterdam. She is a co-author of awarded architectural projects (Multifamily Housing on Miedzyborska St. in Warsaw, with Grupa 5 architekci - 1st prize; Wroclaw Main Train Station Refurbishment with Grupa 5 architekci - VII Life in Architecture competition: Best Public Space in Poland 2000-2012 Nominee – shortlisted; Multifamily Housing Clarissenhof, Tilburg with Dok Architecten -1st prize) and a laureate of several scholarship programs, grants and an author of scientific publications and presentations.

Mangesh V. Madurwar is a qualified engineer he completed his doctoral research from VNIT's civil engineering department and has more than 10 years of teaching, research, administrative and field experience in green construction project management. He is a regular conference speaker and has individually published several research papers in prestigious national and international journals. In India, he has been actively involved in the development and deployment of several national professional training programs.

Gladys Mandisvika is passionate about urban planning and development issues.

Shekhar Vishnu Nagargoje is a M.Tech-Urban Planner with specialization in Housing from CEPT (Center for Environmental Planning and Technology), Ahmadabad. Currently, he is a core faculty at National Institute of Construction Management and Research, Pune and is also a Program Coordinator for Real estate and Urban infrastructure management Program. His areas of specialization and research are Urban Resilience, Urban Planning, Smart cities, Investment Trusts,

Asset Management, Project Feasibility and sustainable cities. He has worked as an urban planner for K Raheja Corp, Mumbai. He was actively involved in financial feasibility, business development and market research activities for Inorbit Malls, a leading shopping center development company, covering over 25 cities across India. He has also headed the research and advisory division of IL&FS, western region.

Oluwasola Oni is a lecturer at Pan-Atlantic University, Nigeria. She teaches information systems at the undergraduate and postgraduate levels. She has been involved in research considering challenges involved in the effective diffusion of innovation technologies among individuals and small to medium-sized enterprises (SMEs). In addition to the study of diffusion and adoption of technologies, she is also interested and is currently investigating mobile customer relationship management (mCRMs), digital ecosystems and knowledge sharing.

Anna Teresa Oniszk-Poplawska graduated in 1998 from the Warsaw University of Technology (WUT), department of Environmental Engineering (MSc Eng); and in 1999 from the Central European University in Budapest (CEU) (MSc). From 2000-2006, she worked for the EC Baltic Renewable Energy Center as the manager of the Strategy and Policy section, later employed by the Institute of Renewable Energy (EC BREC IEO). She has worked in numerous EU R&D projects including 5FP, 6FP, 7FP, Norway Funds and Intelligent Energy Europe (IEE). She is an author of over 40 publications, including 5 books. Since 2004, she has participated as the expert evaluator to the following programmes IEE, EU structural funds, in the COST Imitative (7FP) - Domain Transport and Urban Development. Currently she is a PhD candidate at the Department of Architecture, WUT (spatial planning and siting aspects of technical infrastructures).

Jing Pan is a Building Information Modeling (BIM) engineer in the Joint Construction Management (Beijing) Institute. She consults for several public project owners in providing solutions for information processing and management in project delivery for a clear vision of their projects. She is currently working on a commercial property project, the service period since pre-design phase to investigate property users' needs in terms of building services. She is now working with the client to optimize scheduling, and to hold system coordination meetings to assure progressing with as little as interruptions in the project. Her long term career plan is to investigate computing techniques and process optimization for project life cycle, with a special interests in facility resilience. Her research interests and ongoing projects are: pattern recognition for BIM-based quality management; case-base learning chain for undergraduate and graduate education in construction management; augmented reality for facility resilience; and the life cycle approach for work space generation in disaster recovery projects.

Rahul V. Ralegaonkar has a B.E. (Civil-1998), a M.E. (Civil-2000) & a Ph.D. (Energy Efficient Buildings-2005). Apart from academics, he is also the member of several professional bodies of international and national repute. He has over 13 years of professional experience of which around 2 years has been in the industry in a managerial post in GIS industry and civil construction industry. The other 11 years was at an academic and administrative level. He bagged 1 international, 1 national and 2 state level awards. He is presently working as an Associate Professor in Department of Civil Engineering, at Visvesvaraya National Institute of Technology, Nagpur, India with a keen interest in the areas of sustainable construction engineering and disaster management. He has been actively involved in R & D-sponsored projects. Three Indian patents have been filled in the recent past. To date he has around 95 publications in both international and national level journals and conference proceedings. From his journal articles, four articles were sighted in the top 25 articles in Science Direct. Various core and elective subjects have been taught by him to graduate, post graduate, doctoral students and professionals in training. The special software competency in the field of Project Management and energy management has been developed. Presently, he is guiding several doctoral, master's and graduating students apart from already completed projects at doctoral, masters and graduating levels. As far as interactions with other professionals are concerned, lecture notes have been published and presented as part of more than 10 training programs. More than 35 lectures were delivered as an invited speaker on several management and engineering topics like energy management, energy economics, the application of sustainable technologies in the construction industry and disaster management, among others, at various national-level interactive platforms. As an organising committee member, Including principal coordinator, convener and secretary for workshops, training programs and national conferences, he has participated in more than 35 events. He is also actively involved with technical forums as an editorial board member and a reviewer for several journals and conferences at national and international levels.

David Rey is a lecturer in transport engineering at the Research Center for Integrated Transport Innovation (rCITI) in the University of New South Wales (UNSW). He holds a BSc in Electrical Engineering from Université Montpellier 2, France and a MSc in Mathematics from PUC-Rio, Rio de Janeiro, Brazil. After receiving a PhD fellowship from the French Institute for Science and Technology for Transport, Development and Networks (IFSTTAR), Dr. Rey started a joint PhD at the Laboratoire Ingénierie Circulation Transports (LICIT, Lyon, France) and at the G-SCOP laboratory (Grenoble, France) to study conflict detection and resolution strategies in air traffic management. He joined UNSW after receiving his PhD in Operations Research from Université de Grenoble, France. His key research interests lie in resource allocation schemes in transportation networks, combinatorial algorithms, mathematical programming as well as global and fair optimization.

Elzbieta Dagny Rynska has been a Professor, Faculty of Architecture, Warsaw University of Technology, Poland since 2010. Diplomas on BSc and MSc levels, Faculty of Architecture, Warsaw University of Technology 2004-2008. She has supervised numerous graduate students and postdoctoral fellows.

Vishakha V. Sakhare completed PhD (2016) in Low Cost Housing and PG in Construction Technology and Management (2012) at Civil Engineering Department, Visvesvaraya National Institute of Technology, Nagpur, India. She has thirteen research papers published in international and national journals and a conference to her credit and several others in the pipeline. She has been awarded "International Innovation & Research Award-Highly commended" by the CIOB, UK, for her master's dissertation, won first prize for a model in "Colloquium on green building concepts." She has been actively involved in the Fast Track Project, sponsored by the Department of Science & Technology, Government of India.

Pijush Samui is an Associate Professor, Department of Civil Engineering, National Institute of Technology, Patna, Bihar, India. Formerly, he was a professor at Centre for Disaster Mitigation and Management in VIT University, Vellore, India. He obtained his B.E. at Bengal Engineering and Science University; M.Sc. at Indian Institute of Science; Ph.D. at Indian Institute of Science. He worked as a postdoctoral fellow at University of Pittsburgh (USA) and Tampere University of Technology (Finland). He is the recipient of CIMO fellowship from Finland. Dr. Samui worked as a guest editor in "Disaster Advances" journal. He also serves as an editorial board member in several international journals. Dr. Samui is the editor of the International Journal of Geomatics and Geosciences. He is the reviewer of several journal papers. Dr. Samui is a Fellow of the International Congress of Disaster Management and Earth Science India. He is the recipient of Shamsher Prakash Research Award for the year of 2011.

Gaurvav Krishna Somani is a Post Graduate Student - PGP Advanced Construction Management, NICMAR, Pune.

Cheng Sun is a professor, doctoral supervisor and Vice Dean of School of Architecture of Harbin Institute of Technology. He is a candidate of National New

Century Talents Support Program of Ministry of Education of China, a candidate Academic Leader of Heilongjiang Provincial Key Disciplinary of "Architectural Design and Theories", a candidate leader of Heilongjiang Provincial Talents Leader, the principal of Heilongjiang Provincial Architectural Education Research Group, the director of Architecture Post-Doc Station of Harbin Institute of Technology.

Mehul Manjibhai Sutaria is Post Graduate Student - PGP Advanced Construction Management, NICMAR, Pune.

Meng Zhen is a PhD student of Architecture Department in Harbin Institute of Technology. Research area is building energy efficiency.

Index

T

U

W

www.ingramcontent.com/pod-product-compliance
Lightning Source LLC
LaVergne TN
LVHW080601300926
849162LV00005B/1019